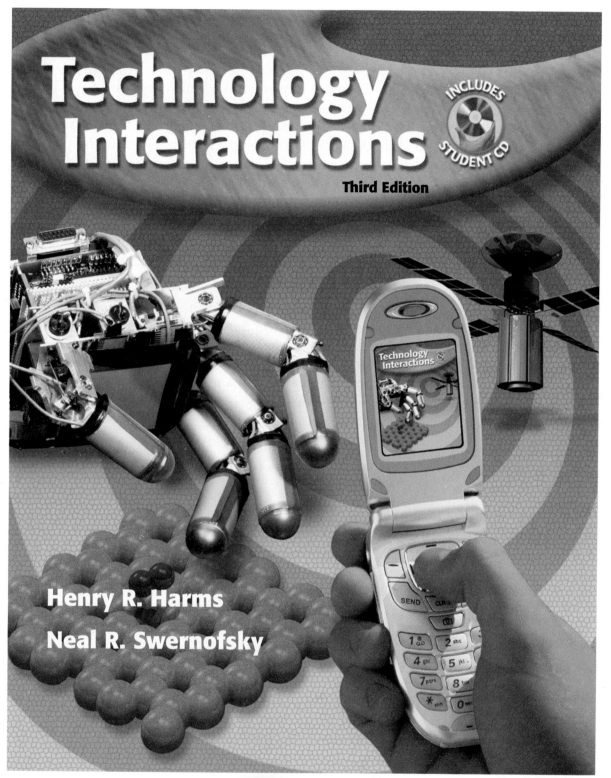

Technology Interactions

Third Edition

INCLUDES STUDENT CD

Henry R. Harms

Neal R. Swernofsky

McGraw Hill Glencoe

New York, New York   Columbus, Ohio   Chicago, Illinois   Peoria, Illinois   Woodland Hills, California

Safety Notice

The reader is expressly advised to consider and use all safety precautions described in *Technology Interactions* or that might also be indicated by undertaking the activities described herein. In addition, common sense should be exercised to help avoid all potential hazards.

Publisher and Author assume no responsibility for the activities of the reader or for the subject matter experts who prepared *Technology Interactions*. Publisher and Author make no representation or warranties of any kind, including but not limited to, the warranties of fitness for particular purpose or merchantability, nor for any implied warranties related thereto, or otherwise. Publisher and Author will not be liable for damages of any type, including any consequential, special or exemplary damages resulting, in whole or in part, from reader's use or reliance upon the information, instructions, warnings or other matter contained in *Technology Interactions*.

Brand Disclaimer

Publisher does not necessarily recommend or endorse any particular company or brand name product that may be discussed or pictured in *Technology Interactions*. Brand name products are used because they are readily available, likely to be known to the reader, and their use may aid in the understanding of the text. Publisher recognizes that other brand name or generic products may be substituted and work as well or better than those featured in the text.

 **Glencoe**

The *McGraw·Hill* Companies

Send all inquiries to:
Glencoe/McGraw-Hill
3008 W. Willow Knolls Drive
Peoria, IL 61614

13-digit ISBN 978-0-07-874172-2
10-digit ISBN 0-07-874172-6

Printed in the United States of America
3 4 5 6 7 8 9 10   027   09 08 07

Contents in Brief

Hi! I'm your Robot Guide.

About the Authors

Henry Harms is a Project Specialist in the Technological Studies Department at The College of New Jersey. He has a special interest in integrating mathematics, science, and technology content. Currently he teaches science methods courses, coaches middle school mathematics teachers, supervises student teachers, and serves as the State Advisor for the New Jersey Technology Student Association. He has co-authored several technology education textbooks and has coordinated the development of pre-engineering curriculum for the New Jersey Department of Education.

Neal Swernofsky has 32 years of middle school technology education teaching experience. Still in the classroom today, Neal enjoys the challenges of developing new activities and curriculum that will excite his students and inspire learning.

Neal is a charter member of New York State Commissioner's Academy for Teaching and Learning, a member of Epsilon Pi Tau, and an ITEA Regional Teacher of the Year recipient.

Contributors & Reviewers

Michael Condurso
Technology Educator
Bordentown Regional High
   School
Bordentown, New Jersey

Peggy Hazelwood
Denver, Colorado

Robert C. Horan
Industrial Education
   Department Head
Erwin Technical Center
Tampa, Florida

David A. Janosz, Jr.
Technology Education Teacher
Northern Valley Regional High
   School
Old Tappan, New Jersey

Robert Knight
Chicago, Illinois

Robert J. Kraushaar
Pinckneyville Middle School
Gwinnett County, Georgia

Ray E. Martin
Technology Education Teacher
RD & Euzelle P. Smith Middle
   School
Chapel Hill, North Carolina

Arlen K. Milne
Technology Education
   Instructor
Park Hill Congress Middle
   School
Kansas City, Missouri

Christopher Napierala
Trade and Technology Teacher
Samuel Morse Middle School
Milwaukee, Wisconsin

David S. Niemierowski
Technology Education Teacher
Colts Neck High School
Colts Neck, New Jersey

Jacquelyn W. Rozman
Technology Education Teacher
Tampa, Florida

Nancy L. Smith
Technology Education Teacher
Riverview High School
Sarasota, Florida

Stuart Soman
Educational Consultant
Wantagh, New York

Eric Thompson
Onalaska High School
Onalaska, Wisconsin

Marlene Weigel
Joliet, Illinois

Raymond D. Wilson
Technology Education
   Instructor
McAlester Middle School
McAlester, Oklahoma

Bill Youngfert
Technology Education Teacher
   (retired)
Herricks High School
New Hyde Park, New York

Contents

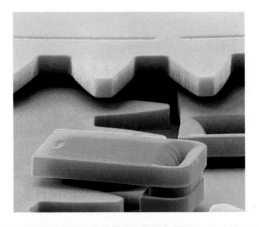

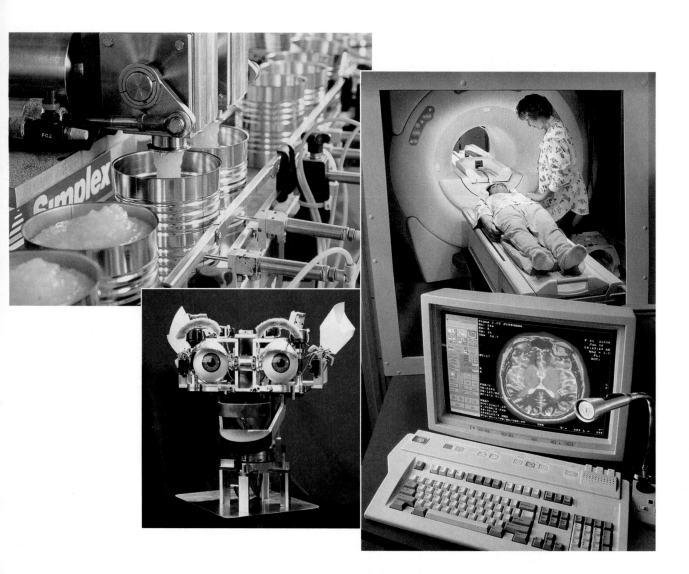

Activities

Activities (continued)

Exploring Careers

Math Link

Science Link

Reading Link

Writing Link

Impact of Technology

Student CD

Interactive Labs

Binary Code
Biometrics
Cartesian Coordinate System
Cochlear Implants
Create a Zoetrope
Designed World
Electromagnetic Spectrum
Elements of Design
Flowcharts
Fluid Power
Forces and Flight
Forces on Structures

Forms of Energy
Four-Color Printing
Gene Splicing
Hand Tools
Internal Combustion Engines
Measurement
Mechanisms
Multimedia Ad
Power Tools and Machines
Primary Processes
Principles of Design
Properties of Materials
Safety Data Sheets
Secondary Processes

Series, Parallel, and
    Combination Circuits
Seven Resources of Technology
Systems Model
Weather Satellites

Chapter Worksheets

Study Guides
Reviewing the Main Ideas
Extending Your Learning

Portfolio Guide

Creating a Portfolio; Design
Notebook; Project Record

Interacting with Technology

Let's tour this textbook.

Technology Interactions takes you on an exciting journey into the world of technology. It will help you understand how technology affects you and the world you live in. You will learn about design and engineering, air and space technologies, communication technologies, manufacturing technologies, and many other exciting areas. You will even learn about the fascinating world of forensics and bio-related technologies.

To be good citizens in today's world it is important to be technologically literate. Many people think that means to be computer literate, but it means much more. To be technologically literate means that you have a good knowledge of technology and how it affects your life. *Technology Interactions* will help you become technologically literate.

The textbook and the Student CD work together.

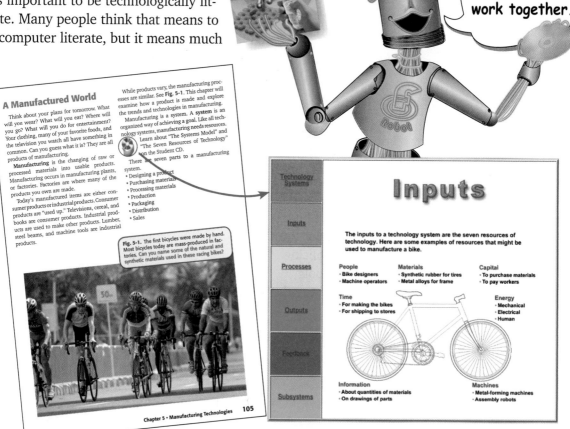

This screen is from the interactive lab "The Systems Model" on the Student CD.

Each chapter provides objectives and vocabulary to guide your reading.

Objectives—List what you will learn from this chapter.

Vocabulary—Terms are bold-faced in the chapter, followed by a definition.

CHAPTER 13

Animation

Objectives

- Discuss the uses of animation.
- Describe the information included on a storyboard.
- Compare and contrast the four types of animation.
- Describe Web animation techniques.
- Discuss how video games are developed.

Vocabulary

- animation
- persistence of vision
- stop-motion animation
- motion-capture animation
- storyboard
- modeling
- key frame
- GIF
- streaming animation

Many animated movies today are three-dimensional. Two of the most recognizable 3D characters today are Shrek and Donkey.

Activities

- Creating a Flip-Book
- Make Your Own GIF
- Presenting Video Games

Design, Build & Evaluate are activities that let you apply what you've learned in the chapter.

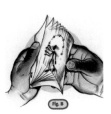

Hands-on activities let you apply what you've learned.

Design, Build & Evaluate

Creating a Flip-Book

Identify a Need/Define the Problem

As you learned in this chapter, each picture in a flip-book is slightly different from the pictures on the page before and the page after. See **Fig. A**. When the pages are flipped the images appear to be moving, as if it were animated. See **Fig. B**. In this activity, you will design and build your own flip-book.

Gather Information

The flip-book, or kineograph as it was originally known, has been around for over a hundred years. It has been made in a variety of shapes and sizes. Before designing your flip-book, you may want to research how a flip-book works and what makes an effective design for a flip-book.

Develop Possible Solutions

Prepare several different designs for your flip-book. You will need to decide on page size as well as the type of paper. Some paper sizes and types may effect how the book flips through your fingers. You will also need to decide on an image and action that the flip-book will show the user.

Materials and Equipment
Select from this list or use your own ideas.

- paper
- pens
- pencils
- computer system with drawing program
- binding material

Model a Solution

1. Once you have chosen the most effective solution, you will need to prepare the pictures. The flip-book pictures can either be drawn by hand or by computer. (It is relatively easy to draw the pictures on a computer. You can save the image as a separate slide and then modify it slightly, save it again, and so on.)
2. Cut the paper to the appropriate size. Make sure that each image is correctly placed in the sequence you designed.
3. Staple the pages together. If the book is too thick you may need another possible solution, such as using brush padding compound along one edge. (Padding compound is a substance used to make pads.) You may also want to put the book on a stiff backing, such as cardboard.

Test and Evaluate the Solution

Thumb through your flip-book to see how the images blend together. Share your design with your classmates and ask them for their feedback.

- Do the images appear to move smoothly when the pages are flipped?
- Are the pages of the flip-book fastened tightly enough to allow repeated use?
- Does the flip-book contain enough pages to effectively communicate the story or action?

Refine the Solution

- If the action in the flip-book wasn't clear, maybe you need to add more pages.
- If the animation didn't work, can the materials be improved to help?
- Revise your flip-book as needed.

Communicate Your Ideas

Create a portfolio of the research and work that you did in order to create your flip-book. Include some facts and information about how flip-books work and are made. Document the solutions you came up with and the steps you took to create it. Also include important information on how you tested your solution and any refinements you made. Present your portfolio to the class.

Fig. A

Fig. B

Math Link

Is It Work? In everyday language, "work" is what people do to earn a living. In physics, work is thought of a bit differently. It is the product of a force times the amount of weight being lifted (or displaced) by that force. The formula usually is written as $W = F \times d$ (work = force times distance).

So, which would take more work: lifting a 25-pound weight up 2 flights of stairs or a 10-pound weight up 5 flights? Be prepared to show your calculations and explain your answer. How much work would be done if you just stood still for an hour after someone placed a 25-pound weight in your hands?

> These really DO link to your other classes and to your world.

Link—This feature links the content of the chapter to other important areas: science, mathematics, reading, and writing. It will help you understand how science, math, reading, and writing are part of technology.

Impact of Technology

Energy Trade-Offs

Energy production is of concern for many people throughout the world. Along with the fear of running out of nonrenewable fuels, there is also worry over the trade-offs that must be made when producing energy. Making a trade-off means giving up one thing to gain another. For example, hydropower is a renewable energy source, and it causes less pollution than fossil fuels. However, the dams and the lakes created behind them may force people to leave their homes or may harm native plants and wildlife. Even wind farms can have some negative effects on the environment.

Investigating the Impact

Research trade-offs in energy and power technologies.
1. Select one of the energy sources described in this chapter. Find out about the benefits and drawbacks of using that energy source. Do you think the positive effects outweigh the negative? Why or why not?
2. What energy sources are used to produce the electricity used by your community? What trade-offs are involved?

Impact of Technology—Draws your attention to some major technological developments and how they affect our daily lives in both positive and negative ways.

Exploring Careers

Commercial Pilot

ENTRY LEVEL | TECHNICAL | PROFESSIONAL

Commercial pilots fly airplanes and helicopters for companies. Most of them transport passengers or cargo, but some dust crops, test airplanes, track criminals, or rescue and evacuate injured people.

Commercial pilots usually work as part of a team that includes a copilot, a flight engineer, and the ground crew. Pilots start the plane's engine, operate controls, monitor instruments, and fly the airplane. Pilots must follow a flight plan as well as Federal Aviation Administration (FAA) regulations and procedures.

Qualifications

A commercial pilot must have a commercial pilot's license with an instrument rating issued by the FAA. Pilots who want to fly helicopters must have a commercial pilot's certificate with a helicopter rating. To qualify for these licenses, an applicant must be at least 18 years old and have at least 250 hours flying experience.

Part of the training may include using a flight simulator, a system that mimics the conditions of flying an airplane. Using the simulator, trainees learn to take off, pilot, and land a plane.

Applicants must pass a strict physical exam. They must have good hearing and 20/20 vision. They may wear glasses, but they cannot have any physical disabilities that might impair them while flying. They also have to pass a written test that covers safe flying, navigation techniques, and FAA regulations. The final test is to demonstrate their flying ability.

Every employer has different guidelines for hiring pilots. Many larger companies require commercial pilots to have a bachelor's degree. Test pilots often need to have an engineering degree.

Outlook for the Future

In a good economy, more people fly; so there is more of a demand for pilots. There is a great deal of competition for the better, more desirable jobs.

Handling Pressure

Good communication skills, the ability to work well on a team, and staying calm under pressure are very important. Pilots must know how to focus on priorities. They must stay alert and be quick to react if something goes wrong.

Researching Careers

Search the Internet to find out about three jobs for commercial pilots. What education and experience are needed? What kind of aircraft will the pilots fly? Write a report about what you find.

174   Technology Interactions

Exploring Careers—Describes the qualifications, outlook, and the responsibilities of a specific career in technology.

> You might find your future career.

This page will help you remember what you learned.

Chapter Review—Lists the key points of the chapter. It will ask you to respond to thought-provoking questions and apply the principles you learned in the chapter.

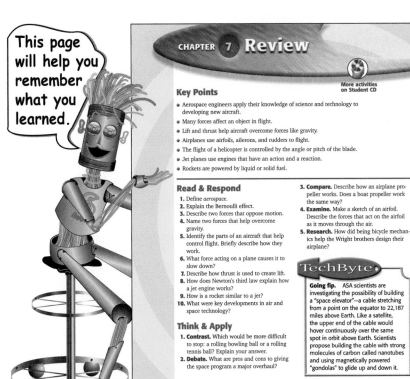

TechByte—This thought-provoking "byte" of information contains new or interesting facts about technology.

Have fun learning!

Student CD—Has more review activities that will expand your understanding of the chapter: Reviewing the Main Ideas, Study Guide, and Extending Your Learning.

Interactive Labs—Each Interactive Lab on the Student CD has a link to the *Technology Interactions* Web site.

CHAPTER 1

Design & Engineering

Objectives

- Define *technology*.
- Describe the connection between the development of technology and human needs and wants.
- Identify the steps of the engineering design process.
- Describe the role of engineers as designers and problem solvers.

Vocabulary

- **technology**
- **nanotechnology**
- **design**
- **invention**
- **innovation**
- **criteria**
- **constraint**
- **brainstorming**
- **prototype**
- **engineer**
- **productivity**

The Walt Disney Concert Hall in Los Angeles, California, presented unique design and engineering challenges.

Activities

- Go with the Flow
- Mapping an Ocean Floor
- Planning Your Dream Room

What Is Technology?

Technology takes place when people use their knowledge to extend their abilities, satisfy needs and wants, and solve problems.

Where does technology come from? The spirit of human creativity has always been the driving force in developing new technologies to solve new problems. We live in an amazing world that at times may resemble science-fiction or an action movie, and technology is the "star of the show."

Right now, the International Space Station is orbiting Earth 250 miles above your head. See **Fig. 1-1**. The ISS is the largest multination science project ever to take place. The station is almost 300 feet long and 400 feet wide. Astronauts are living and working in the station at this very moment, just like in sci-fi movies.

Did you know that robots are performing operations on people? The da Vinci Surgical System allows doctors to sit at a console a few feet away from their patient and guide robotic arms through delicate surgeries. In 2004, an estimated 20,000 procedures were done using this system. See **Fig. 1-2**.

Skyscrapers are often seen in adventure movies. However, the skyscraper Taipei 101

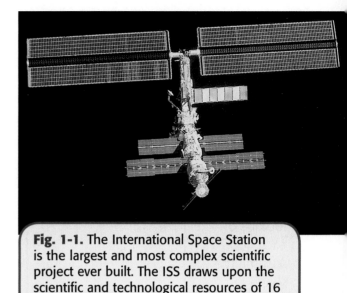

Fig. 1-1. The International Space Station is the largest and most complex scientific project ever built. The ISS draws upon the scientific and technological resources of 16 nations.

in Taipei, Taiwan, was not created by movie special effects. The 1,667-foot structure is the tallest building in the world, rising over 200 feet more than the Sears Tower, which is the tallest building in the United States. Taipei 101 contains more than three million square feet of office, hotel, and shopping space.

Can you imagine machines being inside you? **Nanotechnology** is an engineering field that

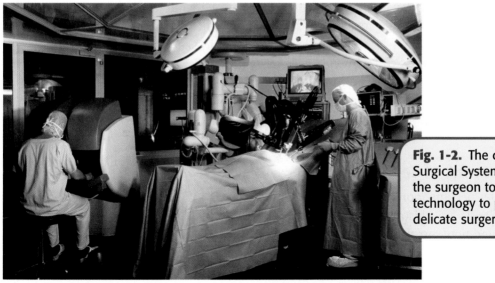

Fig. 1-2. The da Vinci Surgical System allows the surgeon to use robotic technology to perform very delicate surgery.

Math Link

Area and Square Root. Office buildings are often described by the square feet of space available. The area of any one floor can be determined by multiplying its length by its width. If Taipei 101 is rated at 3 million square feet contained on 101 floors, what is the approximate square footage of each floor? Assume the shape of each floor is square. What would its width and length be?

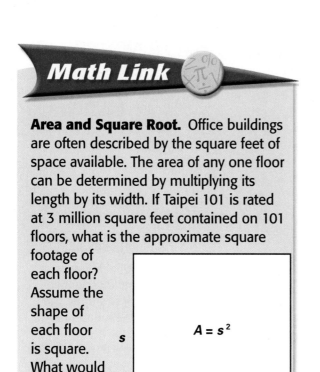

s $A = s^2$

s

For example, when people abandoned a nomadic life of hunting and gathering to settle into villages, their new lifestyle started to focus on farming and raising livestock. Agriculture changed how people lived and worked. New technologies were developed to meet these new needs. The plow and sickle were tools of the Agricultural Age. Their design helped to make farm work easier and more productive.

In the 1700s, new technologies and products exploded onto the scene. This marked the beginning of the Industrial Age. The steam engine replaced muscle power with machine power. Factories harnessed this new energy resource and increased their output. As the demand for more products increased, new production technologies and processes answered the demand.

Transportation systems had to be improved to rapidly carry products and people around the world. Train, automobile, and air transportation began.

involves manipulating materials on an atomic or molecular level. See **Fig. 1-3**. Nanotechnology has produced machines that are smaller than 1/1000 of the diameter of a human hair. Such machines could flow through your veins like red blood cells. They could be moved to specific parts of the body through human tissue and organ ducts and release specialized medicines where needed.

These great accomplishments of technology did not happen overnight. They are the result of human creativity changing simple technologies into more complex technologies over thousands of years.

The Development of Technology

As our daily lives became more complex, satisfying our needs and wants also becomes more complex. By tracing how our wants and needs have developed over time, we can see how technology has developed. See **Fig. 1-4**.

Writing Link

Technology Time Line. Technology changes fast. Just during your lifetime, many new technologies have emerged. The Segway Human Transporter is one example. Research the Internet or other

sources to find out about technology that has been developed since your birth date. Present your findings as a time line.

Fig. 1-3. This computer model of nanotechnology shows balls inside a nanotube. The nanotube is a cylinder of carbon atoms less than a billionth of the width of a human hair.

Today we live in the Information Age, which began around the mid-1800s. It was brought about by the need to gather, store, and share large amounts of information. The Information Age was accelerated by the development of the transistor and the computer.

The technologies of the Information Age touch all aspects of modern life, from advanced medicines used in health care to playing an MP3 player for recreation. Visit "The Designed World" interactive lab on the Student CD to learn more about the technologies in our lives.

Agricultural Age	Industrial Age	Information Age
—began about 10,000 years ago	—began around 1750	—began around 1850
Economy was based on growing crops and raising livestock.	Economy was based on manufacturing.	Economy is based on knowledge.
Some Key Inventions • plow • loom • reading and writing • measurement • money	Some Key Inventions • steam engine • blast furnace • mass production • train, plane, automobile • skyscraper	Some Key Inventions • telephone, radio, TV • transistor • computer • Internet and World Wide Web
Some Key Impacts • growth of towns and cities • specialization; not all people have to work at raising food • most people work where they live, on the farm or in their shop	Some Key Impacts • public schools • people move from farms and villages into cities • greater productivity; lower-priced goods • most people go to a place of work, such as a factory or office	Some Key Impacts • electronic business (e-commerce) • electronic education (e-learning) • globalization • work goes with you—in the office, in the home, on the road

Fig. 1-4. Ages of Technology.

The Design Process

Products of technology that moved people from the Agricultural Age to our Information Age often were the result of design. **Design** is the process of creating things by planning.

The process of designing new products is called **invention**. Thomas Edison was a very successful inventor. His phonograph was the first machine that could record and reproduce sound. See **Fig. 1-5**.

Improving and changing a technology that already exists is called **innovation**. Edison was not only an inventor, but he was also an innovator. His invention of the phonograph actually happened while trying to improve the telegraph. Today, CD players and recorders are innovations based on Edison's design for the phonograph.

Designing is not a simple or quick task. It requires investigating, answering questions, and making decisions. We can say that designing is the step-by-step process of developing solutions to problems. This process can guide you as you create design solutions to problems presented to you.

Although the following steps are listed in order, you might skip around from step to step. For example, if you are modeling a design solution and immediately discover the solution just will not work, you can step back and choose another solution to model.

Identify a Need or Define the Problem

Identifying a need or stating the problem is a good place to start problem solving. Some needs are easy to understand and lead to simple solutions. For example, you may need to keep several sheets of paper together. A paper clip or staple would be a good solution to this simple problem.

Other needs may be more complex and require deeper understanding. If you were asked to design an athletic shoe, many design questions would have to be answered and many requirements, or criteria, would have to be developed.

Fig. 1-5. Thomas Edison with his phonograph, one of the over one thousand inventions he patented in his lifetime.

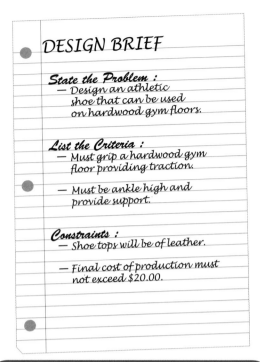

Fig. 1-6. Preparing a design brief will help you clarify the design problem and understand the criteria and constraints.

Criteria are the things that the product must do or include. Defining the criteria is an important part of understanding the need you wish to satisfy. The criteria for an athletic shoe might be that the shoe must grip the hardwood surface of a gym floor and that the shoe must support the ankle and be easy to fasten.

A design also has constraints. **Constraints** are the limits placed on the design and the designer. Constraints are usually related to the resources that can be used. Examples of constraints for an athletic shoe could be the materials that could be used and the maximum cost for producing each pair of shoes. The more the shoes cost to produce, the higher the price will be to the customer.

Identifying the needs and defining the problem ensures that the designer has a good understanding of the criteria and constraints. See **Fig. 1-6.**

Gather Information

The next step in the design process is to start looking for information that will help you design a solution. Have you ever heard the saying "no need to reinvent the wheel"? Research how other people have solved the problem. If you're designing an athletic shoe, visit shoe stores. Look over the different styles of shoes currently available. See **Fig. 1-7.** Try conducting an Internet search to locate information.

Talking to people is also a great way of gathering information. Shoe designers, salespeople, and shoe repair people have a wealth of knowledge they can share with you about shoe design. People who might use the shoes can also help. Feedback from customers is always important information when designing a product.

Fig 1-7. If you are designing a new athletic shoe, first gather information about what other designers have included in their shoes. Then decide how you can improve on their designs.

Develop Possible Solutions

Most problems have more than one possible solution. You should always develop many design ideas even though you may want to grab onto the first idea that comes into your head. The more design ideas you have, the better your chances of having the one solution that will really satisfy the need. That's why there are so many different styles of athletic shoes.

Brainstorming is a technique designers often use to develop ideas. **Brainstorming** is a process in which group members suggest ideas out loud as they think of them. The ideas are usually recorded on a large pad or whiteboard so the whole group can see them. See **Fig. 1-8**. After the brainstorming session is over, the advantages and disadvantages of each idea are reviewed and narrowed down to a few possible solutions. You may find that some of the best solutions originated from ideas that seemed ridiculous when first mentioned.

Principles and Elements of Design. The solution you design should not only work but also look good. The principles of design are guidelines for creating an effective design. Putting those guidelines into practice involves using the elements of design. To learn more, see the interactive labs on "The Elements of Design" and the "The Principles of Design" on the Student CD.

Model the Solution

Model building is a good way to gather additional information and test design ideas. There are many different models you can construct.

- Two-dimensional and pictorial drawings, sketches, and renderings can help you visualize what a design solution would look like in real life. See **Fig. 1-9**.
- Scale models are small but accurate representations of the final product.
- Appearance models are not working models but are used to show what the product will look like when it is produced. The automobile industry uses appearance models to get feedback on possible automobile design ideas.
- Functional models may not look like the final product but they may work like the final product. These models are used to work out mechanical and electronic details.
- A **prototype** is a working model. It looks and functions just like the finished product and is usually made by hand. See **Fig. 1-10**.
- Computer simulations can test newly designed products. For example, the floor-gripping power of an athletic sneaker can be tested on a computer-simulated gym.

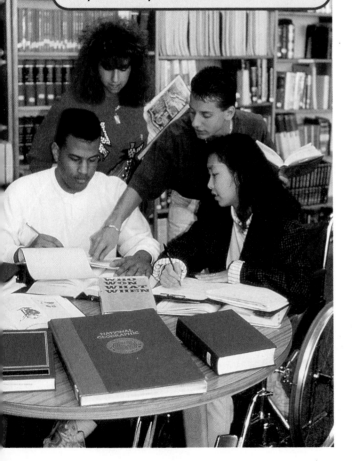

Fig. 1-8. Brainstorming can be used for solving almost any kind of problem. The key to a good brainstorming session is to not exclude any ideas, even though they may seem silly at the time.

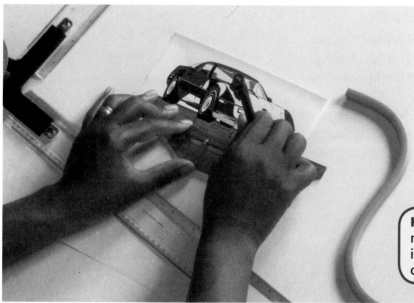

Fig. 1-9. Two-dimensional modeling, called rendering, is used to add realism to a drawing.

Test and Evaluate the Solution

Models of design solutions must be tested, and important questions have to be answered during the evaluation.

• Does it do what it is supposed to do?
• Is it safe to use?
• Will it last as long as it needs to?
• Is the cost within acceptable limits?
• Is it comfortable to use?
• Does it have a pleasing appearance?
• Does it stay within the criteria and constraints?

Refine the Solution

After studying all the test data and evaluating your design solution, you may find that changes are needed. Now is the time to refine (improve) the design, before production begins. For example, you might have to add another shoelace hole to the athletic shoe to make it fit better.

Fig. 1-10. Inside the cab of Toyota's innovative FTX Concept vehicle, the "flying T" console separates the driver from the front passenger and also houses the instrument panel and a large 3D information display panel.

Communicate Your Ideas

Communicating your design ideas to others is just as important as the design itself. See **Fig. 1-11**. Designers often have to prove their designs are worthy of manufacturing. They make presentations to company executives, potential clients, and the people who will be manufacturing the product. Drawings, charts, prototypes, and dialogue are the tools used to sell a design idea. How would you present your athletic shoe design to NIKE, adidas, or Reebok?

Fig. 1-11. Good communication skills are essential in sharing your design ideas with others on your team.

Who Designs Solutions?

Fashion designers create clothing designs. Landscape designers develop outdoor environments using plants, rocks, and water. Industrial designers create products like blow dryers and appliances. Engineers are also designers.

An **engineer** is a person who uses his or her knowledge of science, math, technology, and communication to solve technical design problems. Engineers and the engineering design process have influenced most of the things that make up our human-made world.

Engineering changes as new needs develop. The driving force behind this change is technology. As new technologies emerge, new types of engineering are introduced.

Classic Engineers

Civil, mechanical, mining, chemical, and electrical/electronic engineers are referred to as "classic engineers." See **Fig. 1-12**. Formal programs designed to train these professionals began in the 1700s. The oldest engineering category is civil engineering.

A New Generation of Engineering

With new technologies comes a new generation of engineers. Many of these engineers focus their design efforts on manufacturing, transportation, energy sources, computers, and the environment.

Manufacturing Engineers. Manufacturing engineers try to find the best way to produce a product. Their goal is to increase productivity. **Productivity** is the relationship between how many hours are worked and the quantity of products made in that time period. Manufacturing engineers improve productivity by designing and organizing the placement of machines on the plant's floor. They also develop the sequence of processes used to transform materials into products.

Impact of Technology
New Engineering Challenges

As technology moves forward, new engineering fields will be created. Job opportunities will increase for engineers. What kind of engineering jobs are around the corner?

Structural engineers will design buildings that use cables, motors, and sensors to flex and relax, responding to the earth's movements. These buildings could help the economy by saving millions of dollars in earthquake damage.

Planetary engineers, or macroengineers, think big. As technology develops, they will be able to think even bigger. For example, what would it take to tow an iceberg to a country suffering from drought? Can engineering stop or change tectonic movement to avoid earthquakes? Only time will tell.

Investigating the Impact

If engineers were able to create buildings that were earthquake-proof, what effect might this have on society? Present your ideas to the class.

Fig. 1-12. Classic Engineers.

Civil engineers work with architects and other professionals to design airports, highways, and structures used by the public.		**Chemical engineers** design the chemical reactions that have to take place to produce a product.	
Mechanical engineers design machines and machine parts.		**Electrical/Electronic engineers** design circuits and power supplies for electrical and electronics devices.	
		**Mining engineers** design and manage the process used to remove minerals from the ground.	

Fig. 1-13. Environmental engineers work at the site of a major toxic waste cleanup. Note all the personal protective gear they must wear to protect themselves from being contaminated.

Aerospace Engineers. The people responsible for designing and building new types of aircraft are aerospace engineers. These people focus their work on engine design, the frame of aircraft, and the control systems.

Materials Engineers. New technologies require new materials. Materials engineers are responsible for developing materials with specific properties that meet a manufacturer's needs. Properties such as strength, conductivity, and hardness can be engineered into new materials.

Environmental Engineers. Environmental engineering came about as a result of the rapid growth in manufacturing. The congestion of city life and the reckless use of natural resources increased land, water, and air pollution. Environmental engineers design cleaner production methods and design systems to clean up polluted sites. See **Fig. 1-13**.

Computer Engineers. Computer engineering is a spin-off of electronics engineering. Some computer engineers specialize in hardware or computer design. Others specialize in software design. Companies also hire computer engineers to install and maintain large computer systems.

Going to Work with an Engineer

What would it be like to work with an engineer solving a real engineering design problem? Let's job shadow one and find out! See **Fig. 1-14**.

Jessica is a structural engineer. Her job is to inspect new buildings to ensure that they meet all government requirements for new structures. She also inspects older structures to make sure they remain safe for the occupants.

Jessica has been asked to investigate a problem in a middle school's gym. The gym floor is sagging and some of the concrete walls sitting on it are cracking.

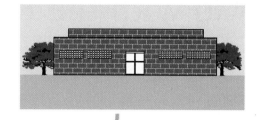

Jessica gathers information by using a laser level to get accurate measurements of how much the floor is sagging. She then hires a company to drill large holes through the floor so a robot video camera can search for the problem.

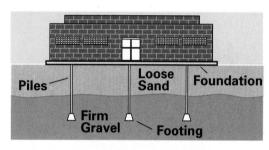

Jessica concludes that the pilings supporting the floor have sunk into the soil, leaving large floor sections unsupported. She calculates the weight that the piles have to support and suggests three different solutions.

Jessica then prepares detailed drawings and cost estimates. After the school board selects a solution, Jessica helps to select contractors who will do the work under her supervision.

Jessica's knowledge of math, science, technology, and communication are all used in a typical day of an engineer.

Fig. 1-14. Follow a structural engineer as she works her way through the various phases of a typical project.

Go with the Flow

Identify a Need/Define the Problem

Flowcharts often aid engineers in the design process. Use a computer to create a flowchart for making an aluminum wheel. The flowchart will need to show the production stages, the work needing to be done, and who will perform each task.

Gather Information

Research the processes involved in manufacturing an aluminum wheel like the one shown in **Fig. A**. Also research flowcharts. For an interactive lab about flowcharts, see the Student CD.

Fig. A

Materials and Equipment

Select from this list or use your own ideas.

• drawings of an aluminum wheel
• computer system

Develop Possible Solutions

Flowcharts can be set up in different ways. Draw several flowcharts and pick the format you think will work best. **Figure B** shows a sample flowchart for a manufacturing process.

Model a Solution

1. Step by step, list your manufacturing process in a flowchart.
2. Make sure that each step in the process has the correct flowchart symbol.

Test and Evaluate the Solution

• Will materials move efficiently through the assembly line?
• Will adjustments have to be made during manufacturing?
• Does your flowchart's sequence make sense?

Refine the Solution

• Now that you have seen your process in flow-chart form, do you see any problems?
• Can you find a way to increase efficiency or reduce costs?
• If necessary, make a new flowchart to show your improved process.

Communicate Your Ideas

Share your flowchart with the class. Walk through each step to see how efficiently the materials will move through production. Select the flowchart in the class that is the most efficient.

STEP NO.	STAGE	DESCRIPTION	LOCATION	WORKER
1	◯	Trace the pattern onto the metal	Bench #1	Worker #1
2	◯	Center punch the areas on the metal to be drilled	Bench #1	Worker #1
3	↓	Move to Bench #2		Worker #1
4	◯	Cut out shape	Bench #2	Worker #2
5	↓	Move to drill press		Worker #2
6	◯	Drill hole in metal	Drill press	Worker #3
7	↓	Move to Bench #3		Worker #3
8	▢	Inspect hole and shape	Bench #3	Worker #4

PROGRAM FLOWCHART FOR _____

Fig. B

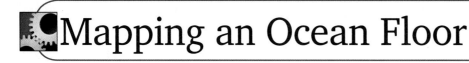

Mapping an Ocean Floor

Identify a Need/Define the Problem

In this chapter, you learned about different kinds of models. Make a model to simulate mapping an ocean floor with sonar.

Gather Information

Gather information on what ocean floors may look like. For example, are they usually flat, or do they have mountains and valleys?

Develop Possible Solutions

Prepare sketches of a model ocean floor you will make from clay. Be sure your model has many of the features actually found on an ocean floor.

Materials and Equipment

Select from this list or use your own ideas.

- shoe box with lid
- clay
- tape
- ¼" graph paper sheets (2)
- ¼" dowel rod, 10" long

Safety Alert

Look up "Safety Data Sheets" on the Student CD and prepare a data sheet for this activity. As you work on the activity, be sure to follow all safety rules.

Model a Solution

1. Using clay, model the ocean floor inside the shoe box. You do not need to cover the entire box bottom. You only need to use a strip of clay 3" or 4" wide down the center of the box, which is where the sensor will be making contact with the floor.
2. Place the lid on the box and tape it closed.
3. Cut and tape a piece of graph paper to fit the top of the lid.
4. Draw a line lengthwise down the center of the graph paper. Divide this line into ¾" segments to represent coordinates.
5. The second sheet of graph paper will become your profile map. On this sheet, draw a straight line lengthwise one inch from the bottom. Divide this line into ¾" segments along its length. Put the profile map to one side.
6. Make a point on one end of the 10" dowel rod. Mark the rod at 1" intervals. Label the mark 1" from the pointed end of the dowel as 8". Label the next 1" mark as 7" and so on. Mark off each 1" interval in ⅛" increments (just like marks on a ruler).
7. Map the ocean floor model by making soundings. (A sounding is a measured depth of a body of water.) For each sounding, push the pointed dowel through the box's lid at each of the coordinates you marked on lid. Keep the dowel straight. Gently lower the dowel until you feel it come in contact with the clay. See **Fig. A**.
8. Read the measurement on the dowel. Plot this number on the second sheet of graph paper. Make soundings on each of these

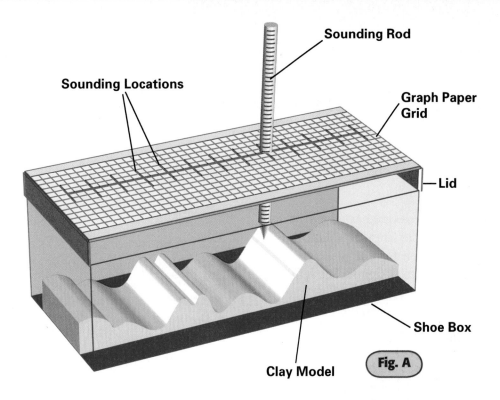

Sounding Rod

Sounding Locations

Graph Paper Grid

Lid

Shoe Box

Clay Model

Fig. A

coordinates. Plot these points on the second piece of graph paper.

9. Connect the points on the graph paper. You now have a profile map of the ocean floor model in the box.

Test and Evaluate the Solution

• Does the profile map look like the model you made?
• How could you show even more detail of the ocean floor model?

Refine the Solution

• If your profile map doesn't mirror the clay, redo the experiment.
• Does this device give you an idea of how sonar really works?

Communicate Your Ideas

Share the results of your project with the class. Be sure to let them know what you found out about how sonar works.

Science Link

Underwater Treasure. Often, scientists use sonar to find treasure or wreckages at the bottom of the ocean. How difficult would it be to find something in a vast ocean? Remember that some ocean floors or trenches are over six miles deep!

Have a classmate put a small object at the bottom of your shoe box without you seeing where the object is at. Have the student then place the lid back on the shoebox. Try to find the object by using the dowel. How many times did you poke a hole through your shoebox lid before you found the object? What could sonar do that the dowel couldn't?

Planning Your Dream Room

Identify a Need/Define the Problem

Your parents have decided to construct an addition to your house. They are planning on adding a new bedroom for you as part of the addition, and you get to design it! The first step required in planning a room is a floor plan. A floor plan is a view of a house or building looking down from the ceiling.

Create a floor plan of your dream room. The scale of the drawing should be 1" = 1'-0". Your bedroom must have the following:

- Floor area of 168 sq. ft.
- One door (standard 30 inches wide)
- Two outside walls
- Two windows, 12 sq. ft. each
- One desk
- One desk chair
- One bed
- One dresser
- At least two additional pieces of furniture, such as a bookcase or nightstand. Be creative!

Safety Alert

Look up "Safety Data Sheets" on the Student CD and prepare a data sheet for this activity. As you work on the activity, be sure to follow all safety rules.

Math Link

Finding Square Feet. Rooms and other living quarters are often described in terms of square feet. How are square feet determined? Why would the number of square feet a room has be important to you?

If a two-story house was 30 x 45 feet in width and length, how would the total square footage (all rooms, all floors) be calculated?

Materials and Equipment

Select from this list or use your own ideas.

- CAD program or graph paper
- T-square
- standard triangles
- pencil
- eraser
- common patterns for furniture (circles, squares, etc.)

Gather Information

In designing your dream room, you must consider your house. Do all the bedrooms have the same layout? How important is the placement of the windows and doors?

You must also research typical sizes of furniture. Measure the furniture in your own bedroom to give you an idea of dimensions. **Figure A** shows an example of a floor plan. Can you find examples of others?

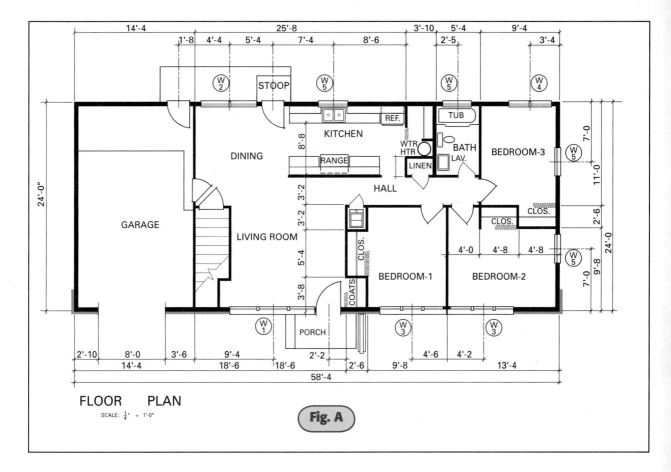

FLOOR PLAN

SCALE: $\frac{1}{4}$" = 1'-0"

Fig. A

Develop Possible Solutions

Sketch several floor plans for the bedroom. Remember that the room has 168 sq. ft. The floor plans will vary, depending on the length of the walls and where the doors and windows are placed.

Model a Solution

1. Cut out paper patterns to represent your furniture. Each piece of furniture should be to scale. Label all the patterns.
2. Place the furniture patterns on the floor plan. Move them around until you have a pleasing and efficient layout.
3. Draw a final floor plan with the furniture in place. You can either tape the patterns down or trace them directly onto the floor plan.

Test and Evaluate the Solution

- Does all of the furniture that you need fit into the room?
- Is the floor plan neatly drawn?

Refine the Solution

- Is there more furniture you want to include?
- Will the room's layout meet your needs?
- Revise the floor plan if needed until you are happy with your new dream room.

Communicate Your Ideas

Present your completed floor plan to the class. Be sure to share the reasons for the layout of doors and windows as well as the furniture.

Exploring Careers

Architect

ENTRY LEVEL | TECHNICAL | **PROFESSIONAL**

Architects plan, design, and oversee the construction of many types of buildings. They work closely with clients, engineers, and building contractors.

First, architects meet with the client to learn about the client's wants and needs. Architects then prepare information on the structure's design and specifications, materials and equipment, estimated costs, safety, and construction time. They also prepare scale drawings and ensure the project meets the building codes and laws.

Architects work mainly in an office setting but may need to visit construction sites. Most architects work for architectural firms, but about 25% are self-employed. Some work for construction-related businesses or for government agencies.

Qualifications

Architects must have a solid math and science background as well as good drafting and problem-solving skills. A bachelor's degree from an accredited architectural program is required to become licensed or registered. In addition, architects must have three years' experience in an architect's office and pass an exam before they can call themselves architects. Architects who will be responsible for a building must be registered or licensed by the state.

Students interested in architecture should take high school classes in mathematics—specifically geometry and trigonometry—and science, such as chemistry and physics. Because

the building's appearance is important, students should also enjoy more creative fields, such as design, fine arts, and history and archeology.

Outlook for the Future

The job outlook for architects is good. As with all jobs, both local and national conditions affect the outlook. Prospects will vary depending on the construction industry.

Planning for Change

To be successful, architects must be on the lookout for new developments in the field. More opportunities may result because of changes in technology. Employers want workers who ask for advice and support coworkers when changes occur.

Researching Careers

Find out about jobs for architects. What areas of the country have the best employment opportunities? Write a brief summary of what you find.

CHAPTER 1 Review

More activities on Student CD

Key Points

- Human creativity is a driving force in creating technology.
- Design is the process of creating things by planning.
- The design process is what engineers use to solve problems.
- Engineers specialize in many different fields, but all engineers are designers.

Read & Respond

1. Define *technology*.
2. Describe the link between our needs and wants and the development of technology.
3. What need led to the Information Age?
4. Describe the difference between inventions and innovations.
5. What are criteria and constraints?
6. Define *brainstorming*.
7. List the steps in the design process.
8. What knowledge does an engineer use to design solutions to problems?
9. What are the five categories in classic engineering?
10. What is the difference between a manufacturing engineer and a materials engineer?

Think & Apply

1. **Decide.** Technology is developed to meet human needs and wants. Does technology also create needs and wants? Explain your answer.
2. **Design.** Make a sketch to show how a product you use could be improved. Then explain how the improved design would meet a specific need.

3. **Research.** Investigate the work of a famous designer, inventor, or engineer. Prepare a poster showing that person's accomplishments.
4. **Design.** Using the design process, explain how you would proceed to design a new kind of book bag. Be specific.
5. **Convince.** You are the engineer who designed a new ballpoint pen that writes upside down and under water. Prepare a presentation to a potential client who might manufacture the product.

TechByte

A Progressive World. Fifty years ago, humans hadn't yet traveled into space. Twenty-five years ago, the Internet still wasn't available to the general public. Fifteen years ago, DVDs hadn't been invented yet. We are creating technology at an amazing rate. Futurist Ray Kurzweil predicts that human progress in the 21st century will be equivalent to progress made by humankind in the last 20,000 years.

CHAPTER 2
Computer-Aided Design

Objectives

- Identify three advantages of CAD.
- Describe three uses for 2D CAD.
- Compare and contrast three kinds of 3D CAD models.
- Explain four uses for 3D CAD.

Vocabulary

- **drafting**
- **computer-aided design (CAD)**
- **solid modeling**
- **mechanical design**
- **rendering**
- **rapid prototyping**

A 3D CAD drawing of a Formula One race car.

Activities

- Making a CAD Cube
- Planning in CAD
- Designing a Portable Music Player

What Is Computer-Aided Design?

Have you ever sketched plans before making something? **Drafting** is the process of creating drawings needed so that a part or product can be manufactured or built. For many years, drafting was done by hand directly on paper. The tools used were pencils and pens, rulers, triangles, T-squares, erasers, and compasses.

Today, many engineers do drafting on a computer. However, they can do more on computers than just create drawings. **Computer-aided design (CAD)** is the process of designing on a computer. CAD helps designers make quick and accurate decisions about how a product should look and how it should work. CAD drawings help them find out if their ideas will work. The design might be for something as small as an electronic circuit used in a hearing aid or as large as a skyscraper.

Many kinds of jobs involve CAD. Architects use CAD to design buildings and building systems. Interior designers use CAD to show clients redecorating plans. Automobile designers use CAD to create their next concept car. Just about anyone who designs can become more efficient by using CAD. See **Fig. 2-1**.

Math Link

Angles. CAD operators, manual drafters, and people in many other jobs are often faced with drawing decisions that involve the understanding of angles. Traditionally, drafters used two plastic or metal triangles to help them in their drawing: the 30-60-90 degree triangle and the 45-45-90 triangle. They are shown below.

Using the two triangles individually and in combination with each other (see example below), how many ways can you divide a circle, which has 360°?

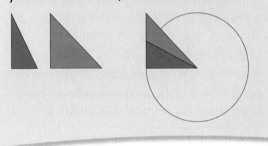

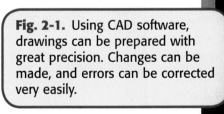

Fig. 2-1. Using CAD software, drawings can be prepared with great precision. Changes can be made, and errors can be corrected very easily.

CAD Advantages

CAD has a number of advantages over traditional drafting. Time is saved because revisions can be made during the design process. For example, suppose an architect is preparing plans for a new home. If the client suddenly wants a different type of window throughout the home, the changes can be made in a matter of seconds.

CAD also produces more accurate drawings than those produced by hand. How are these drawings more accurate?

CAD systems produce accurate drawings because the user can select exact points, such as the ends of lines or the centers of circles. CAD systems use the Cartesian coordinate system, which allows you to plot points on an imaginary grid. Two-dimensional (2D) CAD uses an X axis and a Y axis. Three-dimensional (3D) CAD has an additional Z axis. The computer can mark the exact same spot on an axis every time, which is harder to do using manual drafting. For more on the Cartesian coordinate system, see the Student CD.

CAD systems can draw some shapes automatically. For example, you can enter a "Circle" command, and a circle will appear on the screen. You can make the circle any size and move it to any location. Once a shape has been drawn, you can copy and reuse it rather than drawing a new shape each time. Learn more about shapes and forms in the Student CD's interactive lab on "Elements of Design."

Science Link

Cartesian Coordinates. Scientists, engineers, and CAD operators all must be able to think in three dimensions. These dimensions are length, height, and depth. In the Cartesian coordinate system, length is measured along the horizontal axis (X axis). Height is measured along the vertical axis (Y axis). The third axis is called the Z axis and is depicted as running into and away from the viewer.

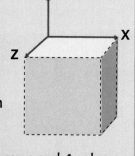

How might you locate a point using this 3-axis scheme? Sketch a block similar to the one shown here, dividing each surface into a grid pattern of perhaps 4 rows and 4 columns to help illustrate your answer.

1. Show your solution to a classmate. Can he or she use your system to correctly position an object within the volume of the box?
2. What would be the coordinates for each of the 8 corners?

Writing Link

Compare and Contrast. Prepare a report about the similarities and differences between CAD and manual drafting. Research drafting books or the Internet as well as this chapter. Are there reasons why someone today would do manual drafting rather than CAD?

Another advantage of CAD is that it improves communication among team members. People working in different locations can view and discuss the design as it takes shape.

Some CAD programs help determine if a design will work as desired. For example, ship designers can use their CAD program to verify that the new hull being designed will float properly. Structural engineers can test the strength of a steel beam before including it in a design. New electronic circuits can also be tested during the design process. See **Fig. 2-2**.

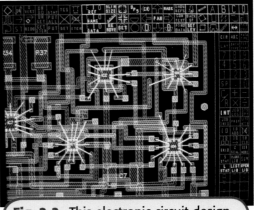

Fig. 2-2. This electronic circuit design is being tested on the computer before it is actually built.

The CAD System

A CAD system consists of hardware and software. The hardware includes the computer itself and devices used to enter information (the input) into the computer, and devices used to display the drawings (output). A keyboard and mouse are common input devices. A monitor is one type of output device. Most designers prefer large monitors.

Printers and plotters are also output devices. Standard laser printers can be used for printing small drawings. For larger drawings, plotters that have ink pens, felt-tip markers, or ink jets are used.

There are many CAD software packages. Most are designed for a specific purpose, such as architectural design or three-dimensional modeling. Add-on programs are also available. For example, an architect may use a standard CAD program for designing the structural elements of a building and an add-on CAD program to design the required electrical systems.

Many of the programs are complex and require training. However a few programs are user-friendly. For example, you could use a simple CAD program to plan an addition to your house.

CAD software can be used for 2D and 3D designing. Three-dimensional design programs have more features than 2D programs. When a 3D design is complete, 2D drawings can be generated very quickly if needed.

Two-Dimensional CAD

While 3D designing is more popular, there are several applications where 2D programs work very well. These 2D programs show width and length, width and height, or length and height. The drawing in **Fig. 2-3** is two-dimensional. Which dimensions does it show?

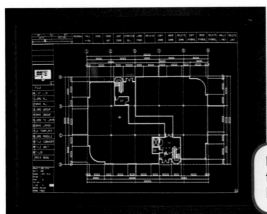

Fig. 2-3. This 2D drawing shows the floor plan for one floor of a large office building.

Fig. 2-4. This vinyl cutter can cut out a very detailed sign using information from 2D CAD drawings.

CAD programs can also create 2D schematic drawings of circuits. The components for the circuit are chosen from a library of parts included with the program. The mouse is used to drop the components in place and wire them up. When the designer thinks the circuit is ready to test, its operation can be simulated on the computer. If the circuit does not work as planned, the designer can make changes and test again. When the circuit is working properly, the design is saved. With a few additional mouse clicks, the layout for the printed circuit board can be produced.

The flat layout of cardboard boxes can also be designed with 2D CAD. These drawings are called developments or nets. Packaging is an important industry since almost everything we buy comes in some kind of package. **Figure 2-5** shows the computer screen image of a package being designed with a program that is used in many schools.

Three-Dimensional CAD

We live in a 3D world, and when we design we usually think in 3D. Three-dimensional CAD programs make it easy to capture ideas in 3D form.

Two-dimensional CAD is used to design packages, signs, and electronic circuits. For each of these tasks, specialized software programs are used.

One common use of 2D CAD is for designing and making vinyl signs. Since the finished products are two-dimensional, it makes sense to design them using a 2D program. Once the sign has been designed, the computer file is sent to the vinyl cutter. Vinyl cutters resemble printers and can produce high-quality products. School versions of vinyl cutters are also available. See **Fig. 2-4**.

Fig. 2-5. This net (development) was drawn as a 2D CAD drawing. Special packaging software allows the designer to add graphics to the package.

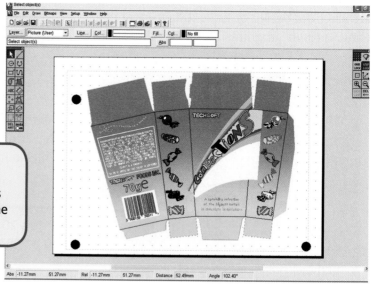

There are three kinds of models in 3D CAD:
• Wireframe models
• Surface models
• Solid models

Wireframe Models

Using wireframes is the simplest form of 3D modeling. You can think of wireframe modeling as "stick-figure" modeling. All you see are lines connected together.

You can use a wireframe to determine the amount of space between objects. However, a collection of lines can often be confusing when seen on the screen. Wireframes also do not provide as much information as surface or solid models. See **Fig. 2-6**.

Surface Models

Imagine taking a piece of paper and shaping it into an object, such as a square box. You have created a surface model. CAD surface models can give more information than wireframes. The actual area of the object can be shown, as well as the CAD coordinates needed for making an object. Surface models can look like the shape of the object when viewed one way. However, since the surface is as thin as paper, viewing the object from another angle might only show a curve or line.

Solid Models

The most realistic kind of modeling uses solid models. **Solid modeling** shows the shape, area, and volume of an object. What's so important about showing volume? The volume might show how much an object can hold of something or how much weight it can endure. If you see a very detailed object in CAD that has actual depth, you are probably looking at a solid model.

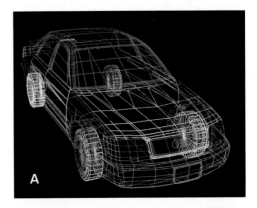

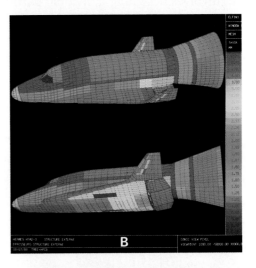

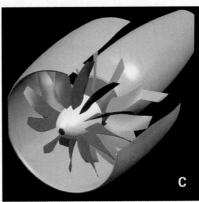

Fig. 2-6. A. Wireframe modeling resembles 3D stick figures. B. Surface modeling is like gluing paper over the wireframe. C. Solid modeling is the most realistic form of 3D modeling.

Using 3D CAD

Now that you've learned about the different kinds of 3D modeling, let's look at how 3D CAD is used in different industries. Three-dimensional CAD can be used in mechanical design, CAM, rapid prototyping, and architectural design.

Mechanical Design

Mechanical design is an engineering activity that involves designing individual parts and assemblies. Three-dimensional CAD has greatly simplified this process. Because of CAD, companies have reduced the time and cost of creating new products.

Fig. 2-7. CAD drawings are printed out with much more accuracy than those that are manually drawn.

In the past, mechanical designers sketched their ideas on paper and then turned their sketches over to drafters. The drafters created the drawings that showed how to make the new part or product. Today most designers begin by making 3D sketches on the computer. As they refine the ideas, they modify the sketches and add details such as dimensions. When the designer is satisfied with the design, the kinds of drawings that the drafter used to prepare manually can be created automatically by the computer. See **Fig. 2-7**.

The design of a part often begins by sketching its length and width. Then depth, the third dimension, is added. Other features of the CAD program, such as the ability to round corners, drill holes, and create smooth curves, are selected from a menu using the mouse. As the part is designed, it appears as a 3D model on the computer. The software saves all of the design information as mathematical data.

The mathematical data created during 3D modeling is very important. When any portion is modified, the program recalculates any other changes that are needed. It works very much like a spreadsheet. If one number is changed, other numbers change, too. The mathematical data also has other important uses, such as making it possible to test the design while it is still in digital form. If changes are needed, they are easy and inexpensive to make.

Individual parts are often combined to make larger parts. These are called assemblies. The parts can be combined on the computer screen to make sure they fit properly. The assembly can be rotated so that it is visible from all angles. When moving parts are involved, the animation feature of the program can simulate the parts in motion. See **Fig. 2-8**.

Rendering is surface shading used to give realism and depth to drawings. The rendering feature included in 3D CAD programs makes it possible to select from hundreds of possible colors, textures, and shading. Attractive and realistic parts may be produced through rendering.

CAD/CAM

CAD/CAM is a process that combines computer-aided design and computer-aided manufacturing. In CAD/CAM, the CAD drawings are input into computers that operate machines. The machines manufacture the actual product. Many different kinds of machines can be controlled this way. Sometimes, CAD/CAM is referred to as CAE, or computer-aided engineering. For more on CAM and its relationship to CAD, see Chapter 5, "Manufacturing Technologies." Also see the interactive lab "Binary Code" on the Student CD to learn about the language computers use to communicate information and control machines.

Rapid Prototyping

Rapid prototyping uses CAD data to create physical models. These models are helpful in sharing ideas. They can also be used to test designs. For example, an aerospace engineer can use rapid prototyping to create an accurate model of a new airfoil (wing). The model airfoil can then be tested in a wind tunnel.

First, a 3D CAD program is used to create a solid model. Then the CAD file is entered into a program that can be used to operate the rapid prototyping machine.

Some rapid prototyping machines use a laser to create solid parts by heating a powdered material such as nylon. Other machines are similar to ink-jet printers. See **Fig. 2-9.** They deposit very thin layers of material and build up layers to create the part. In a technology education classroom, a vinyl cutter is used to cut out layers of material, such as light cardboard, that can be stacked to create the 3D part.

Creating a part using rapid prototyping can take anywhere from several hours to several days. Does this seem like a long time? Actually, rapid prototyping can save time. It helps the designer verify that the part has been properly designed. The prototype can then be tested,

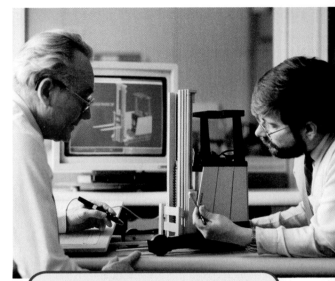

Fig. 2-8. Machines can be tested as 3D models on the computer screen and then compared with the prototype.

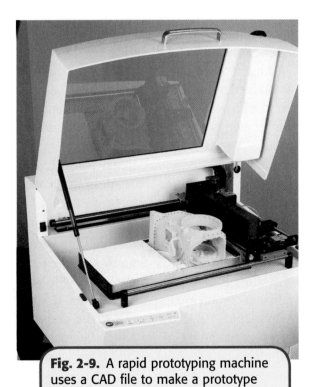

Fig. 2-9. A rapid prototyping machine uses a CAD file to make a prototype layer by layer.

and any flaws can be corrected before producing the actual product.

For some products, the same system used for making prototypes can be used for making the finished product. The process is called "rapid manufacturing." See Chapter 5, "Manufacturing Technologies," for more information.

Architectural Design

For architecture, 3D CAD might be used to design components of a house's interior such as stairs. It might also be used to design the exterior of the house and any landscaping. Architects use CAD both for designing buildings and for designing neighborhoods.

To plan a new building, architects meet with clients to determine their needs. The architect will need to get answers to several important questions:
• Where will the building be constructed?
• How will the building be used?
• How big is the building going to be?
• How many floors does it need?
• What is the client's budget?
The architect will take notes and may make sketches to clarify the client's needs.

Computer-aided design is changing the way all kinds of buildings are designed. In the past, architects spent quite a bit of time preparing hand-drawn sketches. When the client and the architect agreed on a design, drafters created the drawings needed for actual construction. Now many architects use 3D CAD from the very beginning of the design process.

Residential Design. The CAD programs used for residential design are very powerful. Houses can be designed room by room or all at once. CAD is also used by designers who specialize in certain areas, such as kitchen design. Many home improvement stores employ designers who use CAD to design kitchens and bathrooms for new homes and for homeowners interested in remodeling.

Three-dimensional architectural programs have libraries of thousands of related objects.

Rooms are designed by using the mouse to select and arrange walls, windows, and doors. Many additional details such as appliances, cabinets, and fixtures can be added as the design is refined.

Several features of the programs are particularly useful in helping clients visualize their new home. Realistic 3D renderings that resemble full-color photographs show how the interior and exterior of the home will look. See **Fig. 2-10.** Some programs can even show the home in a 3D view without a roof. This helps the architect and client get a feel for the size and relationship of the rooms. Furniture arrangements can also be shown. Clients can know where their furniture will go before the house is even built!

Many 3D CAD programs have a feature that creates animated walkthroughs to let the client experience how it will feel to be in the home. The virtual walkthroughs can be viewed within the program or over the Internet. Visualizing a

Impact of Technology

The Impact of CAD

Until the 1980s, nearly all drafting was done by hand, line by line, object by object. It was, to say the least, a labor intensive effort. Some large companies had CAD programs as early as the 1960s, but they required powerful, expensive computers. The first CAD software for personal computers was introduced in 1982. Today virtually all drafting is done by computer.

Investigating the Impact

CAD software made it possible to produce drawings faster, so a drafter could be more productive. Did this mean that fewer drafters were needed? See if you can answer the following questions:

1. Did many drafters lose their jobs?
2. Were they retrained to use CAD?
3. Did they have to learn other job skills?

design can help save time and money by identifying the need to make changes before final drawings are prepared. For example, clients may realize they need more space in a hall or a different kind of staircase.

When the design of the home is complete, the program can generate everything that a builder will need to obtain a building permit and for construction. The program will create all the working drawings, including floor plans with dimensions, framing diagrams, detail drawings, and drawings needed for the plumbing and electrical systems.

In addition to the drawings, the program can create other documents such as specifications and a bill of materials. This lets the builder know how much material needs to be ordered for framing and other aspects of construction. Builders need this information to estimate construction costs.

Fig. 2-10. Realistic 3D renderings can show interior and exterior views of a house.

Making a CAD Cube

Identify a Need/Define the Problem

Explore CAD commands to create a layout pattern for a paper cube made up of 2-inch squares. The lower left corner of the pattern must be placed at coordinates (1,2).

Gather Information

Most CAD programs have more than one way to create objects such as squares and circles. In some CAD programs, you can create a square using the LINE, POLYGON, or RECTANGLE commands. Find out more about each of these commands. (Hint: Refer to the pull-down menu in your CAD software.) What must you know to use each command?

Develop Possible Solutions

Using a graphing calculator or graph paper, sketch the pattern for the cube. See **Fig. A**. Find the coordinate pairs for each of the six squares.

Materials and Equipment

Select from this list or use your own ideas.

- calculator
- graph paper
- computer with CAD software

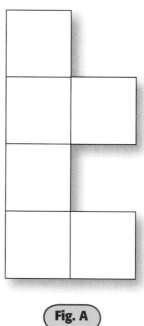

Fig. A

Model a Solution

LINE Command

1. Enter or click on the LINE command.
2. Enter the following coordinate pairs manually, or by using the coordinates on the screen: (1,2), (3,2), (3,4), (1,4), (1,2). (*Hint*: Be sure to do them in order.) In most programs you will not enter the parentheses or spaces.

POLYGON Command

1. Enter or click on the POLYGON command and specify 4 sides.
2. Specify the location of the lower left-hand corner of the polygon by entering the following coordinate pair (1,2).
3. Specify the location of the upper right hand corner of the polygon by entering the following coordinate pair (3,2).

RECTANGLE Command

1. Enter or click on the RECTANGLE command.
2. Enter the coordinate (1,2) for the first corner and (3,4) for the second corner.

Finish entering the rest of the coordinates with the method that you have chosen.

Test and Evaluate the Solution

• Cut out and assemble your cube. Does your pattern form a cube shape like the one in **Fig. B**?

• Which method is the most efficient for drawing the pattern for the cube? Is this the method you thought would be best?

• If you thought another method would work better, explain why. Why did you change your mind?

Refine the Solution

• Did you try all three commands? If not, try the other commands to see how they compare to the command(s) you used.

• Would your cube be easier to assemble if you had tabs? Modify your original pattern by adding tabs to it. Be sure to only add them where they will be needed and not on all sides.

Communicate Your Ideas

Write a brief summary of the process that you used in creating your cube. Describe the benefits of using CAD for this activity.

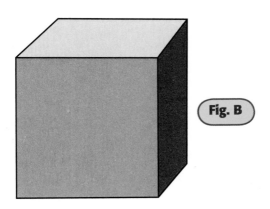

Fig. B

Planning in CAD

Identify a Need/Define the Problem

Architects often create site plans in CAD. Use CAD to create a site plan. The house is to be built on a 35' × 50' lot. The house must sit on the lot exactly as shown in **Fig. A**. The house's upper left corner should be at coordinates (40,25).

Gather Information

What information is usually on a site plan? Your site plan should include such information as dimensions and the total square footage of the lot.

Develop Possible Solutions

Using a calculator and/or graph paper, find the coordinate pairs you will need to create the outline of the house. Prepare a sketch of the property and the site plan for the house. Decide where you will start (with which coordinate pair).

Materials and Equipment

Select from this list or use your own ideas.

• calculator
• graph paper
• computer with CAD software

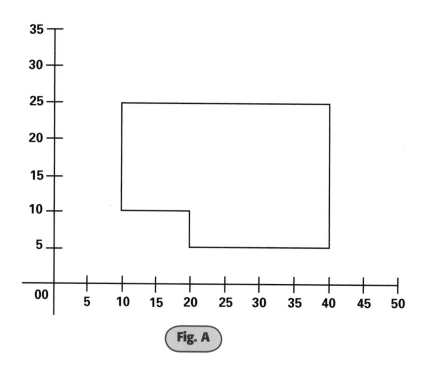

Fig. A

Odd Areas. Calculating the area of a house requires you to multiply the two dimensions, length and width, of the structure. If the building is not a perfect rectangle, you simply divide the shape into smaller rectangles and add the products together. The building pictured below has three rectangles that together equal 750 sq. ft. Can you write a formula that gives you the final square footage?

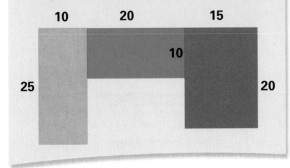

Model a Solution

1. Use the LINE command to create the outline of the house on the site plan. Depending on the CAD software you use, you will have to enter the coordinates for the property itself. Start by entering (0,0) for the lower left hand corner of the property.
2. Next enter (50,35) to set the upper right hand corner.

3. Enter or click on the LINE command. Enter or click the first coordinate pair to start the floor plan.
4. Enter the rest of the coordinate pairs in the order you planned.

Test and Evaluate the Solution

• Does your site plan look like the one shown? If not, how does it differ?
• If you had to create another site plan, what might you do differently? Why?

Refine the Solution

Now that you have created a basic site plan, modify the design of your house to meet the following criteria:
• The house is set back 10' from the road (your X axis).
• There is 5' between the house and the property line.
• The house has a minimum of 800 sq. ft.
• The house must have at least 6 sides.

Communicate Your Ideas

Complete your site plan by adding a border to the drawing as well as a small table that identifies when the drawing was created, who created it, and a title. Be sure to also include such things as dimensions and the total square footage of the structure.

Designing a Portable Music Player

Identify a Need/Define the Problem

Use 3D CAD to design and model a music player that meets the following specifications:
- It must have controls for POWER, PLAY, STOP, PAUSE, FAST FORWARD, REWIND, and VOLUME.
- It must have a 2" × 3" LCD that displays menus and song selections.
- It must have a plug for a USB cable and headphones.

Gather Information

Research other portable music players, such as the Apple iPod™ shown in **Fig. A**. Why are certain design elements used, such as making personal electronics sleek and simple or loud and colorful?

Become familiar with the controls for your 3D CAD software. Many programs will have similar commands and features. Learning the software before you use it is always helpful.

Materials and Equipment
Select from this list or use your own ideas.

- computer with 3D CAD software
- modeling clay
- assorted modeling tools

Develop Possible Solutions

Develop a sketch of what your music player will look like. You will probably need to prepare several design sketches. Be sure to include all the features listed in the specifications.

Fig. A

Model a Solution

1. Start by creating the rough shape of your music player.
2. Add the screen, control buttons, and headphone and USB plugs.
3. Give the music player final touches by adding design elements, such as rounded or squared edges.

Test and Evaluate the Solution

- Does your design have all the specifications required?
- Is it pleasing to look at? Does it have a clear layout?
- Will your player be easy to use?

Refine the Solution

- What other features or design changes can you add to make the product more appealing to consumers?
- If you are satisfied with your 3D design, make a solid model of your design using clay or other materials. This will give you an idea of approximate size and feel. How does the model feel in your hands? Is it the right size? Can the controls be easily reached?

Science Link

Portable Calculators. How is it that a device as small as the iPod can hold hundreds (or even thousands) of songs? The answer is a device you may take for granted—the portable calculator. This device helped jumpstart the continuing need for portability and miniaturization. Research who was responsible for this popular device and how it was incorporated into the iPod. Without the portable calculator, you might not have the ability to listen to so many songs on one device!

Communicate Your Ideas

Create an informational brochure that could go out to potential retailers and customers. Discuss the various features that are included in the device as well as the design and ease of use. Be sure to include your sketches.

Exploring Careers

CAD Drafter

ENTRY LEVEL **TECHNICAL** PROFESSIONAL

CAD drafters prepare technical drawings on a computer. Construction and manufacturing workers use the drawings to build all kinds of products, from houses and offices to toys and machines. CAD drafters create these drawings based on technical information and specifications provided by engineers, architects, or scientists.

Nearly 50% of all CAD drafters work in architectural and engineering fields. The rest work in manufacturing, construction, government, and utilities. Only a few CAD drafters are self-employed.

Qualifications

CAD drafters need knowledge of drafting rules, mathematics, science, and engineering technology. A solid background in computer-aided design and drafting techniques is also needed. Communication and problem-solving skills are important.

An associate's degree is usually required for a job in CAD drafting. Community colleges and technical institutes offer CAD programs that include drafting and mechanical drawing, rules of drafting, math, science, and engineering technology. Specific fields may require more specialized training. Certification for CAD drafters is available from the American Design Drafting Association (ADDA). Drafters must pass a test to become certified.

Drafters should possess mechanical ability and visual aptitude, be able to draw well, perform detailed work neatly and accurately, and enjoy working with people.

Outlook for the Future

The job outlook for CAD drafters is good. Drafters with experience and more advanced education will be more in demand. Most job openings will result from the need to replace drafters who transfer to other jobs or retire.

Organizational Skills

For a CAD drafter, being organized is important for success on the job. Employers want people who can budget their time, keep good records, and store everything in its place. CAD drafters must make neatness a priority and dispose of outdated files and papers.

Researching Careers

Find out about jobs for CAD drafters. Name different job titles for types of drafters. List the salary range for drafters. How much education do drafters need? Do drafters make more money if they have more education? Make a poster using the information you find.

More activities
on Student CD

Key Points

- CAD is the process of designing on a computer.
- CAD has many advantages over traditional drafting.
- While 3D CAD is more popular, 2D CAD is still useful.
- Three kinds of 3D CAD models are wireframe, surface, and solid models.
- CAD is used in mechanical design, CAM, rapid prototyping, and architecture.

Read & Respond

1. What are three advantages of CAD?
2. What is the Cartesian coordinate system?
3. How is 2D CAD still used?
4. What does every CAD system have?
5. Why are wireframe models not used as much as surface or solid models?
6. What might you see by turning a surface model to a certain angle?
7. Define *solid modeling*.
8. What is mechanical design?
9. How long does rapid prototyping take to do?
10. How do architects use CAD?

Think & Apply

1. **Connect.** Describe how CAD can be used in manufacturing and construction.
2. **Extend.** Make a chart that shows the basic drawing commands.
3. **Summarize.** Make a labeled sketch to show the hardware components of a CAD system.

4. **Hypothesize.** If CAD has so many more advantages, why do you think some people continue to draft manually?
5. **Construct.** Create simple 3D solid models of common objects.

TechByte

Coloring Cars with CAD. A team of MIT engineering students working with architect/designer Frank Gehry recently developed a new concept car using a CAD program called CATIA (computer-aided 3D interactive application). Among other features, the car's surface contains thousands of LEDs that enable the driver to change the car's color with the flick of a switch.

CHAPTER 3

Structural Engineering

Objectives

- Explain who is involved in designing structures.
- Name four internal forces that act on structures.
- Describe four different structural materials.
- Explain how tall buildings are designed to resist wind and earthquakes.

Vocabulary

- **structure**
- **civil engineer**
- **structural engineer**
- **architect**
- **force**
- **load**
- **grain**
- **engineered wood**
- **structural member**

Dames Point Bridge in Jacksonville, Florida, was designed by structural engineers.

Activities

- Changing Shapes
- Can It Take the Load?
- Earthquake-Proofing a Structure

What Is a Structure?

Imagine you want to build something. What would you build? A tree house? A dog-house? A shed? Each of these is a structure. A **structure** is something that is constructed, or built. Structures are made by joining parts to meet a certain need or perform a certain task. Structures can be natural or human-made.

Examples of natural structures are a spider web, a bird nest, and a wasp nest. See **Fig. 3-1**. Examples of human-made structures are houses, buildings, and bridges. What similarities do you see between natural structures and human-made structures? See **Fig. 3-2**.

The design of any structure depends on its use. For example, a dam used to control the Colorado River must be strong. A tower that transmits television, radio, or cell phone signals must be tall. Houses are usually designed for comfort and beauty. Factories and office buildings focus more on function.

Fig. 3-1. Intricate structures exist in nature. This spider web provides us with design ideas.

Fig. 3-2. This arch in Arches Natural Park in Moab, Utah, is very similar in design to the human-made St. Louis Arch in St. Louis, Missouri. Can you think of other examples of where human-made structures and natural structures are similar?

Think about why the buildings around your school were built. When they were built, what human needs were they meant to satisfy? Do they still satisfy those needs?

Who Designs Structures?

Civil engineers design and supervise the building of structures that serve the public. Most civil engineers work on roads, water supply systems, sewers, and some public structures. **Structural engineers** are civil engineers whose work focuses more on the mechanics of load-bearing structures. (To learn about a typical work day for a structural engineer, see page 33.) An **architect** is someone who designs buildings and often oversees construction.

On large projects, structural engineers and architects work together. The architect designs a building, and the structural engineer determines what kind of structural system is needed. Structural engineers decide what sizes and types of columns and beams are needed to make the structure stable. They consider the forces that act on a structure. Some questions structural engineers might consider are:

• How many vehicles or pedestrians will travel on a bridge each day? See **Fig. 3-3**.
• How might a skyscraper be affected by high wind?
• What needs to be done to protect a structure being built in an area with frequent earthquakes?

Forces on Structures

Structures must be designed to withstand the forces that will act on them. A **force** is a push or a pull that transfers energy to an object. Forces on a structure can be external or internal. External forces come from outside the structure. They are forces acting *upon* the structure. Internal forces are those that one part of the structure exerts on another part. They are forces acting *within* a structural material.

Some internal forces are tension, compression, shear, and torsion. See **Fig. 3-4**. For example, a material that is being pushed in opposite directions along adjacent planes (bordering surfaces) is being subjected to shear. A material that is being twisted is being subjected to torsion. A material that is being squeezed is compression. Stress from being stretched is tension. For an interactive lab about forces on structures, see the Student CD.

Fig. 3-3. Eight hundred thousand people crowded onto the Golden Gate Bridge in San Francisco, California to celebrate its 50th anniversary. The weight of all those people temporarily flattened the arch of the bridge slightly while they were there.

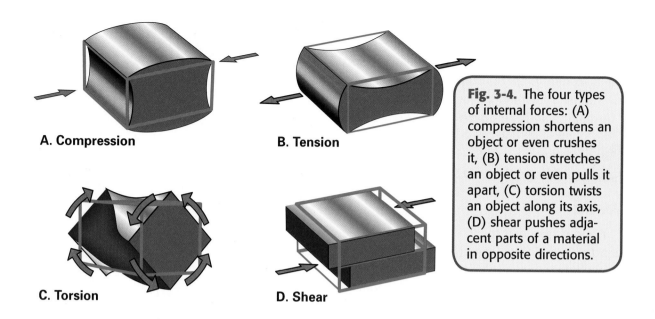

A. Compression

B. Tension

C. Torsion

D. Shear

Fig. 3-4. The four types of internal forces: (A) compression shortens an object or even crushes it, (B) tension stretches an object or even pulls it apart, (C) torsion twists an object along its axis, (D) shear pushes adjacent parts of a material in opposite directions.

A **load** is an external force acting on an object. A load on a structure can be any kind of weight. It can also be force caused by pressure from wind or water. Two types of loads are static and dynamic.

A static load, or dead load, changes slowly or not at all. The materials used to build a structure are part of this kind of load. For example, the bricks in a building are part of its static load.

The twigs in a bird nest are part of its static load. Can you think of other examples?

Dynamic loads, or live loads, move or change. A car crossing a bridge and oil flowing through a pipeline are examples of dynamic loads. See **Fig. 3-5**. Wind blowing on a building and waves pounding on a seashell are also examples of dynamic loads.

Fig. 3-5. A pipeline, such as the Alaskan Pipeline shown here, is sometimes the most economical way to move certain materials.

Structural Materials

Structural engineers have to carefully select the materials to use for a structure. These materials must withstand various loads and forces. The materials will help determine the structure's strength, cost, and appearance. See the "Properties of Materials" interactive lab on the Student CD to learn more.

Wood

Wood was one of the first materials used for construction. In the United States, it is still the primary building material for home construction. Wood is easy to cut, shape, and join to other materials. If properly used and maintained, it can last hundreds of years.

Fig. 3-6. Laminated-veneer wood products include I-joists (left) and beams (right).

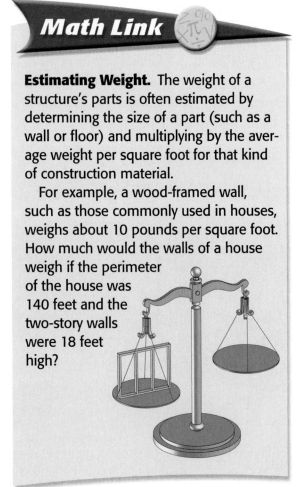

Math Link

Estimating Weight. The weight of a structure's parts is often estimated by determining the size of a part (such as a wall or floor) and multiplying by the average weight per square foot for that kind of construction material.

For example, a wood-framed wall, such as those commonly used in houses, weighs about 10 pounds per square foot. How much would the walls of a house weigh if the perimeter of the house was 140 feet and the two-story walls were 18 feet high?

Wood is different than most other structural materials because it comes in so many varieties. For construction, strong woods are needed. Oak, fir, and certain pines are strong. Balsa wood is very weak. The grain in the wood helps determine its strength. **Grain** is the direction, size, and appearance of wood fibers.

Wood does have some disadvantages. For example, it expands and contracts as temperature and humidity change. Wood can also be damaged by weather and insects.

Engineered Wood

Engineered wood is a composite material. It is made by bonding together wood strands, fibers, or veneers with adhesives. Because it is a manufactured product, engineers can control the material's strength and stability. Engineered wood is formed into panels, laminated beams, and I-joists. See **Fig. 3-6**.

Structural panels, such as plywood, are the most widely used engineered wood products.

Plywood was the first engineered wood product. It is made by gluing together veneers (thin layers of wood). Usually plywood has an odd number of layers, with the grain running in alternating directions. Alternating the grain of the layers helps to give plywood desirable qualities such as strength and dimensional stability. Dimensional stability means that the material does not change size or shape readily because of changes in temperature or humidity. Plywood is much less likely than regular wood to shrink or swell.

A newer kind of structural panel is oriented strand board (OSB). It consists of strands of wood three to four inches in length bonded with adhesives and put together in alternating layers similar to plywood. OSB is less expensive than plywood and is most often used to cover the outside of buildings.

OSB can also be made into I-joists. See **Fig. 3-7**. Typically, these lightweight joists are used for floor construction in homes. They are also available in lengths up to 60 feet for use in larger buildings.

Fig. 3-7. One advantage of I-joists is that they are lightweight and can be made longer than traditional joists.

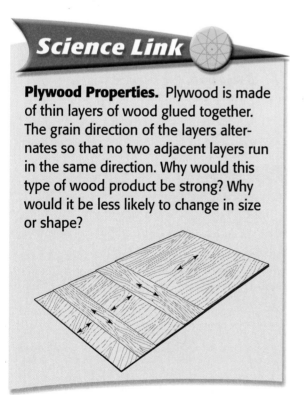

Science Link

Plywood Properties. Plywood is made of thin layers of wood glued together. The grain direction of the layers alternates so that no two adjacent layers run in the same direction. Why would this type of wood product be strong? Why would it be less likely to change in size or shape?

Laminated beams are made by gluing together thin strips of wood to produce large beams that are then cut to desired length. See **Fig. 3-8**. Engineers value laminated beams because they are consistently strong. Such beams can be used to span great distances. You may have seen them in churches or other buildings with large, open areas.

Fig. 3-8. This laminated beam is so strong that it can support the second-story framing.

Fig. 3-9. After the concrete is poured over the reinforcing bars, this bridge pier will be very strong.

Steel

Steel is an important alloy. (You can learn more about alloys in Chapter 4, "Materials Science.") Steel is made from iron and carbon. Other elements are also added to give the steel certain properties. For example, chromium and nickel may be added to resist rust.

Steel can be made in many shapes, including I-beams, pipes, and wires. The steel's type, size, and shape help to determine its strength and where it can be used. It can be joined by riveting, bolting, or welding. Steel is also used in the form of reinforcing bars or wires hidden within concrete. See **Fig. 3-9**.

Concrete

Concrete is made by mixing cement, sand, gravel, and water. The mixture hardens into a strong, stonelike material. We see concrete every day. Roads, building foundations, sidewalks, and patios are all examples. In structural engineering, concrete can be used by itself or along with other materials.

Major advantages of concrete are that it is very strong in compression and that it can be poured into forms to create almost any shape. A disadvantage of concrete is that it is relatively weak in tension. To overcome this weakness, steel bars may be placed within the forms before concrete is poured. The resulting product is called reinforced concrete.

Pre-stressed concrete is different from reinforced concrete. Pre-stressed concrete contains wires that are under tension all the time. The wires can be put under tension before or after the concrete is poured. See **Fig. 3-10**.

Structural Members

Wood, engineered wood, steel, and concrete can be used as structural members. **Structural members** are building materials connected to make the structure's frame. Wood studs, joists, and rafters are structural members typically used to frame houses and small commercial buildings. See **Fig. 3-11**. Steel beams and columns are used in towers, bridges, and large buildings.

Fig. 3-10. Pre-stressed panels were used in building this garage deck.

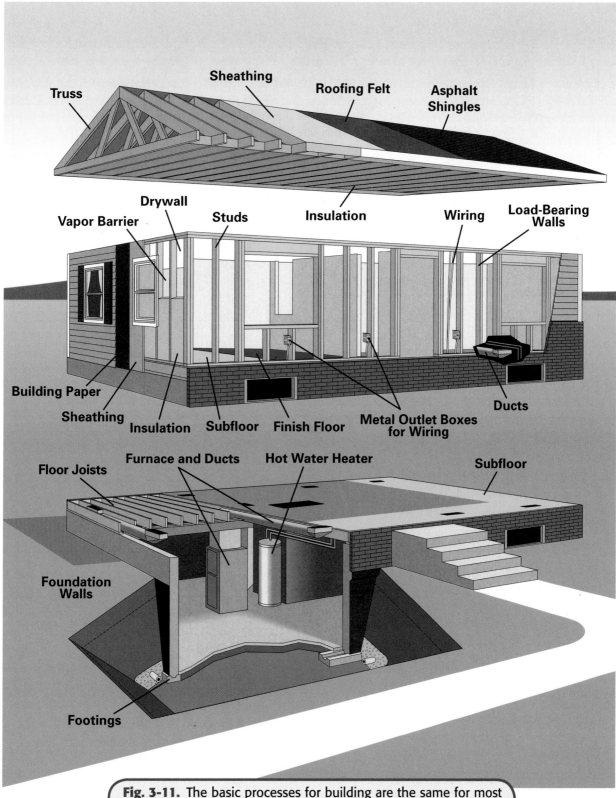

Fig. 3-11. The basic processes for building are the same for most types of structures. Note the parts of this building.

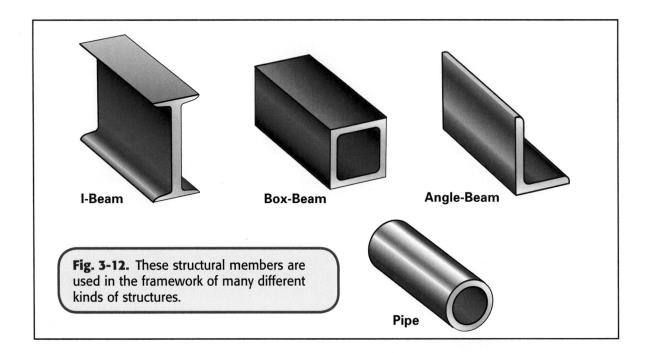

Fig. 3-12. These structural members are used in the framework of many different kinds of structures.

I-Beam

Box-Beam

Angle-Beam

Pipe

Beams are horizontal structural members. The top and bottom surfaces of a beam are subjected to the greatest internal forces. Beams can be strengthened by giving them shapes such as those in **Fig. 3-12**. What other shapes would add strength?

Columns are vertical structural members. Columns must have high compression strength to support the weight of a structure.

Designing Large Structures

Do you often travel over a bridge? Have you ever been in a skyscraper? These are two of the most familiar large structures. Other structures are even larger. One large human-made structure isn't even on Earth—it's in outer space! It is the International Space Station.

Bridges

A bridge extends a roadway across a land obstacle or over water. The design of a bridge depends on the obstacle being crossed and the load the bridge will carry. Planning and carrying out the construction of a bridge is a complex process.

Before construction begins, a great deal of information must be gathered. Soil samples will be taken, and records of wind speed and direction will be studied. Water levels and the speed of the moving water will be considered. Models of the proposed bridge may be tested in a laboratory or by using computer simulations. In many communities, there will be hearings to discuss how the bridge will affect people and the environment. Planning will usually take several years and may cost millions of dollars. See **Fig. 3-13**.

Reading Link

Gathering Information. Read about the types of bridges described on pages 70–71. Prepare a table with one column for each type of bridge. Within each column, list all the information you find about that bridge. Be sure to look at the pictures as well as the words.

Impact of Technology

Bridges: Connecting or Dividing?

Building a bridge sounds like a "no-lose" situation. A bridge can help traffic move efficiently. If there is advance planning, a walkway can be included for walkers and joggers. Construction of such a large structure provides many jobs. Finally, if people do not want to use a bridge, they do not have to.

But there are also drawbacks. Construction will disrupt people's lives, perhaps for a long time. Traffic will no doubt increase through the neighborhood near the bridge (although this could be good for some people, such as owners of restaurants or gas stations). Taxes might be increased or a toll might be charged to help pay for construction or maintenance.

Investigating the Impact

Discuss how a community planning board might analyze the pros and cons of a bridge proposal.

1. What kinds of information would the board need?
2. Whose opinions should be sought?
3. If some people want the bridge and others don't, how should the board determine which action to take?

Fig. 3-13. This civil engineer is planning and testing a bridge by using a computer simulation.

Once approved, construction begins with excavation and the building of a foundation. Keeping water out and preventing collapse of the excavation can be a major challenge, especially in rivers with fast moving currents.

Construction details vary according to the type of bridge being built. See **Fig. 3-14**. For example, arch bridges must be supported temporarily until construction is complete. Beam bridges made of reinforced concrete are usually built in sections and then lifted in place by a crane. Suspension bridges are especially complex because they usually span great distances and require building towers in a river or other body of water. Cable-stayed bridges, like the one shown on page 60, can be built out from the towers, although the spans remain quite flexible until the structure is completed.

Fig. 3-14. There are different types of bridges and each requires a different method of construction.

BEAM BRIDGE
This bridge is built from steel or concrete beams, or girders. The beams provide horizontal supports on which the concrete roadway rests.

TRUSS BRIDGE
This bridge uses trusses to carry the load of the roadway. A truss is a framework formed from triangles. The truss may be placed above or below the roadway. A truss bridge is strong and economical to build. A truss may be used in other bridge designs to add strength.

MOVABLE BRIDGE
This bridge is usually used to span canals and rivers that carry heavy boat traffic. Also called a lift bridge, this bridge has a section of roadway that can be raised to allow large ships to pass.

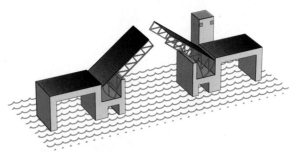

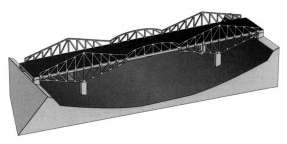

CANTILEVER BRIDGE

A cantilever is a beam that extends from each end of the bridge. A cantilever does not reach all the way across the bridge. Cantilevers are connected in the middle of the bridge by a part called the suspended span.

ARCH BRIDGE

The load on the roadway of this bridge is carried by the arch. The arch is supported at each end by a support called an abutment. Here the roadway is shown above the arch. However, the roadway can also be below the arch.

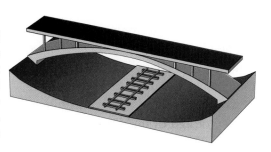

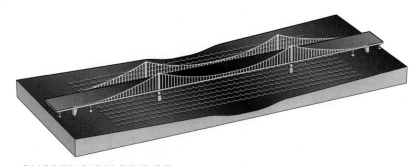

CABLE-STAYED BRIDGE

This bridge supports the roadway by cables that run from towers to the roadway. In this way it is similar to the suspension bridge.

SUSPENSION BRIDGE

Tall towers on both sides of the roadway support the main cables. These cables run the entire length of a suspension bridge. These main cables are anchored in the concrete at each end. Smaller cables are suspended from the main cables. These suspended cables support the roadway.

Skyscrapers

Some of the technologies used to construct modern skyscrapers date back to the time when towers were built in Europe centuries ago. The Leaning Tower of Pisa, completed in 1178, is one of the world's most famous towers. The Eiffel Tower, completed in 1889, is made of iron. It was the world's largest freestanding structure when built and remains a popular tourist attraction today.

In the United States, beginning in the mid-1800s, taller buildings became practical because of the invention of the elevator and the use of metal framing. By the mid-1900s, new materials and structural engineering techniques encouraged the design of buildings with more than 100 floors. The world's tallest skyscraper is currently the Taipei 101 in Taipei, Taiwan. It has 101 stories and is 1,670 feet tall. However, several dozen taller structures are in various stages of planning.

One challenge to constructing tall buildings is gravity. As the height of the building increases, the total force on everything below continues to increase. One of the most important developments for skyscraper construction was the production of high-quality steel beams and columns.

Beams and columns are welded, bolted, or riveted together at each floor. This continues until the finished height is reached and the superstructure is formed. See **Fig. 3-15**. Almost all the weight is transferred to the vertical columns. This force is spread out at the base of the building in the substructure.

A large variety of materials are used to complete the skyscraper. Floors are made of poured concrete. The exterior of the building can be made from glass, brick, plastic, or a combination of materials.

After the skyscraper's exterior is complete, the builders must focus on the interior. Elevators will be added. Complex utility systems will be required, as will modern communication systems such as fiber optics. Fire safety is a major concern since tall buildings are difficult to evacuate. Sprinkler systems, alarm systems, and specially designed stairwells are among the things that will need to be installed.

Fig. 3-15. The steel beams for this skyscraper are held together by welding them.

Fig. 3-16. The International Space Station (ISS) is one of the most ambitious building projects ever attempted. It is a tremendous research laboratory for scientists and engineers.

The International Space Station

Structural engineering isn't just for structures on Earth. The International Space Station is the most complex structure ever built. See **Fig. 3-16**. Sixteen countries are participating in its construction. The ISS is also the largest structure in space. It is 356 feet (108 meters) across and 290 feet (73 meters) long. The inside of the ISS is about the size of a large airplane's passenger cabin.

The station is in orbit at an altitude of 250 miles above Earth. This orbit allows the participating nations to reach the station with rocket-powered launch vehicles.

The ISS is equipped with research modules and laboratories enabling engineers and scientists to conduct experiments in a microgravity environment. See **Fig. 3-17**. Experiments conducted on the ISS will explore cell growth, manufacturing in microgravity, and the impacts of extended space stays on the human body. The outcomes of these experiments may lead to cures for diseases, produce new materials with special properties, and improve the quality of life on Earth.

Fig. 3-17. U.S. astronauts and Russian cosmonauts work side-by-side in the International Space Station.

Structurally Sound

You have already read how forces like gravity can impact how a structure is designed. The larger the structure, the more factors affect it. To make a structure sound, or stable, structural engineers must consider the surface the structure rests on, wind resistance, and earthquake resistance.

Soil

Soil is a major concern to engineers designing large structures. They need to know that the soil beneath the foundation of a structure will provide the support needed to prevent sinking or tilting. Engineers will adjust the foundation design according to the kind of soil at the construction site. Sometimes soil at the site needs to be replaced.

Wind Resistance

A tall building may sway several feet on a windy day. It's important that the movement not harm the structure or the people inside. See **Fig. 3-18.**

Fig. 3-18. On a windy day in Chicago, the Sears Tower may sway as much as one foot. Tall buildings like this are designed to withstand high winds.

Fig. 3-19. A shake table is used to simulate the effect of an earthquake on a scale model of an adobe house.

One good way to control the impact of wind is make sure that the structure is tightly constructed. Sturdy concrete and securely bolted steel cores help reduce swaying. Some modern skyscrapers use a computer-controlled system to monitor how wind sways the building. The information gathered by the computer system is used to shift a large concrete weight to compensate for the movement created by the wind.

Earthquake Resistance

When the ground shakes below a tall building, the structure will shake. Whether it will break depends on a number of factors, such as the magnitude and location of the earthquake, the height of the structure, and the flexibility of the structure.

To reduce the damage caused by earthquakes, communities now ban construction along known fault lines. In addition, engineers have learned how to help buildings survive earthquakes. An effective way to do this is to isolate the motion of the building from the motion of the ground. This can be done by installing flexible layers of rubber and other materials on the foundation. The base can also be isolated by creating a sliding surface so that the building does not break apart as the ground moves. See **Fig. 3-19**.

Although it is not possible to create an earthquake-proof building, much progress has been made in designing buildings that will suffer minimal damage during major earthquakes. Other structures, such as bridges and power plants, have also been modified to reduce earthquake damage.

Changing Shapes

Identify a Need/Define the Problem

Structural engineers must determine the best design for their project. An important step is to test and understand how various structural shapes work under loads. For this activity, determine how the strength of a material can be affected by changing its shape.

Gather Information

Research various structural shapes and take notes on their various designs. Some examples have been provided for you. See **Fig. A**.

Materials and Equipment

Select from this list or use your own ideas.

- 4" x 6" index cards
- masking tape
- set of standard weights
- ruler
- string

Develop Possible Solutions

Consider what shapes you will make. Which shapes seem the strongest? Why?

Model a Solution

1. Prepare a table that will allow you to record the results of your tests. Decide on the items that should be represented in the table.
2. Create several different structural shapes using the index cards and masking tape. Make at least two of each shape.

Box-Beam

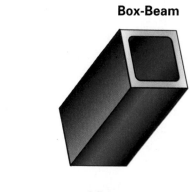

Fig. A

I-Beam   **Angle-Beam**   **Pipe**

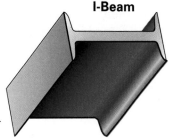   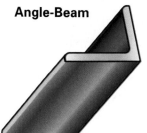   

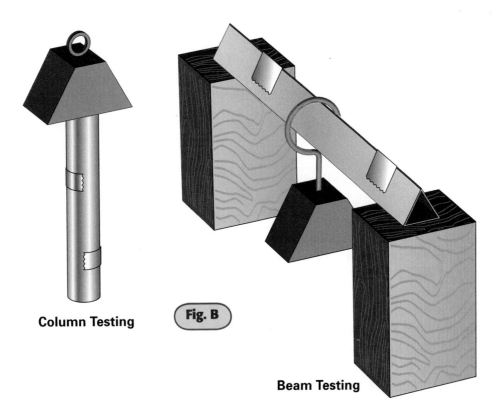

Column Testing (Fig. B) **Beam Testing**

Test and Evaluate the Solution

Figure B shows setups for beam and column testing.

- Do beam testing on one set of shapes. Find the force required to buckle the beams. Record your results.
- Use similar tests for columns using the second set of shapes. Record your results.
- Which shapes were the strongest during the beam testing? Which shapes were the strongest during the column testing? Which shapes performed well in both tests?

Refine the Solution

- Make smaller structural shapes.
- Perform the same beam and column tests.
- Did reducing the size of the beams and columns affect strength?

Communicate Your Ideas

Create a handout of your testing that includes the steps of your testing procedure and the data recorded in your table.

Math Link

Geometric Shapes. In geometry, you study many different shapes, such as circles, squares, triangles, spheres, prisms, trapezoids, and cubes. Research these shapes. Identify which geometric shapes might work better in designing structures and explain why this is so.

Can It Take the Load?

Identify a Need/Define the Problem

Design a small structure capable of supporting as much weight as possible. The structure must meet the following specifications:
* The structure cannot exceed the weight limit set by your teacher (e.g., 18 ounces).
* The structure should be no larger than the size specified by your instructor (e.g., must fit within a 6" cube).
* The structure should have a flat surface on top and bottom to allow for easy testing.

Gather Information

Research the kinds of weight that a structure must be able to hold. Examples include the weight of its own building materials, people, furniture, etc. Also research the kinds of forces acting against a structure, such as wind and gravity.

Develop Possible Solutions

Design and draw small structures that could be made from the materials listed. You will probably need to prepare several design sketches. One possible design is shown in **Fig. A**. This design is presented only to give you a general idea of what will be needed. Don't copy this design.

Materials and Equipment

Select from this list or use your own ideas.

* thin strips of wood
* glue
* set of weights
* video camera

Safety Alert

Look up "Safety Data Sheets" on the Student CD and prepare a data sheet for this activity. As you work on the activity, be sure to follow all safety rules.

Model a Solution

1. Select the design you think will be most effective.
2. Cut the wood pieces to length and assemble the structure.
3. Weigh the structure.

Test and Evaluate the Solution

Test the structure to determine how much weight it will hold. Your teacher will assist you. Compute the weight-to-strength ratio. The more weight a structure can hold (per unit of weight of the structure), the stronger and more efficient its design.

After you finish testing the structure, evaluate its design.

* Is the structure within the specified weight limit?
* Did the structure have a weight-to-strength ratio as good as or better than the average of the other structures built by the class?

* Was it possible to determine exactly where the structure failed as the load was applied? (Recording the test with a video camera and playing the scene back frame by frame will help provide the answer.)

Refine the Solution

* Did your design perform as expected? If it did not, create a new design for the structure and test again.
* If you did a second test, where did the structure fail this time?
* Was the previous failure point better or worse?

Communicate Your Ideas

Create a diagram of your structure as it appeared before testing. Identify the areas of your structure where your design failed. Be sure to include the weight-to-strength ratio, as well as any refinements you have developed.

Weight

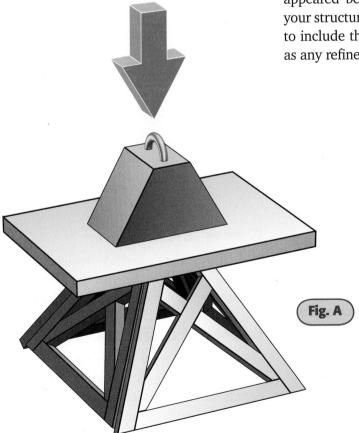

Fig. A

Earthquake-Proofing a Structure

Identify a Need/Define the Problem

Design a small structure capable of resisting the force of a simulated earthquake. The structure must meet the following specifications:

- The structure cannot exceed the maximum amount of material (e.g., 5 sheets of copy paper, 10 strips of wood).
- The structure must meet or exceed the minimum height limit set by your teacher (e.g., 18 inches, 6 inches per floor).
- The base of the structure should be no larger than the size specified by your teacher.
- The structure must hold a minimum amount of weight (e.g., 12 ounces).

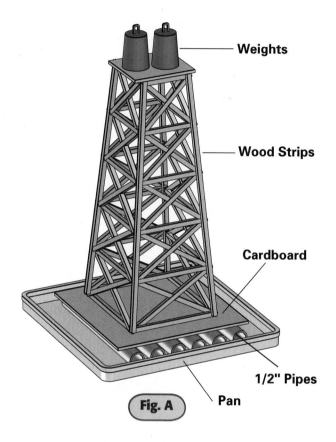

Weights

Wood Strips

Cardboard

1/2" Pipes

Pan

Fig. A

Materials and Equipment

Select from this list or use your own ideas.

- thin strips of wood (balsa or bass)
- copy paper
- glue
- cardboard
- pieces of ½" pipe
- set of standard weights
- video camera

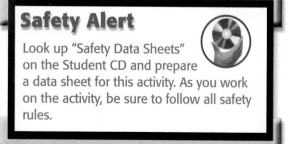

Safety Alert

Look up "Safety Data Sheets" on the Student CD and prepare a data sheet for this activity. As you work on the activity, be sure to follow all safety rules.

Gather Information

Research some earthquake-resistant structures and take notes on how they work. Be sure to observe the forces acting on structures.

Develop Possible Solutions

Design and draw structures that will meet the required criteria and constraints. Include details such as the design of the frame (made from wood strips) and placement of the exterior walls (copy paper). You will probably need to prepare several design sketches. See **Fig. A**.

Model a Solution

1. Select the design you think will be most effective.
2. Cut the wood pieces to length and assemble the structure.
3. Measure the height and the base of your structure.
4. Attach your structure to a piece of cardboard that will simulate the earth's crust.
5. Place your structure and cardboard on pieces of ½" pipe inside a tray to create a rolling surface. Shaking the pan will simulate the earthquake.

Test and Evaluate the Solution

- Test your structure to determine if it will withstand the vibrations. Your instructor will assist you.
- Add a selected amount of weight (e.g., 12 ounces) and test again. Keep adding weight and testing until your structure fails.
- Was your structure able to withstand the initial vibrations?
- Was it possible to determine how the structure failed as the earthquake was simulated? (Recording the test with a video camera and playing the scene back frame by frame will help provide the answer.)

Refine the Solution

- Was your building able to withstand the strength of the vibrations?
- What happened as you increased the weight on the structure?
- Design elements that could better absorb the vibrations caused by the earthquake. If time permits, build another structure that incorporates these redesigns.

Science Link

House of Cards. Try building a house out of playing cards. First, try to make a house without bending any of the cards. If the first story of your house holds, add a second story. How high can you go? Place a small object on the top of your house. Does the house hold the weight?

Make another house of cards using cards that are bent to add support. Can you build a taller house? Does the house hold a larger object? The forces acting on a house of playing cards are the same forces that act on a building. Without earthquake-proofing a structure, a tall building can collapse as easily as a house of cards.

Communicate Your Ideas

Create a portfolio of your work. Include information from all the steps of this activity. Include your redesigned work and a summary of your findings. Also be sure to include all your sketches and designs. Present your portfolio to the class.

Exploring Careers

Steelworker

ENTRY LEVEL | TECHNICAL | PROFESSIONAL

Steelworkers place and install steel girders, columns, and other construction materials to build bridges, buildings, and other structures. Sometimes they are called ironworkers because iron, as well as steel, is used.

Steelworkers read drawings and follow instructions from structural engineers who design the structure. The instructions tell how to connect the steel pieces. Steelworkers may be either reinforcing iron and rebar workers or ornamental workers. Reinforcing iron and rebar workers set reinforcing bars (rebar) in forms. They fasten the bars together using wire. Concrete is then poured over the bars. Ornamental workers install elevator shafts, nonstructural walls and window frames, and stairs after the structural part of a building is finished. They fit pieces together using bolts or by welding or soldering.

Qualifications

A steelworker usually works on the job as an apprentice for three or four years while taking evening classes. The classes are in subjects such as blueprint reading and the basics of constructing and assembling structures. Workers must be at least 18 years old, and most employers want workers to have a high school diploma. High school classes that would be helpful include general math, mechanical drawing, and metalworking or construction.

Because materials used on the job are heavy, workers must be strong and healthy. They also need good eyesight, depth perception, agility, and balance to work safely at heights. Workers should not be afraid of heights.

Outlook for the Future

The job outlook for steel and iron workers is good. The need for workers will depend on the health of the construction industry. New workers will be needed to replace workers who take other jobs or retire.

Teamwork

It is important to have good teamwork skills. Members of a successful team respect each other, keep an open mind, and tolerate others' opinions, even if they are different from their own. They are considerate of others' feelings, compliment team members, and encourage those who may not seem as committed.

Researching Careers

Interview a steelworker who helps build structures. How long has this person been a steelworker? What types of buildings has he or she helped build? Ask the steelworker to describe a typical day on the job. Prepare a five-minute talk to give in class.

**More activities
on Student CD**

Key Points

- A structure is something that is constructed, or built.
- Structures are affected by both internal and external forces.
- Engineers must decide what structural materials to use in construction based on the materials' properties.
- Bridges and skyscrapers are examples of large structures that structural engineers design.
- The International Space Station is our most complex structure.
- Structures can be designed to resist forces such as wind and earthquakes.

Read & Respond

1. Who is involved in designing structures?
2. Name four structural materials.
3. Identify the seven types of bridges.
4. How can steel's properties be changed?
5. How is engineered wood made?
6. What are the ingredients in concrete?
7. What was one of the most important developments for skyscraper construction?
8. Name two ways to help structures resist wind.
9. How can buildings be isolated from the ground for protection against damaging earthquakes?
10. Describe the difference between a static load and a dynamic load.

Think & Apply

1. **Hypothesize.** Why do you think the Leaning Tower of Pisa did not collapse once it started leaning? Research the structure and form a hypothesis.
2. **Propose.** Communities often have hearings before approving construction of a new bridge. Create a proposal for a new bridge, addressing social, economical, and environmental concerns.

3. **Relate.** Describe functions for five different kinds of structures in your community.
4. **Summarize.** Describe the forces affecting structures.
5. **Assess.** Some structures need continual maintenance. Determine why some buildings are updated every few years, while others stay the same for long periods of time.

TechByte

What Can Concrete Do? Engineers have developed concrete embedded with optical fibers. The concrete glows on each side when daylight filters through it. Concrete walls can now allow sunlight to shine through them. Other engineers have developed a self-cleaning concrete that prevents buildings from turning darker over time. This kind of concrete can even help clean the surrounding air!

CHAPTER 4

Materials Science

Objectives

- Define *materials science*.
- Describe five material properties.
- Identify four groups of materials.
- Explain the difference between renewable and nonrenewable resources.

Vocabulary

- **materials science**
- **matter**
- **element**
- **material property**
- **alloy**
- **ferrous**
- **polymer**
- **composite**

Strands of special fiber are drawn into this production line to make synthetic fabric.

Activities

- Fighting Fatigue
- Protecting Your Food
- Recycling Materials

What Is Materials Science?

What does your shirt look like after a hard day at school? Is it all crumpled from the day's activities? Thanks to durable press fabrics, the answer could be no.

Durable press fabrics are made to retain their shape and resist wrinkling. See **Fig. 4-1**. Durable press is sometimes referred to as permanent press. These fabrics are often a blend of polyester and cotton threads. The threads are heat-treated to activate the chemicals applied to them.

Durable press fabrics were created to satisfy a want. People's busy lives no longer provide time for the chore of constantly ironing wrinkled clothing. So where did durable press fabrics come from?

These fabrics are a result of materials science. **Materials science** is the study of the properties and applications of materials. Materials science is important to many engineering fields, such as electronic, aerospace, mechanical, and structural engineering. We now see evidence of materials science every day. See **Fig. 4-2**.

How did we begin studying materials? During earlier times, the selection of materials was limited by what nature made available.

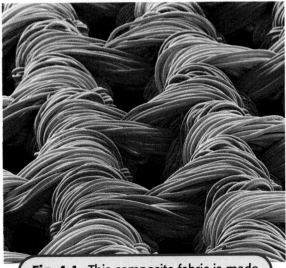

Fig. 4-1. This composite fabric is made from knitted nylon and polyester. The result is a highly elastic material.

As improved products were developed to meet new needs, the materials used to create these products became more complex. See **Fig. 4-3** on page 86. Steel, plastic, and particleboard are materials that have been developed because their special properties meet special needs.

As people experimented with combining materials to create new and more useful resources, the field of materials science was born.

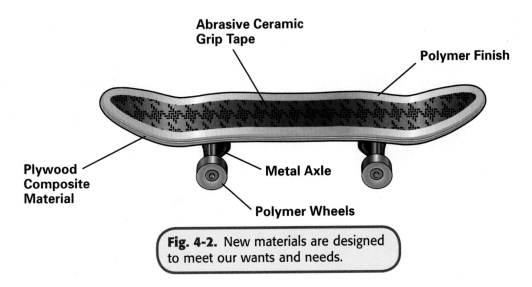

Abrasive Ceramic Grip Tape

Polymer Finish

Plywood Composite Material

Metal Axle

Polymer Wheels

Fig. 4-2. New materials are designed to meet our wants and needs.

Fig. 4-3. As more resources became available, more complex tools were made.

Materials science engineers are responsible for developing and testing new materials. See **Fig. 4-4.** These materials meet specific manufacturing and construction requirements. Wrinkle-free fabrics are one example. To understand how to create these materials, engineers must first know about matter.

Matter

All materials are made of matter. **Matter** is anything that occupies space and has mass. Most matter is a solid, a liquid, or a gas. Solid matter has a definite size, shape, and volume. Wood is an example of a solid material. Liquid matter doesn't have a specific shape but still has volume. The most common liquid on the planet is water. Matter in a gaseous state has no shape or volume of its own. A good example of a gas would be the air you breathe.

All matter is made of atoms. Matter made from only one kind of atom is an **element.** There are over a hundred identified elements. Most of them occur in nature, but a few are made in laboratories. Common elements include gold, silver, iron, and hydrogen.

Most atoms are made up of protons, neutrons, and electrons. See **Fig. 4-5.** The hydrogen atom is an exception. It has only a proton and an electron.

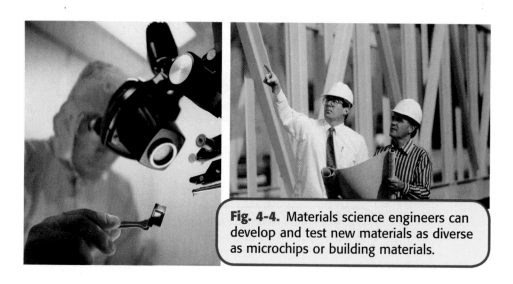

Fig. 4-4. Materials science engineers can develop and test new materials as diverse as microchips or building materials.

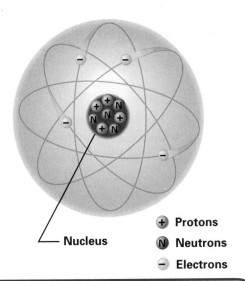

Fig. 4-5. Atoms are the building blocks of all matter. Atoms have a nucleus that contains protons and neutrons. Negatively charged particles called electrons move around the nucleus.

Protons
Neutrons
Electrons
Nucleus

Both protons and neutrons are found in an atom's nucleus. Protons are positively charged. Neutrons have no charge. The number of protons determines an element's identity. All elements have a different number of protons. The Periodic Table of Elements lists the elements according to their number of protons. A portion of the Periodic Table is shown in **Fig. 4-6**.

The third kind of particle in an atom, the electron, has a negative charge and is responsible for different atoms bonding together. Electrons are arranged around the nucleus in levels. Electrons with the lowest energy are found in levels closer to the nucleus. Electrons that are very active are found in the outer higher-energy levels.

Atoms will bond with each other by losing, gaining, or sharing electrons in their outer shell. This process is called chemical bonding.

Fig. 4-6. The Periodic Table gives a great deal of information about the elements. Why is mercury a different color than the other elements shown?

	8B			1B	2B
	8	9	10	11	12
	Iron 26 **Fe** 55.847	Cobalt 27 **Co** 58.933	Nickel 28 **Ni** 58.693	Copper 29 **Cu** 63.546	Zinc 30 **Zn** 65.39
	Rumthenium 44 **Ru** 101.07	Rhodium 45 **Rh** 102.906	Palladium 46 **Pd** 106.42	Silver 47 **Ag** 107.868	Cadmium 48 **Cd** 112.411
	Osmium 76 **Os** 190.2	Iridium 77 **Ir** 192.22	Platinum 78 **Pt** 195.08	Gold 79 **Au** 196.967	Mercury 80 **HG** 200.59

When the positively charged nucleus of one atom attracts the negatively charged electrons of another atom or when two atoms share electrons, tiny particles of a material, or molecules, are formed. If the atoms are from different elements, the resulting molecule is a compound. See **Fig. 4-7**.

You are probably familiar with many everyday compounds. Water, salt, and sugar are compounds created when different kinds of atoms join. Have you ever heard water called H_2O? This refers to the two hydrogen (H) atoms and the one oxygen (O) atom that make up water.

Not all matter combines by forming compounds. When two or more substances mix but do not bond chemically, they are mixtures. Saltwater is a mixture of salt and water. If you boil the water, salt will still be left over.

A solution is a mixture in which one substance dissolves into another. When you mix cocoa powder into hot water, the powder dissolves in the water and a solution forms.

Math Link

Atomic Size. How small is an atom? Here's a way to find out! If you start with a strip of paper about 11 inches long (the size of copy paper) and cut it in half 31 times, you will end up with a speck of paper about the size of an atom. Try to do it mathematically by completing the list below.

① $11 \div 2 = 5.5$
② $5.5 \div 2 = 2.75$
③ $2.75 \div 2 = 1.375$

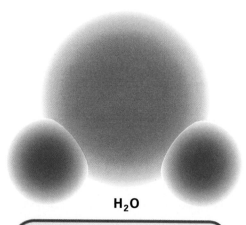

H_2O

Fig. 4-7. A water molecule is composed of two atoms of hydrogen and one atom of oxygen.

Material Properties

A **material property** describes how a material reacts under certain conditions, such as when it is heated or weight is put on it. Product designers must select materials with the right properties. **Figure 4-8** shows some characteristics that should be considered when selecting a material for a product. See the Student CD for more about "Properties of Materials."

Materials can possess many other properties, such as being magnetic or being unaffected by chemicals. What do you think a material with excellent optical properties would look like?

Engineers often must overcome a material's properties for a specific use. For example, concrete is a very heavy material. Can you imagine developing concrete that floats? Civil engineering students from around the country participate in an annual concrete canoe race. The students are challenged to develop concrete that is shaped into a canoe but will float. The students create a composite material consisting mainly of cement, fly ash, silica, and glass air bubbles. When well designed, the canoes are sleek and float without problems. See **Fig. 4-9**.

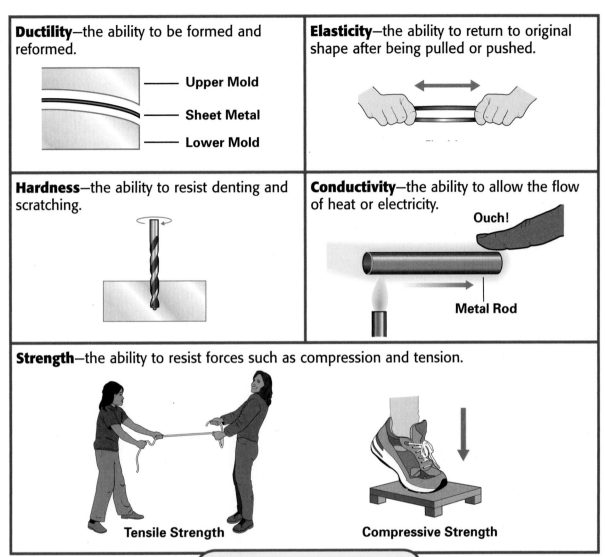

Ductility—the ability to be formed and reformed.

Upper Mold
Sheet Metal
Lower Mold

Elasticity—the ability to return to original shape after being pulled or pushed.

Hardness—the ability to resist denting and scratching.

Conductivity—the ability to allow the flow of heat or electricity.

Ouch!

Metal Rod

Strength—the ability to resist forces such as compression and tension.

Tensile Strength

Compressive Strength

Fig. 4-8. Properties of Materials.

Fig. 4-9 These students are in a unique type of canoe race. All of the canoes in this race must be made out of concrete.

Density. Density is a measure of how much mass (material) is contained in a given volume (amount of space). When scientists try to calculate density, they use the formula:

$$density = mass \div volume$$

The answer is usually expressed as weight per cubic unit, such as pounds per cubic foot or kilograms per cubic meter.

Put simply, if mass is a measure of how much "stuff" there is in an object, density is a measure of how tightly that "stuff" is packed together.

ALUMINUM FOIL

An object placed in liquid will float if it is less dense than the liquid it displaces. If it is denser, it will sink. Using a piece of aluminum foil and a container of water, design a demonstration to prove this.

Fig. 4-10. Alloys are created when metals are combined. Why is bronze a better choice for this statue?

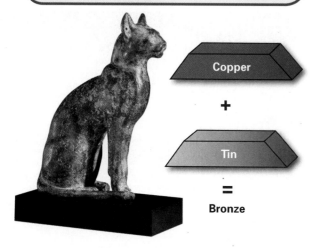

Copper

+

Tin

=

Bronze

Classification of Materials

Most materials can be arranged into four groups:

• Metals
• Ceramics
• Polymers
• Composites

Metals

Metals are often used in products where strength and stiffness are needed. The majority of elements fall into the metals category. Iron, copper, and tin are elemental metals.

The combination of two or more elements, at least one of which is metal, forms an **alloy**. For example, when the elements copper and tin are combined, the alloy bronze is created. Bronze is stronger and has a higher melting point than either of the two elements from which it is made. See **Fig. 4-10**. Materials engineers create alloys to combine the best properties of many metals into one material. So far, more than 70,000 alloys of metal have been created.

Metals can be divided into two families: ferrous and nonferrous metals. The word **ferrous** means "containing iron." Ferrous comes from "ferrum," which is the Latin name for iron.

Steel is an example of a ferrous alloy. It is made of the elements carbon, iron, and oxygen. By changing the amount and type of each element, we can change the properties of the steel. See **Fig. 4-11**. For example, if we increase the percentage of carbon, the steel gets harder. When the amount of carbon is increased to one percent, we create high-carbon steel. High-carbon steel is used to make cutting tools such as saw blades. If we add chromium and nickel to the formula, we create stainless steel.

Nonferrous metals and alloys contain no iron. Aluminum, copper, and tin are nonferrous metallic elements. When we mix the elements copper and zinc, we get a new nonferrous alloy

Fig. 4-11. This skyscraper is being built from steel beams and girders because it will make the structure very strong.

Fig. 4-12. The ceramic refractory bricks inside this pottery kiln are used because they will resist tremendously high temperatures without melting or cracking.

called brass. Brass is harder than copper or zinc and does not rust. Since it does not rust, brass is used where the times in question will be exposed to water, such as sink handles, faucets, and objects intended for boating.

Ceramics

When the word "ceramics" is mentioned, most people think of clay pots and bricks. However, ceramic materials have many other applications. Did you know that glass, porcelain, and plaster are ceramics?

Ceramics are made from silicates. Silica, bauxite, clay, feldspar, and talc are silicate-based minerals used to make ceramic products. Some of these products include glass, cement, plaster, porcelain, abrasives, and refractory bricks. Refractory bricks can resist extremely high temperatures without melting or cracking. See **Fig. 4-12**.

Ceramic materials have many special properties. They resist being eaten away by harsh chemicals. They also resist the flow of electricity and can absorb a great amount of heat.

Materials engineers have created a variety of new uses for this ancient material. Ceramic semiconductors, acting as electronic switches, are used to control small amounts of electricity in computer circuits. These switches allow computers to work at very high speeds. See **Fig. 4-13**.

Fig. 4-13. This silicon chip will resist the flow of electricity where it is not intended, and it will also absorb the heat generated within the chip.

Impact of Technology

Plastics

The use of plastics in our society is growing year by year. Why? Plastic is lightweight and relatively inexpensive. It is easy to fabricate into various forms and products. Plastic is moisture resistant and durable, and it has good thermal and electrical resistance properties. You can probably think of other reasons.

Yet all technology has both good and bad effects. One of our most pressing worries is what to do with plastic after it has been used. Most types of plastic (there are dozens) do not easily biodegrade, or break down, in nature. The unwanted plastic poses a serious disposal problem. What should be done with it?

Investigating the Impact

Currently only about a quarter of all plastics are recycled. Research the problem and explore solutions. Prepare a PowerPoint presentation of your findings.

1. What might encourage people to do more recycling?
2. Are there any biodegradable plastics?
3. What materials could be used in place of plastics? What impacts might those materials have?

Polymers

Molecules can sometimes form a chain. Each link in the chain is a monomer. The whole chain of molecules is called a **polymer**. Polymers can be strong, lightweight, durable, and very flexible. These properties result from the chainlike chemical structure of the material.

Natural Polymers. Most of the polymers we use today are synthetic (made by people), but polymers do exist in nature. Cotton, silk, wool, and wood are all natural polymers.

Synthetic Polymers. Plastic is a synthetic polymer. You cannot find most plastics in nature, although some plastics are made from plant fibers. Monomers are combined in different chain formations to create plastics with different properties. See **Fig. 4-14**.

Plastics are used in adhesives, food containers, furniture, rugs, bullet-proof vests, self-

Fig. 4-14. These different kinds of plastics are created by combining the monomers in different chainlike formations.

sealing tires, and thousands of other products. Plastics are even used as substitutes for human tissue, bones, and arteries.

Synthetic polymers are made from petrochemicals. Petrochemicals come from crude oil or natural gas. See **Fig. 4-15**. Oil is used to make gasoline, but it is also used to make plastics. Petrochemicals are used to produce fibers like nylon or polyester. Check your shirt label. Are any of these fibers used in the fabric?

Composites

When two or more separate materials are combined or mixed, a new material called a **composite** is formed. Many times composites are a mix of two or more families of materials. For example, to make graphite reinforced plastic (GRP), graphite and a polymer are combined. The result is a super-strong lightweight material used in fishing rods, skis, tennis rackets, and many other products.

Writing Link

Classification. Make a list of 20 items in your home. Next to each item, write whether it is made of metal, ceramics, polymers, composites, or several of these materials. Below are some items to help you get started.
Table
Dishes
Pillows
Faucets
Lamp

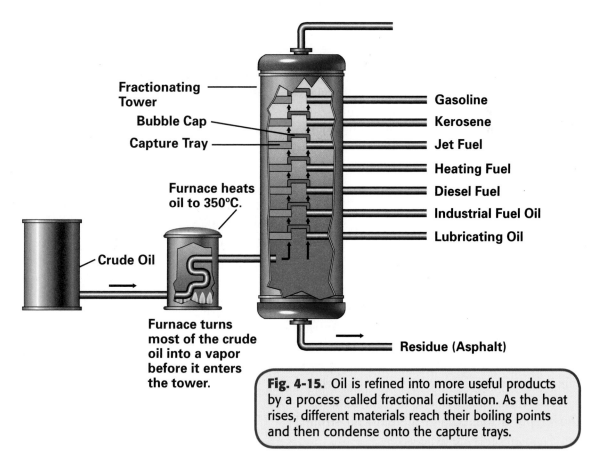

Fractionating Tower
Bubble Cap
Capture Tray

Gasoline
Kerosene
Jet Fuel
Heating Fuel
Diesel Fuel
Industrial Fuel Oil
Lubricating Oil

Furnace heats oil to 350°C.

Crude Oil

Furnace turns most of the crude oil into a vapor before it enters the tower.

Residue (Asphalt)

Fig. 4-15. Oil is refined into more useful products by a process called fractional distillation. As the heat rises, different materials reach their boiling points and then condense onto the capture trays.

Fig. 4-16. Composites such as fiber-glass can be used to make many products, such as this surfboard.

Fiberglass is another composite material. It combines glass cloth and plastic resin. Fiberglass has been used to make furniture, bathtubs, cars, and boats both super-strong and lightweight. See **Fig. 4-16**.

Using Materials Science

How is materials science used to produce new materials? Let's look at diamonds. Diamonds are a form of carbon. Carbon has four electrons in its outer shell and easily combines with other atoms to create many different compounds. Carbon is released from melting rocks in the earth's mantle. Under tremendous heat and pressure, the carbon atoms in the melting rock bond to form diamond crystals. It takes thousands of years for a diamond to form, and, since the supply is limited and the demand strong, natural diamonds are expensive.

Diamonds are valued for their beauty, but they have many other important properties. Diamonds are nature's hardest material and nature's best conductor of heat. These properties have made diamonds a focus of material engineers for half a century.

In the late 1950s, scientists and materials engineers looked to improve the strength and life of cutting tools used in industry. They decided to coat the tools with synthetic diamonds. The diamonds would give the machine tools the hardness and wear resistance that steel alone could not provide.

Reading Link

KWL. A technique called KWL can help you study each chapter. For example, make a table with three columns. In the KNOW column, write what you already know about materials. In the WANT column, write what you would like to know about materials. After reading the chapter, complete the LEARNED column.

Next, look back at what you wrote in the KNOW column. Were any of your statements incorrect? If so, rewrite them so they are accurate. Now look at what you wrote in the WANT column. Were all of your questions answered by reading the chapter? If not, discuss those questions in class.

WHAT I KNOW	WHAT I WANT TO KNOW	WHAT I LEARNED

The process developed by materials engineers used graphite, another form of carbon, to create synthetic diamonds. A machine was designed to provide the high pressure and temperature needed to break the graphite's atomic bonds and form diamond crystals. The machine applies hundreds of thousands of pounds of pressure to the graphite and at the same time heats it to 2,500 degrees Celsius. After an hour of the intense heat and pressure, diamond crystals are formed. Today, industrial diamond cutters are widely used. For example, they are used to drill through rock in the search for oil. See **Fig. 4-17**.

Fig. 4-17. This diamond drill bit is used to drill through very hard rock and soil to tap oil reservoirs many thousands of feet underground.

Nanotechnology

In nature, carbon is a good example of how the arrangement of molecules can affect properties. Graphite, coal, and diamonds are all carbon, but they have very different properties. Today's engineers can manipulate molecules to change the properties of many materials. Working with materials at the molecular or atomic level is called nanotechnology.

Nanotechnology has opened up new possibilities in the world of materials science. Nanoparticles behave differently than larger particles of the same material. Engineers are learning how to put the new properties to practical use. For example, nanoparticles can be used to make very sensitive sensors. Such sensors can detect toxins, even if only a few molecules of the toxin are present.

Managing Material Resources

Many of the materials we use come from the earth. Some of these materials, like trees, are plentiful. These raw materials are called renewable resources. See **Fig. 4-18**.

Unfortunately, we cannot easily replace many raw materials. Coal, oil, and uranium are examples of limited nonrenewable resources. When these resources are gone, they are gone forever.

Materials engineers will have to conserve these nonrenewable resources. For example, most plastic is made from nonrenewable resources. Engineers are working to develop new plastics from renewable resources.

Fig. 4-18. Some materials, like trees, are renewable resources. However, conservation must be considered because the renewal process may take many years.

Fighting Fatigue

Identify a Need/Define the Problem

Metals are often used by materials engineers. Designers must know which kinds of materials are the strongest and which ones demonstrate the most fatigue. Fatigue is weakening or failure of a material after prolonged stress. In this activity, you are going to test wires to see which ones handle fatigue the best.

Gather Information

Research the properties of metal and different kinds of wire. Examples of wires can be shown in **Fig. A**.

Develop Possible Solutions

Decide on the items that should be represented in a table as well as the number of times the test should be repeated to obtain a precise result. Prepare some designs for the table.

Fig. A

Materials and Equipment

Select from this list or use your own ideas.

- assorted wires of equal diameter
- vise
- pliers

 For information about these tools, see "Hand Tools" on the Student CD.

Safety Alert

Look up "Safety Data Sheets" on the Student CD and prepare a data sheet for this activity. As you work on the activity, be sure to follow all safety rules.

Model a Solution

1. Choose the table design that will provide the best test record.
2. Prepare samples of different types of wire. Each sample should be of equal size. Be sure to prepare enough samples for the number of tests you decided to run.
3. Create one additional sample of each type of wire. Set these aside.
4. Set up the testing mechanism as shown in **Fig. B**.

Test and Evaluate the Solution

- Find the fatigue strength of each wire by bending it 90° in one direction, and then bending it back in the opposite direction 180°. Repeat until you have broken the wire. Record your results.
- Repeat similar tests for each kind of wire to test the fatigue strength. Record your results in the table.
- Once you have finished testing the samples, find the average number of bends it took to break the samples.
- Which samples had the best fatigue strength in the test?
- Why is it important to test a material before deciding to use it for a product?

Science Link

Heat and Fatigue. When you performed the material fatigue tests, did you notice whether or not the wire heated up at the point where you repeatedly bent it? (If you try it again, you will probably be able to feel the heat that is generated.) If heat is generated after repeated bending, is the reverse true? Will applying heat to a metal wire cause metal fatigue? Research and see if you can answer the question.

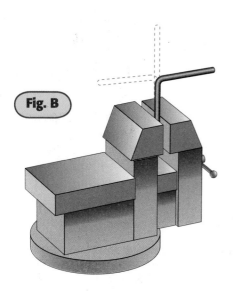

Fig. B

Refine the Solution

- Did each sample of one kind of wire yield the same results test after test?
- Develop an alternate fatigue strength test that can be used on the extra set of wires you set aside.
- Were the results similar to the results from your first testing procedure?

Communicate Your Ideas

Create a handout of your results that can be shared with other students. Make sure that a table is included that has information about what was tested and the number of tests that were conducted. A diagram explaining the testing procedure would also be helpful.

Protecting Your Food

Identify a Need/Define the Problem

Food containers come in many shapes and sizes, and they are all made of different materials. Test the properties of materials used for various food containers. The following properties should be tested: hardness, ductility, elasticity, and strength.

 See the "Properties of Materials" lab on the Student CD to learn about these properties and others.

Gather Information

Research how to test for each property. Also research how materials for containers are chosen based on the containers' contents. (An example would be egg cartons. Egg cartons made from certain types of foam are impact resistant to protect the eggs.)

Materials and Equipment

Select from this list or use your own ideas.

- 5 clean food containers made of various materials
- vise or other clamping devices
- file
- pliers
- power supply
- multimeter
- standardized weights

 For information about the tools listed here, see "Hand Tools" on the Student CD.

Safety Alert

Look up "Safety Data Sheets" on the Student CD and prepare a data sheet for this activity. As you work on the activity, be sure to follow all safety rules.

Develop Possible Solutions

After you've gathered your information, list ways you could check your food containers for each property. Also list the different kinds of food containers you could test. **Figure A** shows some examples. Develop a device or several devices you can use to test each container. See **Fig. B**.

Fig. A

Model a Solution

1. Choose five food containers that are made of different materials. Label each container A through E.
2. Develop a table to list the material of each container, the properties you are testing for, and what kind of food was originally in which container.
3. Set up your testing device(s).

Safety Alert

Some materials, such as glass, may break and shatter. Be sure to wrap such materials in cloth before testing.

Test and Evaluate the Solution

- Test each container for hardness using the method you have developed. Place a 1 next to the hardest container, a 2 next to the second hardest, etc.
- Test each container for the remaining properties and record your results. Did the containers all have the same properties?
- Do the material properties found match the purpose of each container?
- Were some containers made from inferior materials?

Refine the Solution

- What other properties might be needed in food containers?
- Identify one additional property for a food container and develop a test for it.

Math Link

Figuring Shelf Space. Some large supermarkets display many thousands of products. Stores have to organize shelf space to display each product in the best way. Imagine you have 20 containers. Ten of the containers are 6 inches wide, five of them are 7 inches wide, and five of them are 4 inches wide. Approximately how long would a row have to be to fit all the containers? Leave one inch between each container so customers can pick them up.

Communicate Your Ideas

Write a brief report, 150–200 words, describing the testing procedure that you developed and the properties of each container. Include a summary as to why you think the material for each container was chosen.

Fig. B

Recycling Materials

Identify a Need/Define the Problem

As you learned in this chapter, engineers are working on ways to conserve materials. One way they do this is through recycling. In this activity, you will design a new building material made from recycled products.

Gather Information

Research the various materials that can be recycled today at either a community level or industrial level. Take notes on the various properties and what they can be used for. For example, rubber tires can be reused for things such as resurfacing athletic tracks and for surfacing playgrounds, both of which require a firm yet forgiving surface. See **Fig. A.**

Develop Possible Solutions

Your building material may be something as simple as a new type of siding or a more refined product, such as a window. You will need to consider several factors. What will the new material be used for? What types of forces and environmental factors will it be exposed to? Once you have answered these questions, develop several sketches; the more ideas that are developed the easier it will be to produce a solution.

Safety Alert

Look up "Safety Data Sheets" on the Student CD and prepare a data sheet for this activity. As you work on the activity, be sure to follow all safety rules.

Fig. A

Materials and Equipment

Select from this list or use your own ideas.

- recycled products
- hand tools, as needed
- machine or power tools, as available (used under the supervision of an adult)
- fasteners (mechanical or chemical)

For information about the tools listed here, see "Hand Tools" and "Power Tools and Machines" on the Student CD.

Model a Solution

1. Choose one solution and gather the recycled product(s) you will use. See **Fig. B**.
2. Trim off any excess material that will not be needed. Be sure to recycle all waste.
3. Assemble your material as described in your sketches. You may want to use adhesives or other fasteners. (Think of how some materials, such as plywood, are produced.)

Test and Evaluate the Solution

- Many companies will create environments that will allow materials engineers to observe the product in use. Develop a real-life scenario that will test your new construction material. Note all observations and make notes on performance.
- Did your recycled building material perform as expected?
- Did the recycled material you selected have the required properties for the specified use?
- Does the building material you created replace an existing building material?

Refine the Solution

- Did your recycled building material work as planned? If not, what can be improved?
- Is there a better choice to use for the recycled product?
- Try using something else recyclable to make your building material. Have you improved your overall product?

Communicate Your Ideas

Create a sales brochure for your new recycled building material. Include information about the recycled products that were used in its manufacturing and the various properties of the building material. Be sure to also include the benefits to its use over other traditional building materials.

Fig. B

Exploring Careers

Materials Engineer

ENTRY LEVEL | **TECHNICAL** | **PROFESSIONAL**

Materials engineers design new or improved materials and innovative ways to make new products. They work with ceramics, metals, plastics, and composites to give these materials different properties.

Materials engineers spend much of their time in laboratories or industrial plants and have to use safe procedures. Materials engineers analyze and test materials, and they write reports. They use many tools on the job. For example, they use computers to monitor quality and test their designs and to prepare specifications.

Materials engineers work for manufacturing and research and testing companies. Engineers with more experience may become supervisors responsible for entire projects.

Qualifications

Materials engineers must have good problem-solving skills and be detail oriented. Because they design new materials and products, creativity is also a good trait to have. Good communication skills and being a team player are also very important.

Materials engineers must have a strong math and science background. A bachelor's degree in materials engineering is required. Classes in a materials engineering program may include thermodynamics and kinetics of materials, metallurgy, and the physical chemistry of materials.

Many employers want engineers who are certified as a Professional Engineer, or PE. A PE registration requires working for a certain number of years to gain experience and then passing a certification test.

Outlook for the Future

Growth in jobs for materials engineers is good, though slower than average. Because many engineers stay in their jobs throughout their careers, there is not as much turnover in this field. New jobs will be available as other engineers retire.

Self-Discipline

Self-discipline is a necessary skill for a materials engineer. Workers with self-discipline concentrate on the task at hand, work without a lot of supervision, are responsible, and work at a pace to get the job done. They maintain quality and follow through on their commitments.

Researching Careers

Find out about training needed to be a materials engineer. List five skills needed to be a materials engineer and provide detailed information about these skills. List ten high school and college classes to take. Make a PowerPoint presentation of your information.

More activities on Student CD

Key Points

- Materials engineers design, test, and develop processes to make new materials.
- All things are made of matter.
- Atoms are made up of protons, neutrons, and electrons.
- Materials can be classified into different groups, or families.
- Materials are designed to have certain properties.
- Synthetic materials are created by people; other materials come from nature.
- Some material resources are limited and must be conserved.

Read & Respond

1. Define *materials science*.
2. Name the three most common forms of matter.
3. What is found in an atom's nucleus?
4. Which particle of an atom is responsible for atomic bonding?
5. What are four groups of materials?
6. Why do materials engineers create alloys?
7. What are four material properties?
8. What kind of material resources will never disappear?
9. Why is fiberglass considered a composite material?
10. Name one renewable and one nonrenewable resource.

Think & Apply

1. **Hypothesize.** A new material is needed to envelop (cover) a space satellite. Describe some of the actions materials engineers would take to solve this problem. What questions would they have to answer?
2. **Examine.** Examine a penny. List three properties the penny has.

3. **Formulate.** Plastics often replace glass containers for food storage. What are the advantages of plastic containers?
4. **Connect.** Research the different types of materials used in the manufacture of an automobile. Create a table that organizes these materials into the four groups: metals, ceramics, polymers, and composites.
5. **Construct.** Make a mixture and describe its properties.

TechByte

The Value of Orange Peels.
Researchers have formed a new polymer by combining a chemical found in orange peel oil with carbon dioxide. The polymer is similar to polystyrene, a glasslike plastic used in many products. If this polymer can be turned into useful plastics, the plastics industry will be able to draw on a renewable resource and participate in recycling (orange peels) at the same time!

CHAPTER 5

Manufacturing Technologies

Objectives

- Define *manufacturing*.
- Identify the seven parts of a manufacturing system.
- Explain five uses for computers in manufacturing.
- Describe four examples of lean manufacturing.

Vocabulary

- **manufacturing**
- **system**
- **raw material**
- **industrial material**
- **primary process**
- **secondary process**
- **custom production**
- **job-lot production**
- **continuous production**
- **mass customization**
- **quality assurance**

This laser beam is cutting through a piece of thick metal.

Activities

- Assembling an Assembly System
- Toying Around with Mass Production
- Package Perfect

A Manufactured World

Think about your plans for tomorrow. What will you wear? What will you eat? Where will you go? What will you do for entertainment? Your clothing, many of your favorite foods, and the television you watch all have something in common. Can you guess what it is? They are all products of manufacturing.

Manufacturing is the changing of raw or processed materials into usable products. Manufacturing occurs in manufacturing plants, or factories. Factories are where many of the products you own are made.

Today's manufactured items are either consumer products or industrial products. Consumer products are "used up." Televisions, cereal, and books are consumer products. Industrial products are used to make other products. Lumber, steel beams, and machine tools are industrial products.

While products vary, the manufacturing processes are similar. See **Fig. 5-1**. This chapter will examine how a product is made and explore the trends and technologies in manufacturing.

Manufacturing is a system. A **system** is an organized way of achieving a goal. Like all technology systems, manufacturing needs resources.

 Learn about "The Systems Model" and "The Seven Resources of Technology" on the Student CD.

There are seven parts to a manufacturing system.
• Designing a product
• Purchasing materials
• Processing materials
• Production
• Packaging
• Distribution
• Sales

Fig. 5-1. The first bicycles were made by hand. Most bicycles today are mass-produced in factories. Can you name some of the natural and synthetic materials used in these racing bikes?

Designing a Product

To sell products, manufacturers must first design products that people will buy. As time passes, the needs and desires of customers change. To keep up with the demands of customers, manufacturers need to work efficiently and save on costs whenever possible.

Engineers have to design the product to meet a specific need or want. They choose what materials are needed and often direct the process used to make the product. You can learn more about engineers and design in Chapter 1, "Design & Engineering."

Designing for Manufacturability

One way engineers save on cost and time is by designing products for manufacturability. What does this mean? It means engineers design products that will be easier to manufacture. Engineers have developed many ways to improve efficiency and keep costs down.

For example, engineers might:
- Use materials that are readily available instead of ones that are harder to obtain.
- Use materials that can eventually be recycled once a product's life cycle has ended. (A product's life cycle is how long a product should last.)
- Design the product to fit into an existing production process instead of having to create a whole new process.

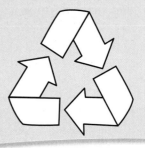
Purchasing Materials

Once a design has been chosen, materials needed to make the product must be ordered. Designers give purchasers a bill of materials, which lists all the materials or parts needed to make the product. Many different materials are used in manufacturing.

Raw materials are materials as they occur in nature. For example, a tree is raw material. See **Fig. 5-2**. Most raw materials cannot be used until after they are processed into industrial materials. **Industrial materials** are materials that are used to make products. Lumber is one example of an industrial material.

What materials might purchasers order? A sawmill needs raw materials such as logs. A

Fig. 5-2. Wood is a raw material. To be usable for manufacturing the wood must be processed into lumber. Trees must be cut down, transported, and sawed into boards.

manufacturer of metal products requires industrial materials such as sheet metal. Automobile manufacturers need finished products such as tires, batteries, and window glass to produce their own finished products—automobiles. Purchasing agents work to obtain the right items at the best price.

Processing Materials

Once materials are purchased and received, they need to be processed. **Primary processes** are processes that change raw materials into industrial materials. There are three kinds of primary processes: mechanical, thermal, and chemical. The following are examples.

- Metal ore is mined and crushed. These are mechanical processes.
- The ore is heated in a furnace. This is a thermal process.
- Materials are added to the ore to give it different properties. This is a chemical process.

 You can find more information about primary processes on the Student CD.

Secondary processes are processes that turn industrial materials into finished products. These processes include forming, separating, combining, and conditioning.

Forming

Forming changes the shape of a material. See **Fig. 5-3**. Nothing is added to or taken away from the material during this process. Forming is done in several ways:

- Rolling squeezes the material between rollers. Sheet metal is made in this way.
- Casting involves pouring or forcing softened material into a hollow mold.
- Forging hammers or squeezes material into shape. Most wrenches are shaped by forging.
- Stamping squeezes sheet metal between dies to give it shape. Stamping is used to shape some automobile body parts.
- Extrusion pushes the material into shape. The shape of the opening in the dies determines the shape of the material.

Separating

Separating is the cutting of materials to size and shape. Some material is usually lost. Sawing, sanding, and filing are traditional forms of separating. However, heat, light, chemicals, and even water can also be used to separate certain materials.

Shearing is separating part of a solid material from the rest of the material. No material is destroyed. Thin materials like paper and sheet metal can be separated by shearing.

Material can also be separated by cutting. Usually, cutting a material involves chip removal. Common cutting processes are shown in **Fig. 5-4**.

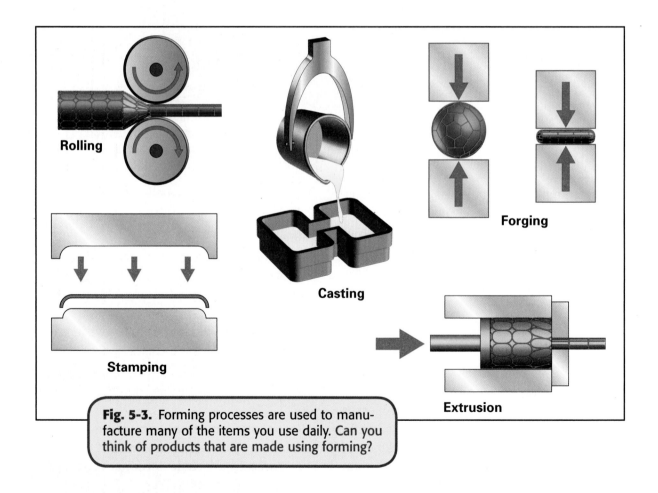

Rolling

Casting

Forging

Stamping

Extrusion

Fig. 5-3. Forming processes are used to manufacture many of the items you use daily. Can you think of products that are made using forming?

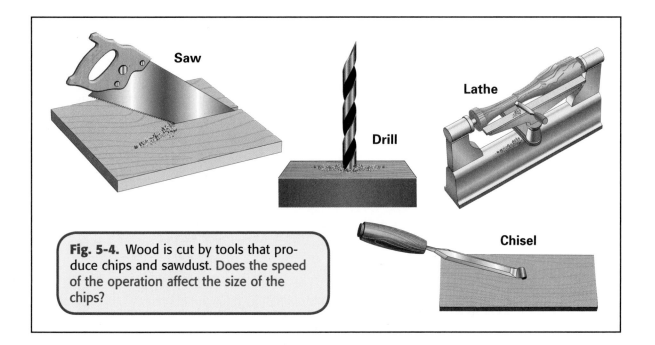

Fig. 5-4. Wood is cut by tools that produce chips and sawdust. Does the speed of the operation affect the size of the chips?

Saw

Drill

Lathe

Chisel

Combining

Combining is the process of joining materials. Combining is done in a variety of ways:

- **Mixing.** Materials may be mixed together to form new materials. Food products such as cake mixes and soup mixes are made in this way.
- **Mechanical fasteners.** These include nails, screws, staples, and nuts and bolts. Automobiles and many other products are assembled using removable mechanical fasteners. Doing this makes it possible to replace parts.
- **Soldering, brazing, and welding.** These are processes that involve heat. In soldering and brazing, a filler material is melted along a joint between pieces of metal. The metal pieces themselves do not melt. In welding, the heat melts the two pieces together.
- **Coating.** This process uses one material to cover another. It may be done to decorate or protect the covered material. Some coatings do both. Painting is the most common coating process. See **Fig. 5-5.**

Fig. 5-5. Coating is one of the final processes in manufacturing. The paint job on this car will be carefully inspected in quality assurance.

Conditioning

Conditioning is a process that changes the internal structure of a material. The three types of conditioning are thermal, chemical, and mechanical.

• Thermal conditioning uses heat. Hardening, tempering, and annealing are thermal processes used to condition metals. Hardening makes metal products more wear-resistant. However, hardened metals are usually brittle.

Fig. 5-6. Production is a key component when manufacturing baseball bats.

In tempering, hardened metals are reheated to remove brittleness and make them tougher. Tempering helps steel-cutting tools keep their sharp edges. Annealing is a softening process. Steel must often be annealed. It can become too hard during processing. Annealing also makes metal less brittle.

- Chemical conditioning uses a chemical reaction. Vulcanization (vul-cuh-na-ZAY-shun) is a chemical conditioning process used to make rubber durable. A substance such as sulfur is added to the rubber. Heat and pressure are then applied.
- Mechanical conditioning uses force. Forging is an example of mechanical conditioning. Hammering a piece of metal will cause it to change. Usually this is done to make the metal harder.

 See the Student CD for more information about secondary processes and about properties of materials.

Production

Both primary and secondary processes are used in manufacturing. Many different processing techniques may be needed to produce one type of product. Products with a number of different parts, such as appliances and computers, must also be assembled.

Production is a key component of any manufacturing system. See **Fig. 5-6**. There are four types of production. The kind of production used depends upon the kind of product and the number of products needed.

In **custom production**, products are made to order. For example, a cake made according to a buyer's choice of size, flavor, shape, and icing is a custom-made product. A custom-made product need not be a small item. A cruise ship that costs $300 million and is designed for 2,500 passengers is also a custom-made product.

In **job-lot production**, a specific quantity of a product is made. For example, suppose a manufacturer received an order for exactly 10,000 skateboards. Job-lot production would be used. Many job-lot manufacturers make products on a seasonal basis. A company making skateboards used mostly in the summer might also make skis for use during the winter.

In **continuous production**, products are mass-produced, usually on an assembly line.

Products move from one work station to the next. Some mass-production systems make thousands of identical items each day. See **Fig. 5-7**.

Mass customization is a new approach to manufacturing that combines elements of custom production and mass production. Companies that use mass customization produce standard products that are modified for individual customers. For example, eye glasses are made of standardized parts that may be modified to fit the particular user.

Mass customization has revolutionized the production of made-to-order computer systems. Customers go online to select a particular model computer and then select options. They might specify how much memory they want, processor speed, kind and number of disk drives, the speaker system, and the type and size of monitor. The ordering process takes only a few minutes and generates a work order that is sent to the factory floor. The computer is assembled from a large selection of standard parts and the requested software is installed. The computer is shipped and will probably arrive about a week after it was ordered. See **Fig. 5-8**.

Quality Assurance

A product must meet the quality standards set by the company. The system used to check quality is called **quality assurance**. Quality assurance means making sure the product is

Fig. 5-7. Continuous production can be used to make thousands of identical items every day.

Fig. 5-8. These customized computers are being shipped just a few days after the order was received.

Math Link

Storing Parts. Manufacturers often keep in stock the parts or materials that they need to make their products. How would a manufacturer know how large a storage area or warehouse is needed to store such items? Suppose 2,000 different items are needed. If the container to hold each item was a box about 2 feet long, 2 feet wide, and 1 foot tall, how many linear (straight) feet of shelf space would be required? Suppose the manufacturer is going to buy shelf units. Each unit is 4 feet wide and has 4 shelves. How many shelf units will have to be acquired?

made according to plans and meets all specifications. Quality assurance focuses on prevention and detection.

When a product is designed, engineers will provide standards a product must meet. Then the manufacturing process will be set up to monitor quality. Materials coming into a factory will be checked. Often, inspectors will check a product by hand or with a machine as the product is manufactured. Other times, programmed machines monitor the quality throughout the entire manufacturing process. A finished product will be checked for any defects before being packaged. Even the packaging will be checked to confirm the overall product is of the highest quality.

Packaging

Products must be prepared for shipping and then transported to customers. Industrial materials are usually shipped to other manufacturers without being individually packaged. Most consumer products are packaged. What is the purpose of packaging? It must protect the product. It should also attract attention on a store shelf. Have you ever seen packaging that makes you want to buy a product? If so, the packaging department has done its job!

UPC and RFID

Have you ever looked at a packaged product and wondered what those white and black parallel bars were for? That is a universal product code, or a UPC. See **Fig. 5-9**. It is often called a bar code, and it contains information about the product and its manufacturer.

Universal product codes help out in both the distribution and sales of a product. For example, UPCs are used for inventory purposes. A manufacturer or seller has equipment that can read the code. Each item that leaves the factory or store is scanned. The information is sent to a computer system that tracks how many products are in stock. When supplies run low, more products are ordered or manufactured. You have probably been to many stores in which items were scanned at the checkout lane. You were part of the stores' inventory process.

Another method of tracking uses RFID, or radio frequency identification. RFIDs use a "tag-and-tracking" system. The product has a tag on it that contains a transponder. A transponder is a device designed to receive a unique radio signal. When the transponder detects the signal, it responds with its own unique signal.

RFIDs are used to track products or their containers, pallets that hold products, and the trucks and trailers that transport products. For example, when a vehicle needs to be located, a transceiver transmits radio signals to locate the transponder. Since products may be moved in warehouses all the time, an RFID is a good way to find what you need quickly. You have probably seen RFIDs on consumer products such as CDs. See **Fig. 5-10**.

0 71402 72712 3

Fig. 5-9. This UPC contains information about the product that can be used by the manufacturer and the seller.

Fig. 5-10. RFIDs are used to track products in a store or warehouse. It is an easy way for stores to know where their products are located and also to keep an inventory record of what is in the store.

Impact of Technology
Life Cycle Analysis

More and more products are being produced. Some are needed to sustain our lives. Others make our lives more enjoyable. But a downside to this trend is we now have products that are no longer needed. For example, what is to become of all the old and unwanted computers, cell phones, and kitchen appliances?

Some manufacturers consider a product's disposal even before they produce it. A "life cycle analysis" looks at the impacts that a product's manufacturing, distribution, sales, use, and disposal will have on the environment. Such an analysis can help product designers create products that have fewer negative impacts on the environment and that can be reused or recycled.

Investigating the Impact

Research "design for recyclability."
1. What are some products that have been designed to be recyclable?
2. What does the consumer need to do in order to make sure the product really will be recycled?
3. Are there stores or other places in your community that accept used products, such as cell phones, for recycling?

Distribution and Sales

Where does a product go once it is made? It depends on the product. Products can go to another manufacturer, a wholesaler, a retailer, or directly to the customer.

If one factory makes a specific component, it may then ship the part to the manufacturer of the overall product. A wholesaler is a company that purchases large amounts of a product and then sells smaller numbers to a retailer. Retailers are the stores that sell products directly to the consumer. Consumers can sometimes order directly from the manufacturer. For example, computers can be ordered, manufactured, and shipped directly to the consumer.

To promote sales, manufacturers will usually advertise their products. For consumer products, television, newspaper, magazine, and Internet ads are often used. Even the sides of some vehicles, such as trucks and buses, display ads to catch the eye of consumers.

Salespeople are employed by many companies. Salespeople are paid on a commission. This means the more products they sell, the more money they are paid.

Computers in Manufacturing

Computers play an important role in all stages of manufacturing. They are used to design products, to control processes and machines, and to manage businesses. New technologies such as rapid manufacturing and nanotechnology are changing how manufacturing is done.

CAD

CAD is computer-aided design. Engineers use CAD software to show what the product will look like, how it will work, how it should be made, and what materials will be used to make it. CAD is faster to do than manual drafting. Three-dimensional models can also give more accurate details than 2D models. You can learn more about CAD in Chapter 2, "Computer-Aided Design."

CAM

Once CAD drawings are produced, they can be given to people who actually make the product. However, CAD drawings are sometimes input directly into a CAM system. CAM, or computer-aided manufacturing, operates and controls many machines and processes. CAM can be used to tell a machine what specific job to do, or it can be used to set up a specific process to create a product.

CNC. Computer numerical control is an important part of CAM systems. CNC machines are programmed to perform a series of operations over and over. For example, CNC mills and lathes are used for cutting, drilling, and turning. Some of these machines can produce thousands of identical parts each day. When production of a particular part is complete, the CNC machine is reprogrammed to make a different part.

CIM

In computer-integrated manufacturing, computers monitor and control every aspect of manufacturing. Computers link design and production operations with purchasing, accounting, inventory, shipping, sales, and payroll.

PLCs

Programmable logic controllers (PLCs) are small computers that help control machines, such as the ones on an assembly line. See **Fig. 5-11**. PLCs have many possible uses. For example, in a chemical processing plant, they can be used to control the flow of a liquid into a tank. A floating device in the tank sends an electronic signal to the PLC when the liquid drops below the desired level. The PLC opens a valve to add liquid and then closes the valve at the right time. If the system is not working properly, the PLC signals that maintenance is needed. When a change is needed, such as increasing the speed that liquid flows into the tank, the PLC can be reprogrammed to make this happen.

Fig. 5-11. The PLC on this assembly line determines how many bicycle parts should be sent to the automated welding station.

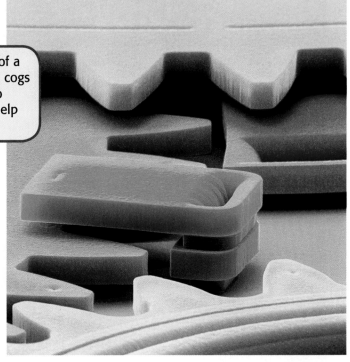

Fig. 5-12. These microgears are part of a very small micromotor. Tiny gears and cogs such as these will someday be able to travel freely in the human body and help combat diseases.

Rapid Manufacturing

In Chapter 2, "Computer-Aided Design," you learned about rapid prototyping. Some companies use rapid prototyping methods to produce final products for sale. In rapid prototyping, a design is sent from a CAD system to a 3D printer. The printer builds the product layer-by-layer. Rapid manufacturing takes the next step. The same machine used to make one prototype can be used over and over to make tens or hundreds of actual products.

Rapid manufacturing exists outside of factories, too. For example, the United States Army has a Mobile Parts Hospital. It is a truck whose interior has individual workstations. The truck also carries standardized metal bars that can be formed into whatever parts are needed. Technicians direct robotic tools and machines to make the parts.

Nanotechnology

In Chapter 1, you learned about nanotechnology. Nanotechnology allows manufacturers to produce materials or products using atoms or molecules as the building blocks. It can be used to create materials or products that are lighter, stronger, and programmable. See **Fig. 5-12**.

With nanotechnology, carbon atoms could be arranged into a perfect diamond. Can nanotechnology be used to make larger items as well? Take those same carbon atoms. With nanotechnology, carbon could be programmed to self-manufacture into larger components used for vehicles or buildings.

Lean Manufacturing

Lean manufacturing systems are designed to produce high-quality products at the lowest possible cost. Lean manufacturing was first used by the Japanese automobile manufacturer Toyota. It has spread throughout the world to industries that produce a wide variety of products. Important elements of lean manufacturing systems include:
• Continuous improvement
• Just-in-time inventory management
• Teamwork
• The use of work cells

Fig. 5-13. This aircraft manufacturer utilizes work cells. The huge planes are hardly moved until they are completed. Can you see individual work cells in this photo?

Continuous Improvement

Continuous improvement is based on the idea that details are important and that it is always possible to make a good system even better. When problems occur, they are viewed as opportunities for improvement. Employees working in a lean manufacturing plant are encouraged to identify small problems and share their ideas about possible solutions.

Just-in-Time Inventory

Keeping large quantities of parts on hand is expensive. Sometimes a large amount of storage space is required and unneeded parts may be wasted. In a just-in-time system, the correct amount of parts is always on hand. For small parts it can be done by having two bins of each part in the assembly area. When one bin is emptied, parts from the second bin are used and another bin is ordered. Other times, parts arrive exactly when they are needed.

Teamwork

Teamwork is also an important component of lean manufacturing. People are hired not only for their technical qualifications, but also because they will fit with a team. In most manufacturing processes, people need to rely on each other. While each person may have his or her own task, that task will affect the tasks of other people.

Work Cells

The importance of being able to work as part of a team carries over to the work cells that are often used in lean manufacturing. The ability and desire to work cooperatively with others is important because work cells usually involve groups of people working together in small areas. The cell contains all of the equipment needed to make the products. Parts are delivered just in time. Communication is easy since everyone is nearby. In many industries, the use of work cells has helped to bring about significant increases in quality. See **Fig. 5-13**.

Automatic Manufacturing

The computer systems mentioned in this chapter have completely changed how a factory operates. Some factories are now operated almost completely by computer. These are called automatic factories.

In an automatic factory, machines are used at all stages of manufacturing. While a few highly skilled technicians are on hand in case a machine breaks down, the machines do most of the work. Materials are delivered by machines. Machines make the products. Quality is checked by machines, and machines package the product. In many areas of an automatic factory, lights aren't even on because machines don't need them. Therefore, automatic manufacturing is often referred to as lights-out manufacturing.

In areas where much of the work is automated, you would probably see industrial robots. Industrial robots can work at a perfectly timed pace, are accurate, and can handle dangerous materials. (For more on industrial robots, see Chapter 17, "Robotics.") If you go past a manufacturing plant that is completely dark, these industrial arms may still be working to help deliver products to consumers like you.

Manufacturing Safety

While industrial robots can often be used to handle dangerous materials, safety is still of the highest importance in manufacturing plants. The Occupational Safety and Health Administration (OSHA) was created by the Federal government and is part of the U.S. Department of Labor. OSHA sets safety standards for the workplace.

To reduce injuries, workers are required to wear safety equipment. Safety glasses, hard hats, and steel-toed shoes are commonly required. Most factories are safe places to work.

Workers who perform the same task all day can develop repetitive motion injuries. To reduce the risk of this happening, workers are often rotated through several different jobs each day.

Figure 5-14 shows colors that signal safety messages. A safe work environment is necessary for an efficient manufacturing system.

Color	Meaning
Red	Danger or emergency
Orange	Be on guard
Yellow	Watch out
White	Storage
Green	First aid
Blue	Information or caution

Fig. 5-14. Safety Colors.

Assembling an Assembly System

Identify a Need/Define the Problem

Most manufactured products are made of many different components. An automobile typically requires the assembly of more than 35,000 parts. A ballpoint pen might have five or more parts. See **Figs. A** and **B**.

Work as a member of a two-person team to develop a system for assembling 10 retractable ballpoint pens.

Gather Information

Research assembly lines. Questions you may ask might include the following:
• What are the assembly methods used in manufacturing?
• Is there a type of chart that can be used to help keep track of the order in which parts are assembled?
• How many individual pieces does a ballpoint pen contain?

Develop Possible Solutions

Brainstorm ways to assemble your pens in a two-person assembly line. Set up your work area so that you and your partner can run an efficient assembly system.

Materials and Equipment
Select from this list or use your own ideas.
• 10 identical retractable ballpoint pens
• stopwatch
• video camera

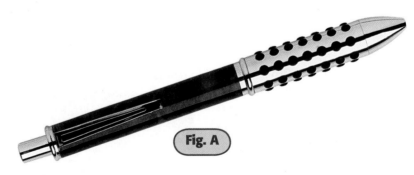
Fig. A

Model a Solution

1. Separate the parts into individual piles. Check to make sure that each pile has the correct number of pieces for the 10 pens. This is the first step in quality assurance.
2. Each team member should assemble and then disassemble one pen. This ensures that each member is familiar with how the product is made before starting the assembly system.
3. Assemble the pens using your assembly system. A stopwatch should be used to time how long it takes to assemble the 10 pens. Have a volunteer time the process.

Test and Evaluate the Solution

- Prepare a production run that will be timed and videotaped.
- How quickly did your team assemble the pens?
- Do the pens work like they are supposed to? Is each pen of equal quality?

Refine the Solution

- After viewing your videotape, discuss the most efficient method for assembling the products.
- If needed, take your pens apart and create a different assembly system.
- As before, time and videotape your production run. Did your new system perform better than the original one?

Communicate Your Ideas

Write a brief script outlining the production process and the team decision-making process. Using this script, narrate the videotape to the class.

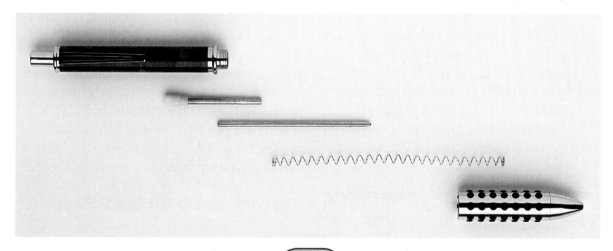

Fig. B

Toying Around with Mass Production

Identify a Need/Define the Problem

Your class has been asked to mass-produce a small wooden toy for a local crisis nursery. The toy must be safe for a child to use. Your team members must be organized efficiently to be able to manufacture the product using mass production. See **Fig. A**.

Gather Information

Research how toys are made safe for children. Identify the specific toy you will be manufacturing. Identify the processes needed to manufacture the toy.

Develop Possible Solutions

Determine the order in which each part of each process is to be performed. Identify the individuals who will perform each operation. Remember that you will also need inspectors to ensure quality control. You may come up with several different plans. Write out each plan so that it is clearly understood.

Materials and Equipment

Select from this list or use your own ideas.

- wood (possibly pre-cut)
- hand tools, as needed
- machine or power tools, as available (used under the supervision of an adult)
- fasteners (chemical or mechanical)

 For information about the tools listed here, see "Hand Tools" and "Power Tools and Machines" on the Student CD.

Model a Solution

1. Select the plan that you think will be the most effective.
2. Make working drawings of your item.
3. Each team member should have a work station with the tools or machines needed to perform his or her assigned task.
4. Make a prototype of your item to make sure the toy you designed will work properly.
5. Begin mass production. Assemble the item.

Test and Evaluate the Solution

- Were the job assignments clearly explained?
- Was the order in which the tasks were done effective?
- Did the pieces fit together as planned? Does the toy work as intended?
- Did the product pass quality assurance?

Safety Alert

Look up "Safety Data Sheets" on the Student CD and prepare a data sheet for this activity. As you work on the activity, be sure to follow all safety rules.

Refine the Solution

- How might this product be improved if it were to be made again?
- If needed, mass-produce another toy.
- During mass production, have team members perform different tasks from the ones they originally did.
- Is the new toy better than the first one? Why or why not?

Communicate Your Ideas

Present your toy to the class. Each team member should describe his or her own role in mass production.

Math Link

Math Assembly. Henry Ford is often credited with creating one of the most efficient mass production systems of its kind in the early 1900s. At one point, Ford's assembly plant in Highland Park, Michigan, was producing a car every 93 minutes! If Ford's plant was open 10 hours a day, 6 days a week, how many cars could that manufacturing plant have produced in one year?

Fig. A

Package Perfect

Identify a Need/Define the Problem

Imagine that a candy manufacturer is willing to donate a large quantity of wrapped mints to your school. In return, you have been asked to design a unique and attractive package that will hold ten mints. Please keep in mind the following:

- The package must offer adequate protection for the product.
- Packaging is used to draw the customers' attention and encourage them to buy the product.
- The package must be easy to store or ship. Packages can be made to hang on pegs or sit on shelves.

Materials and Equipment

Select from this list or use your own ideas.

- individually wrapped circular mints
- paper of assorted colors and weights
- lightweight card-stock of assorted colors and weights
- plastic wrap
- tape or glue
- color markers
- color pencils

Gather Information

Look at packages of similar products to see how they are constructed. You might disassemble a paper clip box, a toothpick box, a candy container, and a fast food carton. Note how they are constructed. A package's materials can add strength. However, packaging material also adds weight. This increases shipping costs. Keep this in mind as you choose packaging materials.

Develop Possible Solutions

Candy packages come in many shapes and sizes. See **Fig. A**. Sketch five package designs. In sketching the designs, remember the reasons for packaging. To provide a protective shape, it might need to be folded or cut into pieces. If the material is to be cut or folded, make sure that you show what will be done on your design.

Fig. A

Also, show how the parts are to be joined together. Be sure to identify the types of polygons used in the package design. Specify what material (i.e., glue or tape) will be used to join the parts. An example of a package pattern is shown in **Fig. B**.

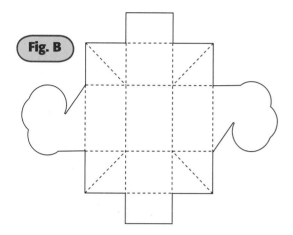

Fig. B

Model a Solution

1. Choose three of your designs and construct prototype packages.
2. Use the markers and pencils to provide written information about the mints and to add color to your packages.
3. Place the mints inside your packages and seal them.

Test and Evaluate the Solution

- Show the three prototype packages to several classmates. Ask them to choose the package they find most attractive and tell you why. Record your results.
- Test the package's ability to protect the contents by dropping it off the top of a desk or by placing something slightly heavier on the package. Open the package to inspect the contents for damage. Record your results.
- Was the package that classmates liked best the same package that provided the best protection?

Refine the Solution

- If needed, create a package that combines the best design with the best protection. If both the design and protection were unsatisfactory, choose a design you did not use.
- Did your new package perform as well as you hoped?

Communicate Your Ideas

Prepare a brief written or oral presentation. Explain why your package is unique. Point out features that will increase sales of the company's mints.

Exploring Careers

Shop Foreman

ENTRY LEVEL | TECHNICAL | PROFESSIONAL

A shop foreman supervises, delegates, and coordinates the activities of workers in manufacturing jobs. He or she also inspects the work to ensure it meets quality standards. A foreman sets the order in which workers will complete projects. The foreman may teach workers and demonstrate how to do tasks.

Foremen order inventory to have parts and materials on hand. They must keep accurate records of inventory, procedures, schedules, and payroll. Shop foremen also keep maintenance logs or records, sometimes using a computer.

Foremen make personnel decisions, such as hiring, transferring, and promoting workers. As leaders, foremen help improve the way work is done, make sure the work is completed safely and on time, and help their teams produce quality products. Shop foremen work in all types of repair facilities and manufacturing plants.

Qualifications

A shop foreman must have good management, organizational, problem-solving, and communication skills. Shop foremen are responsible for producing products or making repairs on time and in a safe manner. Therefore they must be able to motivate and lead workers.

A high school degree is required. Courses to take in high school might include electricity, woodworking, science, math, and computers.

Shop foremen should have several years of experience as laborers in the type of work they will be supervising. They should also have some supervisory experience, usually between two and five years. To gain such experience, a laborer could take on additional tasks that include supervising coworkers.

Outlook for the Future

The job outlook for shop foremen is good. Workers in this field change jobs or leave the field to do other work, and this leaves openings to be filled.

Leadership

Foremen should enjoy leading other people. A leader explains decisions and ideas clearly and shares credit for a job well done. Leaders are supportive of workers and allow them to help make decisions.

Researching Careers

Find out about jobs for foremen. Cut out five want ads for foremen or supervisory jobs. What are some other titles that employers use for these jobs? What skills and experience are needed? Make a poster using your data.

More activities on Student CD

Key Points

- Manufactured products are either consumer products or industrial products.
- There are many stages in a manufacturing system.
- The kind of production used depends on the product and the amount of the product needed.
- Computers play an important role in all stages of manufacturing.
- Nanotechnology can be used to manufacture at a microscopic level.
- Lean manufacturing is producing high-quality products at the lowest possible cost.

Read & Respond

1. Define *manufacturing*.
2. What is the difference between raw materials and industrial materials?
3. What does designing for manufacturability mean?
4. Name the seven parts of a manufacturing system.
5. What are PLCs?
6. What is CAD used for in manufacturing?
7. How is rapid manufacturing different from rapid prototyping?
8. What does CAM stand for?
9. What are four examples of lean manufacturing?
10. What does CNC do?

Think & Apply

1. **Connect.** This chapter discussed nanotechnology in manufacturing. Research and describe other areas of technology in which nanotechnology is used.
2. **Propose.** Imagine you work for a company that makes roller skates during the summer. What kind of product might use a similar manufacturing process but would be for use in the winter?

3. **Summarize.** Why is teamwork important in manufacturing?
4. **Assess.** What problems might a faulty checkout scanner cause for a store?
5. **Design.** Design a product out of existing materials. Will the materials need to be purchased? Where will the materials come from?

TechByte

Smarter RFIDs. More sophisticated RFID tags are being developed that could signal faulty or damaged parts before the parts end up on an assembly line. Bad parts could be pulled out of inventory, reducing the rejection rate in the quality assurance stage of production.

CHAPTER 6

Applied Physics

Objectives

- List the steps of the scientific method.
- Define *applied physics*.
- Explain the forces that are involved in motion.
- Describe Newton's three laws of motion.
- Explain how work is made easier with machines.
- Compare and contrast light and sound waves.

Vocabulary

- **applied physics**
- **motion**
- **force**
- **friction**
- **gravity**
- **inertia**
- **work**
- **machine**
- **mechanical advantage**
- **wave**
- **medium**

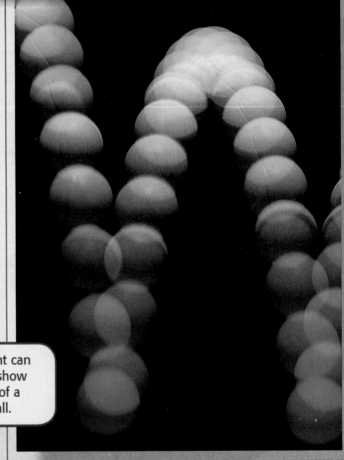

A strobe light can be used to show the motion of a bouncing ball.

Activities

- Giving a Lift
- Racing Reaction Rockets
- Creating a Sound Barrier

What Is Applied Physics?

Do you want to know why things happen and what makes things work? If so, you have the curiosity of a good scientist or engineer. How are science and engineering related? Scientists and engineers use the same kind of problem-solving methods. See **Fig. 6-1**.

Engineers and scientists also both use applied physics. **Applied physics** is the branch of science that applies the principles of science, particularly physics, to solve engineering problems.

Leonardo da Vinci (1452-1519) is famous for applying physics to solve engineering problems. See **Fig. 6-2**. He studied the flight of birds.

Fig. 6-2. Leonardo da Vinci was a great artist, scientist, and engineer.

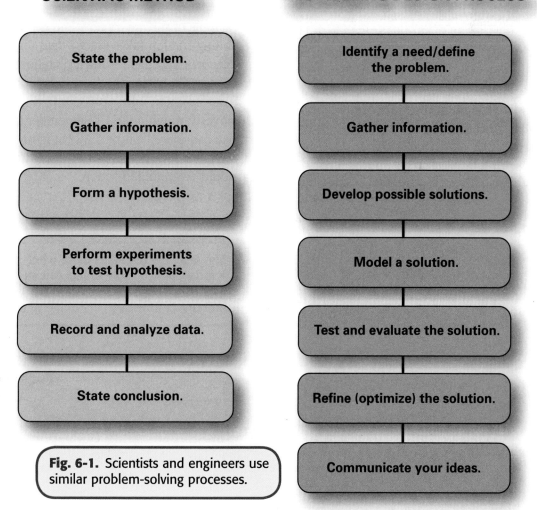

SCIENTIFIC METHOD

- State the problem.
- Gather information.
- Form a hypothesis.
- Perform experiments to test hypothesis.
- Record and analyze data.
- State conclusion.

Fig. 6-1. Scientists and engineers use similar problem-solving processes.

ENGINEERING DESIGN PROCESS

- Identify a need/define the problem.
- Gather information.
- Develop possible solutions.
- Model a solution.
- Test and evaluate the solution.
- Refine (optimize) the solution.
- Communicate your ideas.

He observed and sketched their movements as they flew. He studied their anatomy to make detailed drawings of their muscles, wings, bone structure, and feathers. Using the knowledge he gained about flight, da Vinci created designs for flying machines. He prepared designs for a helicopter, a plane, and a parachute, all this 500 years ago!

Much like da Vinci, today's engineers must have a good understanding of physics in order to design new technologies. Engineers must know about physics concepts such as motion, work, machines, and light and sound waves.

Physics of Motion

In Chapters 7 and 9, you will read about various modes of transportation. Without motion, these technologies would literally stand still. This book also has chapters about manufacturing and robotics. Motion is also critical to these technologies. **Motion** is a change of position in a certain amount of time. See **Fig. 6-3**. Motion involves force, friction, and gravity.

An object's motion is changed by a force. A **force** is a push or pull applied to an object. See **Fig. 6-4**. Once an object is moving, a force can cause it to slow down, speed up, or change direction. Two forces that affect motion are friction and gravity.

To learn what effects friction and gravity have on an airplane, see the interactive lab "Forces and Flight" on the Student CD.

Friction

Friction is a force that opposes motion. It acts in the direction opposite to the direction of the motion. Friction is the force that brings an object to rest.

- Surface friction occurs when one surface slides over another. Using a brake on a bicycle is an example of surface friction. See **Fig. 6-5**.
- Rolling friction takes place when one surface rolls over another. Riding a skateboard would create rolling friction.
- Fluid friction takes place when an object moves through a fluid. (Remember that air is also considered a fluid.) A boat going through water or an airplane moving through air would create fluid friction.

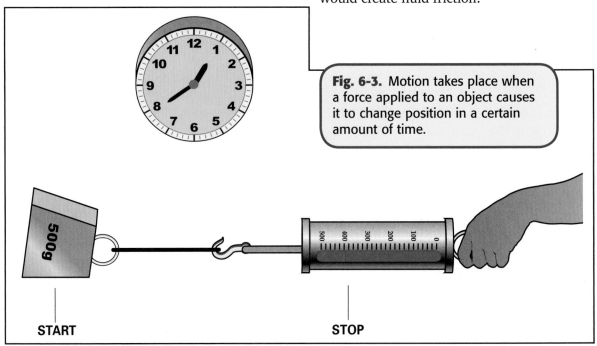

Fig. 6-3. Motion takes place when a force applied to an object causes it to change position in a certain amount of time.

500g

500 400 300 200 100 0

START

STOP

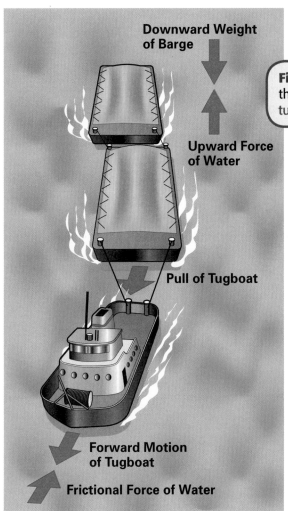

Downward Weight of Barge

Upward Force of Water

Fig. 6-4. Several forces act together on this tug and barge. How do you think a tug boat resists these forces?

Pull of Tugboat

Forward Motion of Tugboat

Frictional Force of Water

Backing Plate

Brake Drum

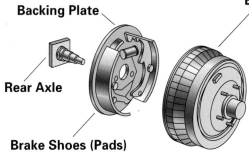

Rear Axle

Brake Shoes (Pads)

Fig. 6-5. Whether you are stopping a bicycle or an automobile, the brake parts rub against the wheel, causing friction. The friction slows down the motion of the wheel and forces it to stop.

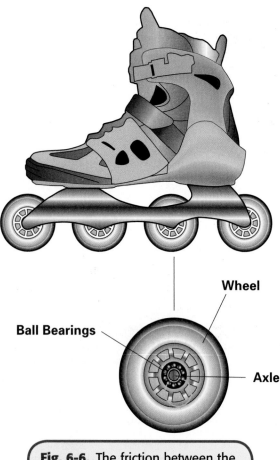

Wheel

Ball Bearings

Axle

Fig. 6-6. The friction between the ball bearings and the wheel is very slight. The spin of the wheel is fast and smooth.

Sometimes, engineers have to design products to overcome friction. To minimize drag, a form of fluid friction, engineers design products to be aerodynamic. See **Fig. 6-6**. (For more on aerodynamics, see Chapter 7, "Air & Space Technologies.")

Gravity

Gravity is the force of attraction that exists between two objects. The amount of gravitational force depends on the size of the objects and the distance between them. For example, the gravitational force between you and the

classmate sitting next to you is almost nonexistent. This is because your roughly equal masses are small. On the other hand, the gravitational force between you and the planet Earth is quite great. In fact, the force of gravity actually holds you to the surface of the planet.

Your weight is a measure of the amount of force placed on you by gravity. You would weigh less on the surface of the moon because the mass of the moon is one-sixth that of Earth. Therefore, the moon's gravitational force is also one-sixth of the force on Earth. See **Fig. 6-7**.

Gravity pulls all objects on our planet towards the center of Earth. You may have heard the old saying, "What goes up, must come down." If that is true, how can a rocket travel beyond the reach of Earth's gravitational pull? To do this, the rocket must have engines powerful enough to produce enough forward thrust, or force, to reach escape velocity.

Escape velocity is the minimum speed at which an object must travel to escape the gravitational pull of a planet. The escape velocity needed to overcome Earth's gravitational pull is 40,000 km/h (17,000 mph). The rocket must reach this speed while moving in the direction opposite that of the gravitational force.

Newton's Laws of Motion

Isaac Newton, an English scientist who lived from 1642 to 1727, came up with an explanation for how force and motion work. His laws of motion are used in designing many technologies today.

First Law of Motion

Newton's first law of motion states that an object at rest will stay at rest until a force acts on it. Once the object is in motion, it will stay in motion until an opposing force acts on it. See **Fig. 6-8**. This law holds true for any object. The tendency of an object to stay at rest or to continue to move is called **inertia**.

On Earth, an astronaut has a weight of about 30 newtons (N).

On Jupiter, she would weigh 80 N. Jupiter's gravitational force is 2.65 times stronger than Earth's.

On the moon, she weighs 5 newtons (N). The moon's gravitational force is 1/6 as strong as Earth's.

Fig. 6-7. Scientists use the newton (N) as a unit to measure force. Weight is a force created by gravitational pull. The newton is used to express the force of weight.

Fig. 6-8. Newton's first law of motion states that a body stays at rest unless a force has enough strength to cause the object to move or change direction.

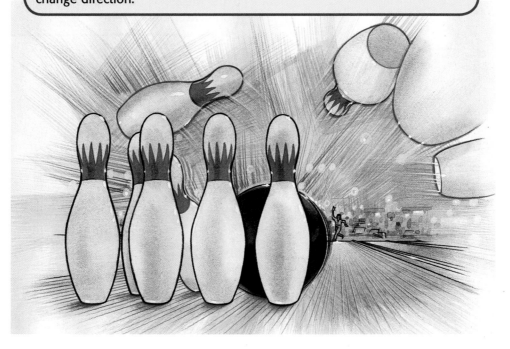

Newton's first law can be used to explain the effectiveness of a car's seat belt system. A car moving down a highway has inertia. Its tendency is to continue to move in a straight line. If the driver slams on the brakes, the car begins to slow down because the brakes oppose the car's motion. However, inertia will cause the passengers inside the vehicle to continue moving forward until something stops them. If the passengers are wearing seat belts, the belts will stop them and help prevent injury. See **Fig. 6-9**.

Second Law of Motion

Newton's second law of motion explains the relationship between force, mass, and acceleration. The rate at which an object moves depends on two things: the size of the force acting on it and the mass of the object. This relationship is stated in the following formula:

$$\text{Force} = \text{mass} \times \text{acceleration}$$

or

$$F = m \times a$$

Suppose you applied the same force to an empty wagon and to a wagon of the same size filled with bricks. Which wagon would move with greater acceleration? Because the mass of

Math Link

Calculating Force. How much force might be generated by a tractor-trailer with a mass of 40 tons and a final acceleration speed of 50 miles per hour? Would that be more force than that generated by a car with a mass of 2 tons being raced at 100 mph?

Apply the formula $F = m \times a$ to solve these problems.

Fig. 6-9. The brake slows the car, but the driver continues forward because of inertia. The safety belt stops the passenger's forward motion.

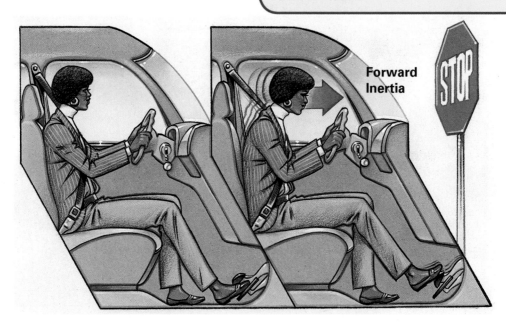

Forward Inertia

the empty wagon is less, it would move with greater acceleration. The wagon filled with bricks would require greater force to accelerate at the same rate. That force can be calculated by multiplying the wagon's mass by the desired change in velocity.

Engineers must often factor in Newton's second law. For example, what if you were asked to design a vehicle powered by a rubber band? Should the vehicle be as light as possible or as heavy as possible? What does Newton's law tell you? See **Fig. 6-10**.

Third Law of Motion

All forces occur in pairs. Newton's third law of motion states that for every action there is an equal and opposite reaction. Suppose you were standing absolutely still on a skateboard. If you were to push a basketball away from you, in what direction would the skateboard move? Remember that the forward motion of the ball would be the action. What would be the direction of the reaction force? See **Fig. 6-11**.

Fig. 6-10. This motorcycle was designed to go fast. Will a motorcycle accelerate faster if it is heavier or lighter in weight?

Fig. 6-11. Newton's third law describes forces as always being in pairs. While such forces work in opposite directions, they have equal strength.

Action

Reaction

A tennis racket is designed to take advantage of this law. The racket has an oval shape. This shape allows the racket frame to withstand the force placed on it by the stretched strings. Why are the strings stretched so tightly?

Imagine that you are on a tennis court. Your opponent serves the tennis ball to your side of the court. The ball is moving at a high velocity. If you simply hold out your racket to meet the ball, the ball will return at the same force applied to it by your opponent. This is action and reaction.

If you swing the racket towards the oncoming ball, you will meet it with increased force. This will cause the ball to stretch the flexed strings even further. See **Fig. 6-12**. The stretched strings try to return to their original position.

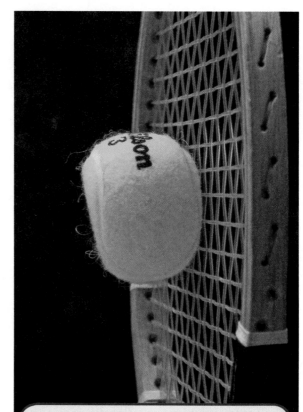

Fig. 6-12. This tennis ball moves away from the racket after it meets an increased force.

Reading Link

Decoding Formulas. Think of a math formula as a procedure written in code. Abbreviations and symbols are used in place of words. If you know what the abbreviations and symbols mean, you can decode the formula and understand the procedure. For example, the formula

$$A = lw$$

means "Area equals length times width." The procedure for finding the area of a shape is to multiply its length times its width.

Write the following formulas in words. Look in a math book to find the meanings of any symbols or abbreviations you don't know.

Area of a circle: $A = \pi r^2$
Volume of a sphere: $V = \frac{4}{3}\pi r^3$
Area of a triangle: $A = \frac{1}{2}bh$
Density: $D = m/V$

In doing so, they multiply the force applied to the ball. This causes the ball to return across the net at an even higher velocity.

How would you use this law to design a small ice racer that would travel across a frozen pond powered only by a balloon? Here's a hint: rockets use a similar system of escaping gases to create forward motion.

Physics of Work

What does the word "work" mean to you? You might think of it as cutting the grass or cleaning your room. However, work has a different meaning to scientists or engineers. To them, **work** is the application of force to make

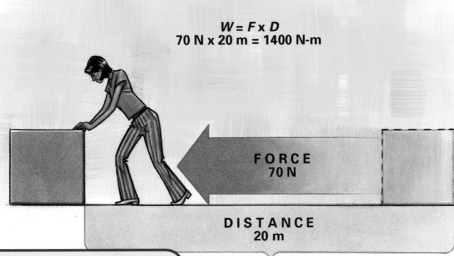

$$W = F \times D$$
$$70 \text{ N} \times 20 \text{ m} = 1400 \text{ N-m}$$

FORCE
70 N

DISTANCE
20 m

Fig. 6-13. Work is accomplished when an object is moved a certain distance by an applied force. To calculate the work accomplished, multiply the force by the distance.

an object move in the direction of the force. If you push a heavy box 20 meters across the floor, you are doing work. See **Fig. 6-13.** The amount of work accomplished can be measured using the formula:

$$\text{Work} = \text{force} \times \text{distance}$$
or
$$W = F \times d$$

In the metric system, the unit of measure for force is the newton (N), and distance is measured in meters (m). The metric unit of measure for work is the newton-meter (N-m). The force of one newton pushing or pulling an object a distance of one meter equals one newton-meter (N-m) of work. Scientists sometimes use another unit, the joule, to describe work. (Joule is pronounced "jewel.") One newton-meter equals one joule (J).

Assume that the box you pushed along the floor weighed 70 N. As mentioned, you moved the box 20 meters. How much work have you accomplished? See again **Fig. 6-13.**

Writing Link

Metric Abbreviations. In the metric system, units that were named for a person are symbolized by capital letters. Thus the symbol for newton is N, and the symbol for joule is J. Note that, while the symbol is capitalized, the complete word is in lowercase letters. An exception is the degree Celsius, which is always capitalized.

Find out what the following symbols for metric units mean. Find out the full names of the persons for whom the units were named. Write the information on a separate sheet of paper.

Hz	K	Pa
T	A	V

Suppose a crane lifts a steel beam of 1,400 N to a height of 20 m. How much work did the crane perform?

$$W = F \times d$$
$$W = 1,400 \text{ N} \times 20 \text{ m}$$
$$W = 28,000 \text{ N-m}$$

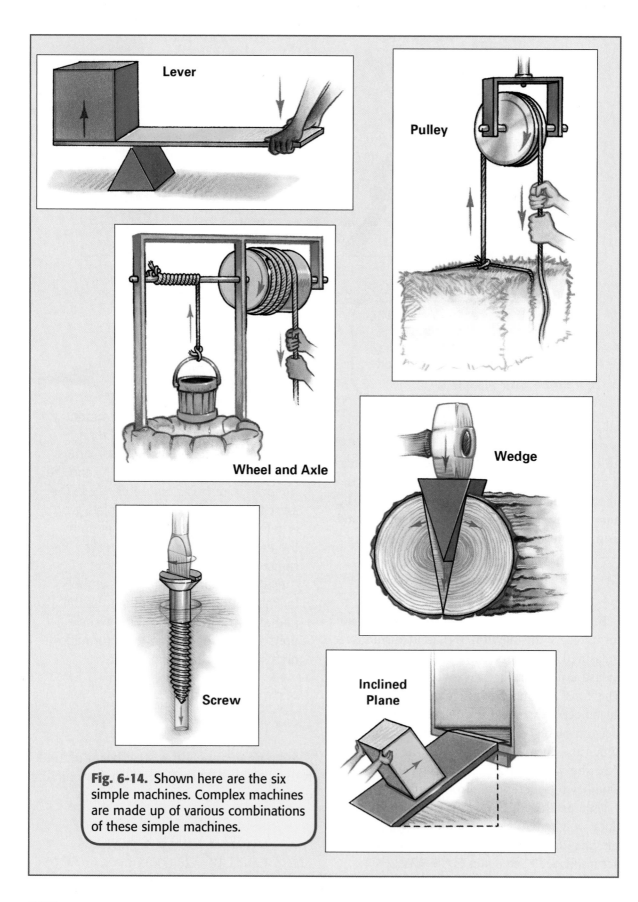

Lever

Pulley

Wheel and Axle

Wedge

Screw

Inclined Plane

Fig. 6-14. Shown here are the six simple machines. Complex machines are made up of various combinations of these simple machines.

Work and Machines

How can work be made easier and faster? You can use a machine. A **machine** is a device designed to obtain the greatest amount of force from the energy used. For example, you can open a metal container in seconds using a can opener. A car jack can help you easily lift a car weighing thousands of pounds. Tasks are accomplished more quickly and easily when we use machines.

Machines help us accomplish work by increasing a force or changing its direction. A machine may look very complicated. However, a complex machine really consists of a combination of simple machines. See **Fig. 6-14**.

Some machines multiply a force more effectively than others. The number of times a machine multiplies a force is called the machine's **mechanical advantage**.

A zipper is a good example of a simple machine. It is made up of two rows of teeth that lock into each other. Have you ever tried to push the teeth together by hand and get them to lock? The force needed is too great.

Fig. 6-16. When a pebble hits the water, it transfers its energy to the water. A wave pattern spreads out from the point of contact, as particles of water are pushed away. Light and sound waves move in a similar way.

Carefully look at the part of the zipper that moves up and down. It is actually three tiny wedges. The wedges increase our mechanical advantage and apply the needed force to lock the zipper teeth together. See **Fig. 6-15**.

 To learn more about how machines accomplish work, see the "Mechanisms" lab on the Student CD.

Physics of Waves

What happens when you drop a pebble into a bowl of water? Can you describe what you see? The pebble creates waves. These waves move outward from the point where the pebble hits the water.

When you drop a pebble into a bowl of water, the energy from the falling pebble is transferred to the water. The particles of water are disturbed. The energy places them in motion. See **Fig. 6-16**. Each particle of water bumps into a neighboring particle of water, and a wave is formed. A **wave** is a disturbance that transfers energy from one place to another through matter or space.

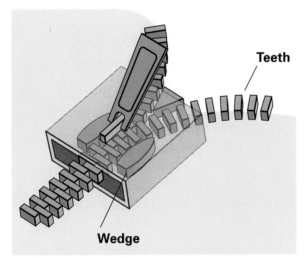

Teeth

Wedge

Fig. 6-15. Simple machines increase our mechanical advantage and help us apply forces we could not provide with just our own strength.

Fig. 6-17. It is possible to hear more than one echo from your voice. How can that happen?

We know that waves can travel through liquids such as water. They can also travel through solids and gases such as metal and air. A liquid, solid, or gas that allows waves to pass through it is known as a **medium**. Water is the medium for ocean waves. Air is a good medium for sound waves. Energy is transferred by a medium because the particles that make up a medium vibrate. Did you know that sound and light also move through the air in waves?

Sound Waves

Sound is a form of energy. It causes the molecules within a medium to vibrate. As the molecules vibrate, they compress neighboring particles and a sound wave is formed. The sound energy is transferred through the medium as a wave.

Some mediums transfer sound better than others. The molecules within a solid material transmit sound waves more easily than liquids and gases. The particles are packed closer together, so the transfer of the wave energy is more efficient.

Have you ever shouted in a large, empty room and heard an echo? Echoes are heard because sound energy bounces off a hard, flat surface back to our ears. The repeated sound is an echo. See **Fig. 6-17**.

Sonar is a device used on ships to measure the depth of water and locate underwater objects. Sonar equipment on the ship sends pulses of sound energy through the water in the form of sound waves. These sound waves strike the ocean floor or an underwater object and echo back to the ship. The time it takes the sound wave to travel down and back is measured. Sound travels through water at a speed of 1,500 meters per second. By knowing the speed of sound and the time it takes the sound wave to bounce back to the ship, we can calculate the distance the wave traveled. See **Fig. 6-18**.

Science Link

Wave or Particle? Light is often described as being a form of wave energy, but light also shows characteristics of particles. For years, scientists argued about whether light was particles or waves. Can both sides be correct? Is it possible for light to be both wave and particle energy? Research how the controversy began.

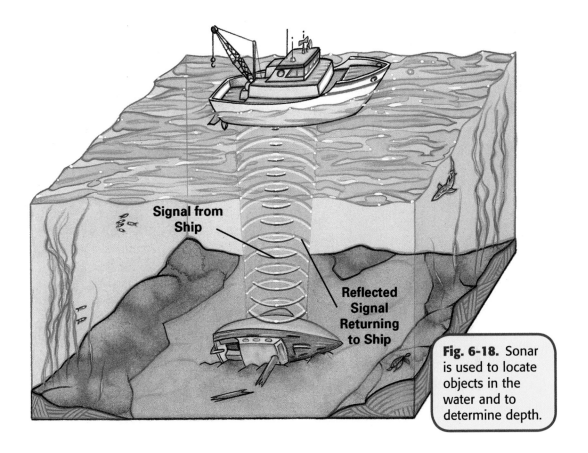

Signal from
Ship

Reflected
Signal
Returning
to Ship

Fig. 6-18. Sonar
is used to locate
objects in the
water and to
determine depth.

Ultrasound is sound that is so high-pitched people can't hear it. Ultrasound machines are used in the medical field. For example, they can produce images of babies before birth. The ultrasound machine works in the same way as sonar. It sends sound waves that bounce off the baby inside the mother. The echo is measured by the machine, which produces an image of the baby on the display screen. See **Fig. 6-19**.

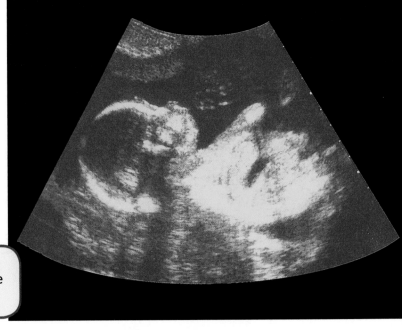

Fig. 6-19. This sonogram from the ultrasound machine is the first time parents can see their child.

Light Waves

Like sound, light also travels in waves. However, light waves and sound waves are very different. Unlike sound energy, light energy can be transmitted without a medium. For example, sound waves cannot travel in a vacuum (the absence of air). Light waves can. That's why we can see the light (which has passed through the vacuum of outer space) from the stars.

The source of all light energy is the atom. An atom consists of a central nucleus surrounded by a cloud of negatively charged particles called electrons. Electrons position themselves around the nucleus in energy levels. At times, electrons absorb additional energy. This change is only temporary. The electron soon loses its extra energy, releasing photons.

Photons can be thought of as packets or bundles of light energy. Photons move in waves. As the waves of photons move, they produce electrical and magnetic fields around them. For this reason light waves are called electromagnetic waves. Electromagnetic waves do not

need a medium in which to travel. To learn more about these waves, see the "Electromagnetic Spectrum" lab on the Student CD.

Natural light is all around you. The most common form is sunlight. Other forms of natural light are the light from a lightning bug or the light from fire.

Light can be artificial, too. The most common sources are incandescent and fluorescent lamps. See **Fig. 6-20.** Lasers are another source of artificial light. A laser is a narrow, high-energy beam of parallel light rays. Laser light is a very powerful source of single-color light. To learn more about artificial sources of light, see Chapter 15, "Lasers & Lights."

Did you know that microwaves are a form of electromagnetic waves, or light waves? The food in a microwave oven absorbs the energy of microwaves. The water or sugar molecules in the food vibrate. This vibration produces heat, causing the food to cook. See **Fig. 6-21.** Without light waves, some of us might starve!

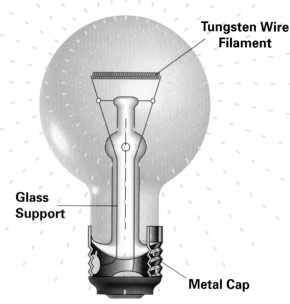

Photons of Visible Light

Tungsten Wire Filament

Glass Support

Metal Cap

Fig. 6-20. The filament of a lightbulb is made of tungsten wire. This metal wire creates heat by resisting the flow of electricity. When the filament is white hot, it gives off photons of visible light.

Impact of Technology

Microwave Ovens

Microwave ovens are in millions of homes and restaurants. They save time and energy and are easy to use. However, there are some safety issues. For example, some containers should not be used for heating food in a microwave oven. When the containers become hot, they release harmful chemicals into the food. Some people believe that the microwaves themselves have negative effects on food and that it is less nutritious than food cooked by traditional methods.

Investigating the Impact

Research the controversies over microwave ovens.
1. What safety rules should be observed when cooking or heating food in a microwave oven?
2. What claims are made about the dangers of microwaved food? What evidence has been found to support these claims?

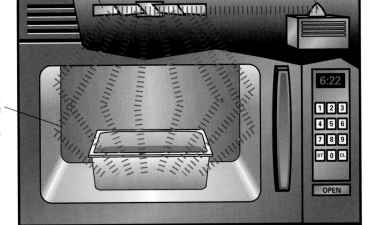

Magnetron tube creates the microwave.

Stirrer Blade

Microwaves bounce off oven walls into food.

6:22

MICROWAVE OVEN

Fig. 6-21. Energy from microwaves is absorbed into the food more quickly than in traditional methods, so the food cooks in less time.

Giving a Lift

Identify a Need/Define the Problem

You and your team must design and build a tabletop crane that can be used to lift panels into place. The crane must increase your mechanical advantage and ability to do work.

Gather Information

Have you ever seen a crane used to lift a heavy object? An example is shown in **Fig. A**. Gather images that you can find of cranes at work. Also research how to increase mechanical advantage. Find out the types of mechanisms that can be used to construct a model crane.

Develop Possible Solutions

Determine the size of the crane you will build. Sketch possible design solutions. Include the frame and body of the crane. Sketch the pulley combination(s) you will use. Show how it will connect to the frame of the crane. An example of a model crane with a pulley system is shown in **Fig. B**.

Safety Alert

Look up "Safety Data Sheets" on the Student CD and prepare a data sheet for this activity. As you work on the activity, be sure to follow all safety rules.

Materials and Equipment

Select from this list or use your own ideas.

- ⅛" × ⅛" balsa strips
- string
- pulleys
- cardstock paper
- cardboard
- dowels of various diameters
- glue
- modeling materials
- material processing tools and machines
- 1,000-gram weight

 For information about tools and machines, see "Hand Tools" and "Power Tools and Machines" on the Student CD.

Model a Solution

1. With your team, select the design that you think will be the most effective.
2. Build the crane using the materials you have chosen.

Test and Evaluate the Solution

- Lift a 1,000-gram weight 10" off the table.
- Calculate the work accomplished by your crane.
- Was the crane able to lift the weight?
- Was the crane able to support the weight once it had been lifted?

Fig. A

Fig. B

Refine the Solution

If necessary, create a new crane model. Keep in mind the following questions:
- How could the crane be improved to lift and hold more weight, based on the knowledge that you have gained from testing it?
- Could the crane be created using cheaper materials but still carry the same weight?
- Run the lifting test again with the new crane. Did your results improve? Why or why not?

Communicate Your Ideas

Demonstrate the operation of the crane to the class. Be sure to point out any modifications that you made after the initial testing.

Math Link

Crane Away. In addition to lifting things, cranes are also used to help dig holes. For example, cranes could be used to help excavate earth in preparation for construction, or to remove ore for mining purposes. These cranes often have a large scoop or shovel at the end. Let's say we have a shovel crane capable of scooping up 20 cubic yards at once. How many scoops would it take to excavate a football field 3 feet deep? (Assume the playing surface of a football field is about 300 feet by 150 feet.) Show your work and be prepared to explain your answer.

Racing Reaction Rockets

Identify a Need/Define the Problem

Rocket and jet engines are "reaction" engines. Newton's third law of motion explains how these engines work. Hot gases escaping from the rear nozzle of the engine are the "action" force. The "reaction" force, called thrust, propels the rocket or jet in the opposite direction. **Figure A** shows a real car with a rocket engine.

In this activity, you will design and build a reaction rocket racer that will travel a measured distance.

Gather Information

Research how reaction engines work. What factors influence the speed at which an object will move across the floor when powered by air escaping out of a balloon? List the factors.

Materials and Equipment

Select from this list or use your own ideas.

- plastic foam meat tray (use only new trays) or foam core board
- pins
- masking tape
- thin cardboard
- flexible straw
- small round balloon
- ruler
- scissors
- compass
- tape measure
- stopwatch
- scale

 If you're not sure how to use measuring tools, see the "Measurement" lab on the Student CD.

Fig. A

Develop Possible Solutions

Use the list of factors to help you prepare several designs for a rocket racer. One possible design is shown in **Fig. B**.

Model a Solution

1. After choosing the design you think will be the most effective, construct your rocket racer.
2. Make a 20' test track and lay it out on the floor of the classroom.

Test and Evaluate the Solution

- Test the rocket racer. Time how long it takes for your vehicle to travel down the track.
- Make at least five test runs. Average the run times.
- Compare the success of your vehicle with the success of other vehicles in your class. What additional modifications might you make to increase speed?

Refine the Solution

- Make changes based on the testing and evaluation. For example, your car might need to be lighter or heavier, depending on its performance.
- Test the racer again and compare the results. Did you notice any improvement?

Communicate Your Ideas

Discuss your final results with the class. Be sure to point out any modifications that you made after testing.

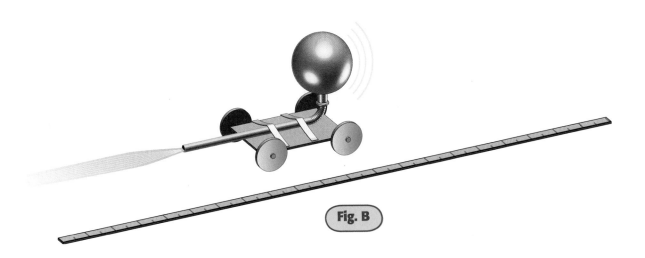

Fig. B

Creating a Sound Barrier

Identify a Need/Define the Problem

As the population of the country continues to grow, more highways are constructed. The cars and trucks on the highways create a lot of noise as they rush down the road. Many of these highways are close to homes and businesses, where a quieter environment is preferred. The solution to this noise problem is to create sound barriers to keep the highway noise from spreading to the surrounding area. See **Fig. A** for an example. In this activity, you will create your own sound barrier.

Gather Information

In order to create sound barriers you will have to have some knowledge of sound waves. It will also help to research sound barriers to see how they are constructed and installed.

Fig. A

Materials and Equipment

Select from this list or use your own ideas.

- balsa wood
- compass
- decibel meter
- flexible straw
- foam core board
- hot glue gun
- masking tape
- noise source (to simulate traffic)
- ruler
- scale
- scissors
- stopwatch
- plastic foam meat tray (new trays only)
- tape measure
- thin cardboard

 To learn how to use measuring tools, see the "Measurement" lab on the Student CD.

Develop Possible Solutions

Sketch at least three different design solutions for sound barriers. Be sure to show how the sound barrier will be constructed and what it will be made out of. The following questions should be answered in your design solutions

- How long will the sound barrier be?
- How high will the sound barrier be?
- How far from the noise source will the barrier be?
- Will the sound barrier be nice to look at?

Model a Solution

1. Choose the solution that you think will be the most effective.
2. Construct your sound barrier according to your design plans.

Test and Evaluate the Solution

- Set the decibel meter 18 inches from the noise source.
- Place the sound barrier between the noise source and decibel meter. See **Fig. B**. Record the decibel reading.
- Change the distance between the sound barrier and noise source. Record the new distance and decibel readings.

Refine the Solution

- Did the sound barrier work?
- Where was it most effective?
- What could be done to improve it?
- If necessary, create a new sound barrier based on your recorded data.

Communicate Your Ideas

Write a proposal to a committee about your sound barrier. Your proposal should include your plans and your recorded data.

Science Link

Sounding Off on Acoustics. The science of sound is called acoustics. Acoustic engineers test and design all sorts of things to make our lives better.

Imagine you are an acoustical engineer. Provide answers for the following three questions. Be prepared to explain and defend your answers.

1. Why does sound seem to travel further at night?
2. Why can you only sometimes hear an echo?
3. How could people tell if a train was a few miles away by putting their ears close to a railroad track?

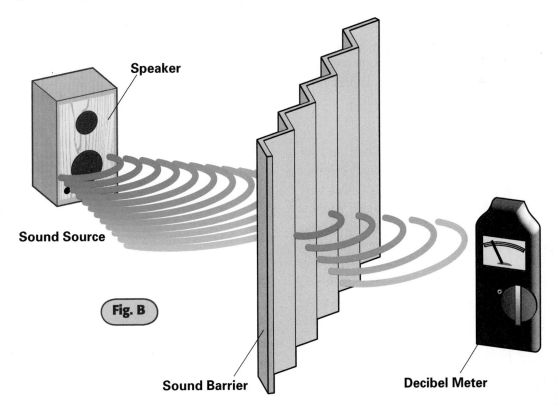

Speaker

Sound Source

Fig. B

Sound Barrier

Decibel Meter

Exploring Careers

Physicist

ENTRY LEVEL | TECHNICAL | **PROFESSIONAL**

Physicists research and study physics, the laws of matter and energy. They work to understand the nature of matter and energy and how they relate to each other. Physicists perform research in a laboratory. They develop a hypothesis and then design and run scientific tests and experiments. They analyze the results using math principles and special equipment.

There are many specialties in physics, such as elementary particle physics, optics, acoustics, and the physics of fluids. Most physicists specialize in one area, but these areas may overlap with each other and with other disciplines, such as chemistry, biology, and engineering.

About 30% of physicists work for research and development companies and 30% for the federal government. Some physicists teach and perform research at colleges and universities. A small number of physicists work in testing, inspection, quality control, and other production-related jobs for companies.

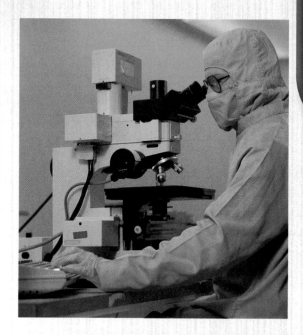

Qualifications

Physicists must have a broad math and science background. A doctoral degree is required for most jobs in research. Physicists with a master's or bachelor's degree may work in manufacturing, in applied research and development, in industry, or in high schools or community colleges.

Students working toward a bachelor's degree in physics take classes in such subjects as electromagnetism, quantum mechanics, and thermodynamics. After receiving a bachelor's degree, students can begin studying for an advanced degree.

Outlook for the Future

Jobs for physicists are increasing, though the rate of growth is slower than average. New job openings will depend on budgets, such as in the defense industry.

Self-Esteem

As with any job, it is important to know that you will make mistakes, so you need to focus on your abilities and successes. A positive step is to set goals and then work hard to achieve them. In addition, doing things for others can help you feel good about yourself.

Researching Careers

Learn more about jobs for physicists. Choose a famous physicist and study him or her. In what area of physics did this person work? What are his or her achievements? Give a presentation to the class.

More activities on Student CD

Key Points

- Engineers must understand physics concepts in order to design new technologies.
- Motion describes the movement of all objects.
- Newton's three laws help explain how motion works.
- People use machines to make work easier.
- Sound and light travel in waves.

Read & Respond

1. Define *applied physics*.
2. What are the steps in the scientific method?
3. What is force?
4. Describe the two main forces associated with motion.
5. Explain Newton's three laws.
6. How do machines help do work?
7. What is a machine's mechanical advantage?
8. What is the formula for work?
9. How does a light wave travel differently than a sound wave?
10. What creates echoes?

Think & Apply

1. **Connect.** What forces are acting on a person parachuting from a plane?
2. **Extend.** Using the knowledge you gained from this chapter, explain how we hear the sounds from a guitar.

3. **Relate.** Sports can be a good way to observe motion. Choose a sport and describe the forces present in a game.
4. **Design.** Design three devices to demonstrate the three kinds of friction.
5. **Plan.** Imagine you have to move a heavy computer still in its box up three stairs and place it on a high shelf. Create a plan to show what simple machines you could use.

TechByte

The Sound of Silence. Audio companies have begun to market active noise cancellation devices. Such a device doesn't just block sound; it "listens" to incoming sounds and instantly produces the inverse (upside-down) version of the sound wave. The inverse wave signal collides with the incoming noise wave, and the two waves cancel each other out. Goodbye, noise!

CHAPTER 7

Air & Space Technologies

Objectives

- Identify key developments in air and space technology.
- Define *aerospace*.
- Describe how airplanes use force to help control flight.
- Explain how jet engines work.

Vocabulary

- **entrepreneur**
- **aerospace**
- **force**
- **gravity**
- **momentum**
- **friction**
- **drag**
- **inertia**
- **aerodynamics**
- **Bernoulli effect**
- **airfoil**
- **aileron**

These pilots are in a Boeing 767 cockpit simulator.

Activities

- Foiling Gravity
- Creating a Propeller
- Blasting Off with Rockets

To Fly Like a Bird

The desire to fly like a bird is as old as human-kind itself. Did our ancient ancestors wonder how birds managed to get off the ground? Maybe they even tried to fly by flapping their arms. We will never know.

How are birds able to fly? Birds have many traits that help them fly. They have wings, powerful muscles that help them rise in the air, hollow bones that make them lighter, and thin bodies covered in feathers that improve air flow. See **Fig. 7-1**.

Fig. 7-1. This cardinal must move its wings many times to overcome gravity.

Aviation Entrepreneurs

In a free enterprise system, most of the resources used to make products or provide services are owned by individuals and groups, not by the government. People decide for themselves what to buy, where to work, and what businesses (enterprises) to develop. In a free enterprise system, almost anyone can become an **entrepreneur**, a person who starts a business. Here are some of aviation's major entrepreneurs.

- William E. Boeing graduated from Yale University in 1903, the same year that the Wright brothers made their first flight. Boeing went on to start an airplane manu-facturing company, which evolved into the huge company that built the Boeing 727, 747, and today's 777.

- In the 1930s, Donald Douglas needed to do something to compete with Boeing's 10-passenger commercial airliner known as the 247. Douglas developed the DC-1, which flew for TWA in 1933. Then in 1936 he produced the legendary DC-3 (see photo), which became the first passenger plane to make money.

- In 1924, Clyde Cessna, Lloyd Stearman, and Walter Beech started Travel Air Manufacturing Company. After a few years, they split up and each started his own successful aircraft com-pany. Today Cessna is the largest manufac-turer of small private planes.

- For the past 30 years, Burt Rutan's company, Scaled Composites, has averaged more than one new aircraft design per year. The most famous might be *SpaceShipOne,* which was the first private aircraft to fly into suborbital space twice within two weeks.

FLYING THROUGH TIME

1485—Leonardo da Vinci designed the Ornithopter flying machine. While it was never built, its concept would eventually be inspiration for the helicopter.

1783—Joseph Michel and Jacques Montgolfier invented the first hot-air balloon.

1891—Otto Lilienthal designed the first glider that could actually fly a person. A glider has wings like an airplane but has no engine.

1903—Two bicycle engineers, Orville and Wilbur Wright, studied the use of gliders and performed experiments to create their own aircraft. On December 17, 1903, Orville made the first powered, sustained, and controlled flight.

1924—The first flight around the world was made by a team of pilots.

1926—Robert Goddard built and launched the first liquid-fueled rocket.

1939—The first jet was flown. It was developed by Hans Von Ohan and Ernst Heinkel.

1981—The first space shuttle launched into orbit.

Fig. 7-2. Name some achievements in flight technology that have taken place in your lifetime.

Fig. 7-3. This laser-powered airplane has a wingspan of five feet. It is controlled by radio commands and powered by ground-based laser beams.

Reading Link

Prefixes. The word *prefix* comes from a Latin term meaning "to fasten before." A prefix is attached to the beginning of a word, and it affects the word's meaning. Knowing what a prefix means can help you figure out the meaning of an unfamiliar term. For example, the prefixes *aer* and *aero* relate to air, or the atmosphere. When you know that, you know that *aerospace* must have something to do with spaces in the air and not, for example, spaces between words or between teeth. What do you think the following terms mean?

aerate   aerobic
aeronautics   aerosol

At first, humans tried using feathers and wood to construct wings. However, humans don't have the muscle power to fly like birds. Humans had to create technology in order to fly. See **Fig. 7-2**.

Aerospace (AIR-oh-space) is the study of how things fly. Today's aerospace industry is responsible for the production of airplanes, satellites, missiles, helicopters, blimps, and space vehicles. The designers of these flying machines are aerospace engineers. Aerospace engineers apply their knowledge of science and technology to develop new aircraft and spacecraft.

Aerospace engineering has produced some remarkable craft. Did you know that some airplanes today weigh more than a million pounds? The Airbus A380 is one example. When fully loaded, it can weigh 1.2 million pounds. In contrast, NASA's experimental laser-powered aircraft weighs only 11 ounces. It has no engine or motor. Instead, this small model airplane gets its "fuel" from a ground-based laser. The airplane is equipped with photovoltaic cells that convert the energy from the laser beam into electricity that powers the plane. See **Fig. 7-3**.

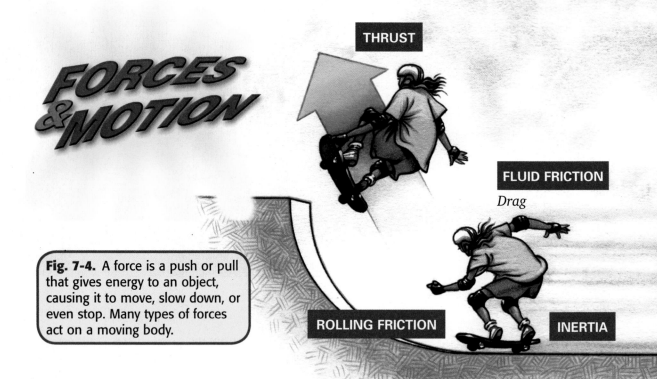

THRUST

FLUID FRICTION
Drag

ROLLING FRICTION

INERTIA

Fig. 7-4. A force is a push or pull that gives energy to an object, causing it to move, slow down, or even stop. Many types of forces act on a moving body.

What Makes Things Fly?

Things move only when force is applied. **Force** is a push or pull that transfers energy. Force can set an object in motion, stop its motion, or change its speed and direction.

Have you ever been skateboarding? Skateboarding can help you understand force. Forces help and hinder you while going up and down a ramp. See **Fig. 7-4**.

The force of gravity pulls you down the ramp. **Gravity** is the force that pulls objects toward the center of the earth.

You gain momentum as speed increases. **Momentum** is the connection between how fast an object is moving and the mass of the object. Objects that are very heavy need a greater force to make them move. When heavy objects are moving, it takes a greater force to stop them or change their direction.

Certain forces oppose, or act against, objects in motion. Friction is one of these forces. **Friction** opposes motion between two surfaces that are touching. Even molecules of air mov-

ing across an object can cause enough friction to slow it down. The force of air on a moving object is called fluid friction, or **drag**. As you move along the ramp, drag slows you down.

After hitting the top of the skateboard ramp, you become airborne. Motion continues because of inertia. **Inertia** can be described as an object in motion resisting any change in its speed and direction unless another force acts on it. See Chapter 6, "Applied Physics," for more about inertia.

Remember how gravity helped you skateboard by creating momentum? That same gravity is the reason you are pulled down to the ground from the air. Everything on the ground and in the air is affected by gravity.

If gravity pulls things toward the ground, how can anything fly? Birds, planes, rockets, and all other things that fly generate forces that are greater than opposing forces such as gravity and drag. See **Fig. 7-5**.

POTENTIAL ENERGY

Energy stored in an object at rest

GRAVITY

KINETIC ENERGY

Energy of motion

GRAVITY FORCE

Weight

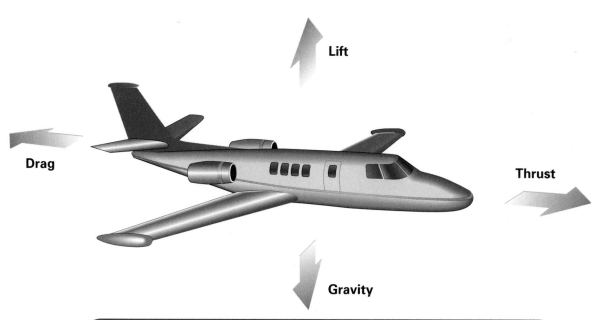

Lift

Drag

Thrust

Gravity

Fig. 7-5. The combined action of these four forces enables a plane to take off, fly, and land. Which of these forces are opposing forces?

Thrust

Thrust is a forward force. On a jet airplane, thrust is created by the plane's engines. On propeller planes, the engines make the propellers spin. The motion of the propellers creates thrust.

Lift

Lift is an upward force. The shape of an airplane's wings—or a bird's—helps provide the lifting force needed to fly. (Wing shape will be discussed later in this chapter.) When the force of lift is greater than the weight of the airplane, the plane will rise into the air.

In order for the plane to remain in the air, it must maintain a minimum speed. This is the speed where lift equals the plane's weight. Below that speed, the plane will start to drop.

Aerodynamics

You read about drag on page 156. Drag opposes forward movement and will slow down an object.

Aerodynamics (air-oh-dy-NAM-iks) deals with the forces of air on an object moving through it. One goal of aerodynamics is to design objects so that fluid friction is reduced.

How does aerodynamics help you skateboard? Helmets are shaped so that air flows over your head more easily, much in the same way air flows easily over a bird's feathers. Many vehicles, such as cars and boats, have aerodynamic designs that reduce fluid friction. Aircraft and rockets are designed with rounded smooth surfaces for the same reasons. See **Fig. 7-6**.

 To see the effects of drag, thrust, gravity, and lift on airplanes, go to the "Forces and Flight" lab on the Student CD.

Propeller Airplanes

The wings of an airplane are used to create lift. Each wing is round in the front, thickest in the middle, and narrow at the back. This shape is designed to speed up the air passing over the top surface. A fast-moving fluid exerts less pressure than a slow-moving fluid. This is the **Bernoulli effect**. Because air moves faster over the top of an airplane's wing than under it, there is less pressure above the wing. The difference in pressure helps create lift.

A shape designed so that air flowing around it produces useful motion is called an **airfoil**. The wings of an airplane are airfoils. See **Fig. 7-7**.

Science Link

The Bernoulli Effect. What does an airplane wing have in common with a shower curtain? When the water is running, a shower curtain seems to "blow in" toward the shower. Something is pushing the curtain inward. What's going on? Here's a clue: A shower spray displaces air and therefore decreases the air pressure within the shower enclosure.

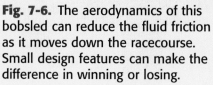

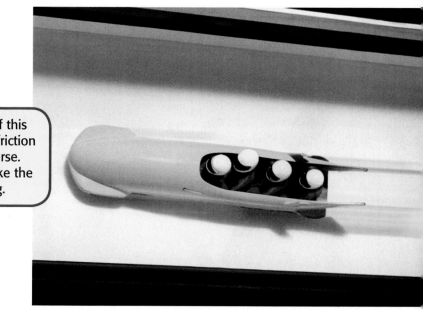

Fig. 7-6. The aerodynamics of this bobsled can reduce the fluid friction as it moves down the racecourse. Small design features can make the difference in winning or losing.

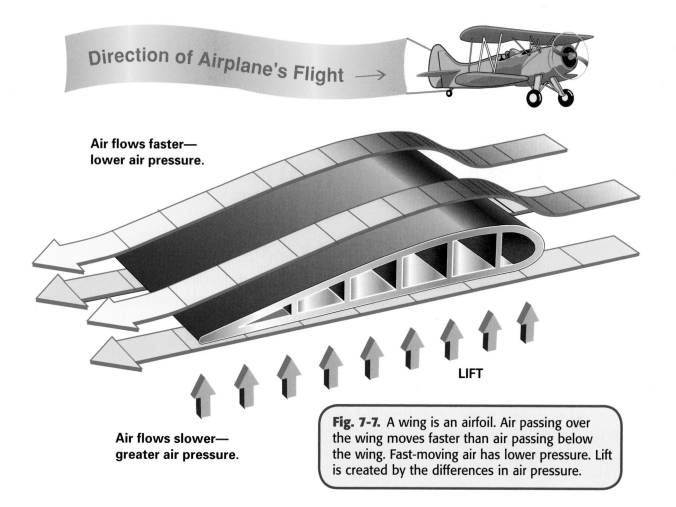

Direction of Airplane's Flight →

Air flows faster— lower air pressure.

LIFT

Air flows slower— greater air pressure.

Fig. 7-7. A wing is an airfoil. Air passing over the wing moves faster than air passing below the wing. Fast-moving air has lower pressure. Lift is created by the differences in air pressure.

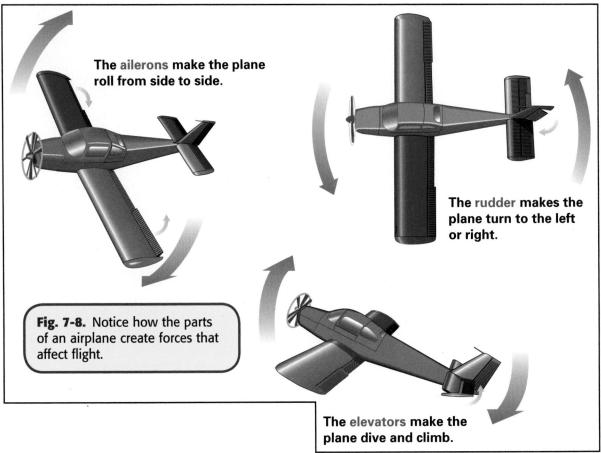

The **ailerons** make the plane roll from side to side.

The **rudder** makes the plane turn to the left or right.

Fig. 7-8. Notice how the parts of an airplane create forces that affect flight.

The **elevators** make the plane dive and climb.

Propellers are airfoils that spin. They create low pressure areas in front of the propeller so that high pressure areas behind the aircraft can push it through the sky. The propeller helps to move the wings through the air so they can create lift.

Controlling Propeller Airplanes

Pilots control the position of an airplane by adjusting specific surfaces located on the wings and tail section of the aircraft. They deflect air and create drag. The increased drag allows the plane to turn.

Ailerons (AY-luh-rons) change the airflow across the wing, increasing and decreasing the amount of lift the wing creates. Flaps at the front and rear edges of the wing also change the airflow and help in takeoffs and landings.

The rudder, located on the tail section of the airplane, is used to turn the aircraft. Elevators are used to help the aircraft climb and dive in

 the air. See **Fig. 7-8**. The "Forces and Flight" lab on the Student CD lets you control an airplane.

Helicopters

How is a helicopter different from an airplane? Helicopters can move in any direction. They can even hover, or float, in midair. The rotors, or blades, of the helicopter control its motion. Rotors are shaped like propellers. They are also airfoils, which provide the lift and thrust that the aircraft needs.

Controlling Helicopters

Helicopters have two to eight rotor blades. The flight of the helicopter is controlled by changing the angle, or pitch, of the blade. The

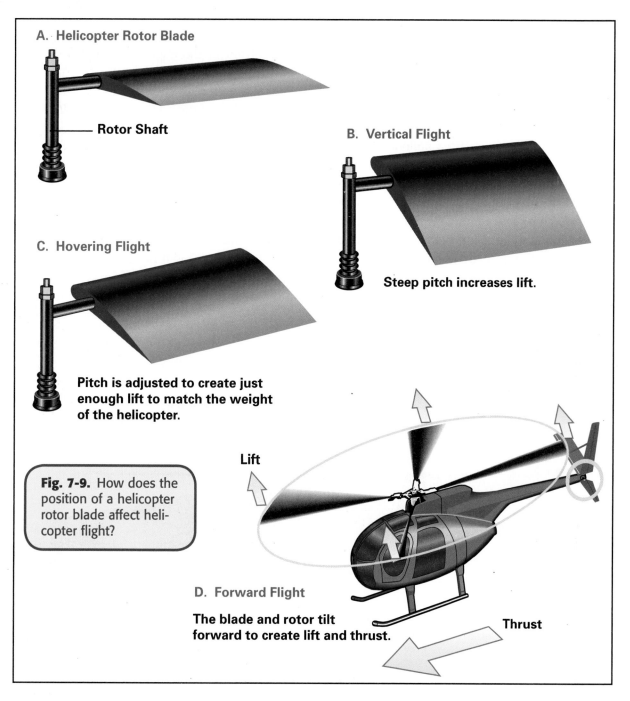

A. Helicopter Rotor Blade

Rotor Shaft

B. Vertical Flight

Steep pitch increases lift.

C. Hovering Flight

Pitch is adjusted to create just enough lift to match the weight of the helicopter.

Fig. 7-9. How does the position of a helicopter rotor blade affect helicopter flight?

Lift

D. Forward Flight

The blade and rotor tilt forward to create lift and thrust.

Thrust

front edge of the rotor blade can be raised or lowered to vary the amount of lift the blade creates.

As the pitch increases, lift increases and the helicopter moves vertically. When a helicopter hovers, the rotors are pitched so that they produce only enough lift to match the weight of the aircraft. See **Fig. 7-9.**

A helicopter moves forward when the rotor blades are moved forward. The combined pitch and forward position splits the force of lift into a raising force and a thrusting force.

A helicopter that has experienced engine failure can sometimes descend safely. Such a helicopter can autorotate to a rough landing. How does autorotation help break the fall of a helicopter?

Jet Planes

Jet planes streak across the sky every day. Have you ever noticed that they have no propeller? How is thrust provided without a propeller to move the aircraft?

Have you ever blown up a balloon and just let it go, allowing the air to escape from the balloon's open end? The balloon reacts by flying around the room in all directions. Newton's third law of motion states that for every action, there is an equal and opposite reaction. In the case of the balloon, the air escaping the open end of the balloon is the action. The balloon moving in the opposite direction is the reaction.

Jet engines work in a similar way. Air is drawn into the front of the engine. It is compressed, heated, and then ejected from the back of the engine at a very high speed. The air streaming out of the back of the engine is the action. The aircraft moving forward is the reaction. The reaction force moves the wings through the air, creating lift. See **Fig. 7-10**. Aircraft powered by jet engines are more powerful than propeller aircraft. They are also more common for commercial use. Aerospace engineers are designing larger and more powerful jet planes.

The Airbus A380 will be the world's largest commercial aircraft, carrying 555 passengers on two separate decks. See **Fig. 7-11**. A third deck

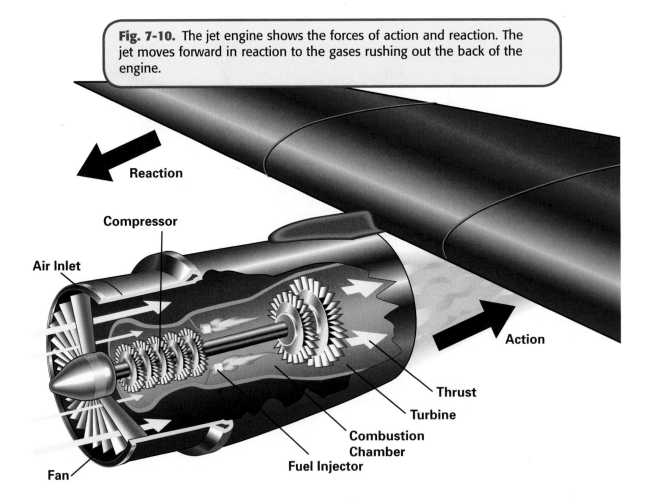

Fig. 7-10. The jet engine shows the forces of action and reaction. The jet moves forward in reaction to the gases rushing out the back of the engine.

Reaction

Compressor

Air Inlet

Action

Thrust

Turbine

Combustion Chamber

Fuel Injector

Fan

Fig. 7-11. The Airbus A380 is the largest and one of the most advanced and efficient commercial airliners ever built.

will be reserved for carrying cargo. Designers boast that the aircraft is more fuel-efficient than your family car. Having a wingspan of over 260 feet, the aircraft weighs 1.2 million pounds. It will take a lot of lift to move this airplane.

Boeing, one of the world's largest manufacturers of commercial airliners, also has a new jetliner. The Boeing 777-200LR is the world's longest-range commercial airplane. The jetliner can fly between any two cities in the world without stopping for refueling. The airplane can carry 301 passengers and fly up to 9,420 nautical miles without stopping. (A nautical

Math Link

Calculating Weight. When empty, the Airbus A380 has a weight of 608,400 pounds. This aircraft can carry 555 passengers. If the average weight of a passenger is 150 pounds, what is the total weight of the aircraft when all passenger seats are filled? (This total will not include the weight of the crew and any cargo that might be carried on the third deck.)

Fig. 7-12. This Boeing 777-200LR can fly further than any other commercial airliner without stopping to refuel. It is capable of flying between any two cities in the world.

mile is an international measurement used for air and water transportation.) See **Fig. 7-12**.

Much like propeller-operated planes, jet planes use the shape of the wings and the rudder to achieve flight. Jet planes also use the thrusters to control speed and direction.

Rockets

Rockets are the largest objects that fly. Like jets, rockets use the forces of action and reaction. Rockets move forward by pushing out powerful streams of hot gases. These gases are made by burning fuel. Rockets are powered by either solid fuel boosters or liquid fuel boosters. See **Fig. 7-13**.

Rockets are unique because they must work in outer space, where there is no oxygen. Rocket engines must carry oxygen with them.

Future space missions will require rockets to travel longer distances. Nuclear rocket engines could replace liquid hydrogen/liquid oxygen engines. Nuclear engines are more efficient and provide more thrust per pound of fuel. A nuclear reactor superheats liquid hydrogen. The hydrogen does not burn but passes through the nozzle of the rocket at high velocity, creating thrust. However, while nuclear engines are more efficient, they are not currently practical to launch. Engineers must determine how to solve safety issues before nuclear rocket engines could be used.

Controlling Rockets

The nozzle at the base of a rocket's engine can swivel. It can direct the burst of hot gases in different directions. The rocket reacts by changing direction. The swivel of the nozzle is used to steer the rocket. The space shuttle orbiter is steered by small rocket engines.

The Space Shuttle Orbiter

The space shuttle orbiter is the moving van of outer space. It is designed to carry astronauts as well as cargo, also called a payload, into space at low orbit around the earth. Carrying up to 65,000 pounds, the orbiter is able to launch communication and military satellites into space from its cargo bays. The United States also involved the shuttle in sending sections of the International Space Station into orbit for assembly. Before the loss of the shuttle *Columbia*,

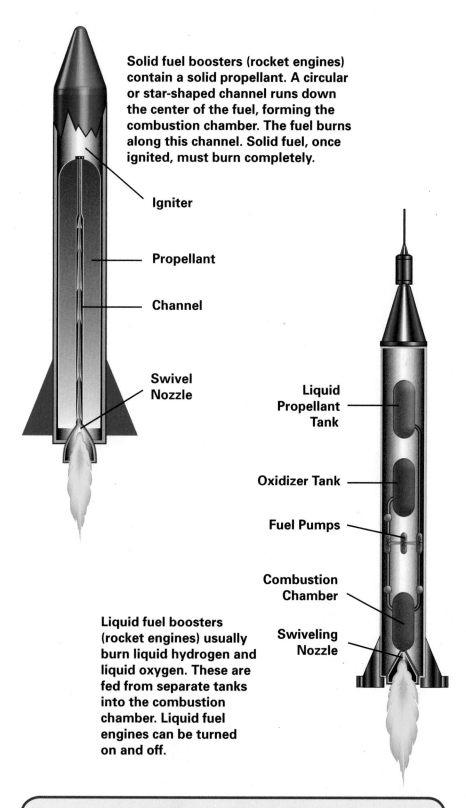

Solid fuel boosters (rocket engines) contain a solid propellant. A circular or star-shaped channel runs down the center of the fuel, forming the combustion chamber. The fuel burns along this channel. Solid fuel, once ignited, must burn completely.

Igniter

Propellant

Channel

Swivel Nozzle

Liquid Propellant Tank

Oxidizer Tank

Fuel Pumps

Combustion Chamber

Swiveling Nozzle

Liquid fuel boosters (rocket engines) usually burn liquid hydrogen and liquid oxygen. These are fed from separate tanks into the combustion chamber. Liquid fuel engines can be turned on and off.

Fig. 7-13. The way a rocket is fueled depends on the purpose of the rocket. What are the advantages and disadvantages of solid fuel and liquid fuel?

Writing Link

Living in Space. Near-zero gravity. Extreme cold. Lack of air. These are just a few of the hazards in outer space. Find out how NASA and other space agencies deal with these challenges so that humans can live and work in outer space. Write a report of your findings. Be sure to list your sources of information.

American orbiters were the main transporters of astronauts and Russian cosmonauts to and from the space station. See **Fig. 7-14**.

Next Generation Spacecraft

Aerospace engineers have been hard at work for years designing the next generation of reusable spacecraft. Engineers are currently working on two designs.

The Orbital Space Plane "Lifeboat" will be used as a rescue vehicle for the International Space Station. The National Aeronautics and Space Administration (NASA) believes that the first Lifeboat will be attached to the ISS by 2010. The Lifeboat will be lifted into space by disposable rockets.

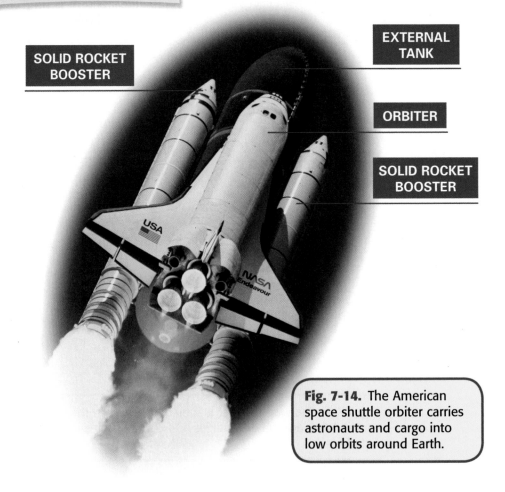

SOLID ROCKET BOOSTER

EXTERNAL TANK

ORBITER

SOLID ROCKET BOOSTER

Fig. 7-14. The American space shuttle orbiter carries astronauts and cargo into low orbits around Earth.

Impact of Technology
Technology Transfer

It is exciting to see astronauts talking to us from a space station or to watch a robot vehicle exploring Mars. Beyond that, though, has the U.S. space program benefited those of us left on the ground? What possible effects, both intended and unintended, has the space program had on Earth since Neil Armstrong landed on the moon in 1969? Has the investment of many millions of dollars yielded valuable results?

Investigating the Impact

Technology developed for one purpose may be used for other purposes. Sometimes this effect is called "technology transfer." Another term is "spinoff."

1. Find out what technology transfers have occurred as a result of the space program.
2. What would you say to someone who asks whether the space program has "paid for itself"?

The Orbital Space Plane will not have the payload space of a shuttle, but will carry astronauts to and from Earth by 2012. See **Fig. 7-15**.

The next generation shuttle will most likely not look anything like the old fleet. Engineers are working on radical designs that will look more like jet fighters. These shuttles, still in the design phase, will take off from runways rather than launch pads. The new shuttles will use a combination of jet engines and ramjet engines that will scoop oxygen from the atmosphere and mix it with a tank of liquid fuel, avoiding having to carry oxygen on board the vehicle.

Fig. 7-15. The Orbital Space Plane will carry people to and from the International Space Station. This space plane will be able to fly with or without a crew.

Foiling Gravity

Identify a Need/Define the Problem

In this activity, you will design and test a model wing section that demonstrates lift when air moves past it.

Gather Information

Airfoils are found in helicopters, planes, and even propellers. Research different airfoils. Be sure to take notes on the various shapes and designs.

Develop Possible Solutions

Using graph paper, draw a variety of patterns for your airfoil design. Look at each design and make notes on which ones you think will demonstrate the most effective lift.

Model a Solution

1. Once you have selected your solution, trace the pattern onto a plastic foam block.
2. Remove material from the block until it takes the shape of your design. See **Fig. A**.
3. Mount the airfoil model as shown, using wire. Use a block of wood for the base as shown in **Fig. B**.

Materials and Equipment

Select from this list or use your own ideas.

- high-density Styrofoam blocks, 3" wide, ¾" thick, and 4" long
- pen
- coat hanger wire
- block of wood (for base)
- graph paper
- wet/dry abrasive paper
- files and rasps
- end-cutting pliers
- blow dryer
- hand drill and twist bits

For information about the tools listed here, see "Hand Tools" and "Power Tools and Machines" on the Student CD.

Safety Alert

Look up "Safety Data Sheets" on the Student CD and prepare a data sheet for this activity. As you work on the activity, be sure to follow all safety rules.

Test and Evaluate the Solution

- Set your blow dryer at a cool temperature and a high speed. Use it to force air past the front edge of the airfoil as shown in **Fig. C**.
- Did the airfoil lift? If not, what corrections do you have to make?

Refine the Solution

- Create several other designs for airfoils. Test each one.
- Which airfoil performed the best? How does the shape of the airfoil affect the lift that is produced?
- Sketch the most successful design. Study your drawings to see what these designs had in common.

Communicate Your Ideas

Create a portfolio of the research and work that you did in order to create your airfoil. Be sure to include background information on the way lift is created, Bernoulli's principle, and other related information. It is important to document your thought process as well, so be sure to include your preliminary designs and notes. Incorporate your conclusions about which airfoil design worked best and why.

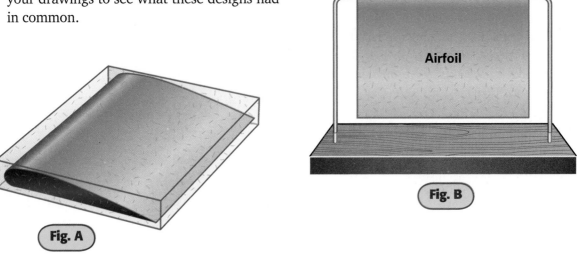

Airfoil

Fig. B

Fig. A

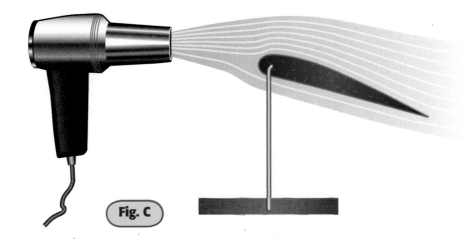

Fig. C

Creating a Propeller

Identify a Need/Define the Problem

In this chapter, you learned about propellers used on airplanes. You will create your own propeller for this activity.

Gather Information

Research and gather information about propellers and how they function. You may want to look at areas of technology where they are used, such as transportation. Gather diagrams and pictures of various air propeller designs that you find. Take specific note of each shape.

Develop Possible Solutions

Using graph paper, draw various propeller designs. Select the best design from your preliminarily ideas and begin to construct a model.

Model a Solution

Shape the Propeller.

1. First, lay out the strip of wood. Locate its center. Create a ¾" center hub. See **Fig. A**.
2. Cut or file away diagonals on the propeller, leaving the center of the hub flat. See **Fig. B**.
3. Round the leading edges. This will help shape the wood strip into an airfoil design. Drill the center hole for the motor shaft. See **Fig. C**.

Construct the Deflection Gauge.

1. Gather a dowel rod, a drinking straw, and a sheet of 8½" × 11" paper.

Materials and Equipment

Select from this list or use your own ideas.

- wood strip 1" wide, ¼" thick, and 4" long
- abrasive paper
- wood support for motor
- wood base for motor support
- DC toy motor
- files and other shaping tools
- deflection gauge
- safety motor mount
- DC battery

 For information about the tools listed here, see "Hand Tools" on the Student CD.

Safety Alert

Look up "Safety Data Sheets" on the Student CD and prepare a data sheet for this activity. As you work on the activity, be sure to follow all safety rules.

---

2. Construct your deflation gauge. See **Fig. D**. Remember that the greater the angle of the paper from the vertical, the greater the force of the propeller.

Assemble Your Propeller System.

1. Mount the propeller to the motor stand.
2. Connect the motor to the battery. See **Fig E**. Be sure to set up a safety net. Place the deflection gauge in front of the motor.

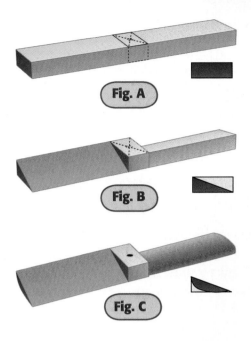

Fig. A

Fig. B

Fig. C

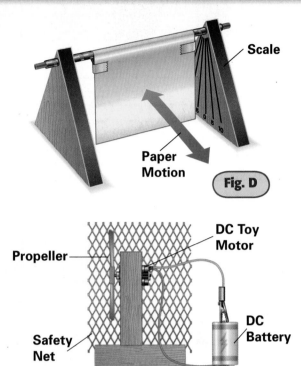

Scale

Paper
Motion

Fig. D

Propeller

DC Toy
Motor

DC
Battery

Safety
Net

Fig. E

Test and Evaluate the Solution

• Test the effectiveness of the propeller.
• Measure the amount of air moved by the propeller. Do this by determining the angle the paper swings from vertical when the propeller is blowing air on it. Plot this amount on a graph. Label this entry "Prop 1."
• Reverse the direction of the motor. How does this affect the output of the propeller?

Math Link

Propeller RPMs. Propellers spin at varying speeds. Speed at which a propeller turns is usually measured in RPMs (revolutions per minute). This means that the number of times that any particular point of the propeller passes a rotation in 60 seconds is how fast it is going.

If a fan is left on for 2 hours and is spinning at 1,500 RPMs, how many revolutions did it spin all together?

Refine the Solution

• Construct a different propeller. Use the data you gained from the first trial. Change the airfoil's shape and size. Test its effectiveness.
• Measure the amount of air moved by the second propeller. Plot this amount on the graph. Label this entry "Prop 2."
• Reverse the direction of the motor. How does this affect the output of the propeller?
• How does the output of the first propeller compare with the output of the second propeller?

Communicate Your Ideas

Write a brief report (150–200 words) regarding the activity. Your report should contain no spelling errors and have appropriate grammar and punctuation. In your report, be sure to mention the difference in the amount of air moved. Explain how the difference relates to propeller design.

Blasting Off with Rockets

Identify a Need/Define the Problem

Produce a model rocket that will reach the highest possible altitude. The rocket will be powered by the gases produced when water and Alka-Seltzer® tablets are combined.

Gather Information

Rockets have been launched into the atmosphere and beyond for well over a hundred years. Research some of the designs that have been created. Gather pictures of various rocket nose and fin designs. Note the difference in shapes.

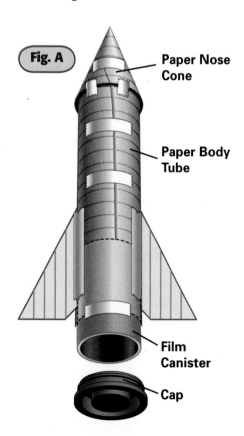

Fig. A

Paper Nose Cone

Paper Body Tube

Film Canister

Cap

Materials and Equipment

Select from this list or use your own ideas.

- Plastic 35-mm film canister. (The canister must have an internal sealing lid.)
- cellophane tape
- sheets of notebook paper or thin cardboard
- Alka-Seltzer® tablets (2)
- water
- graph paper
- pencil
- ruler
- scissors
- bucket for launching

Develop Possible Solutions

Develop a series of designs for a rocket that you think will be able to reach the greatest altitude. Sketch the design on graph paper. See **Fig. A** for an example.

Model a Solution

1. Select the best design from your preliminarily ideas and begin to construct a model.
2. A plastic film canister will become the rocket engine. Produce a body tube by wrapping notebook paper around the film canister.
3. Develop a nose cone.
4. Fasten the nose cone to the top of the body tube.
5. Develop a pattern for the fins. Cut the fins from paper or thin cardboard.
6. Fasten the fins to the rocket. Your rocket is now complete and ready to test.

Test and Evaluate the Solution

• Turn the rocket over. Place about ¼" of water in the canister. Drop one-quarter of an Alka-Seltzer tablet into the canister. Quickly cap the canister. Place the rocket in the bucket for launching. Be sure to stand back. Keep classmates away. *NOTE:* All students should be wearing safety goggles.

• Make note of the height achieved by the rocket.

• What effect did the mass of the body tube, fins, and nose cone have on the height reached by the rocket?

• What effect did the fin size, shape, and area have on the trajectory of the rocket?

• Which of Newton's three laws explains why the rocket lifts off?

Refine the Solution

• Experiment with the body tube length, mass, fin size, and number of fins. Test again for the greater altitude.

• Conduct four more test flights with your designs. Refine the design if needed. Test again. Record the heights reached by each flight.

• Did the rocket perform as expected? Take notes on the effects of any changes in the design of the rocket.

Communicate Your Ideas

Create a poster depicting your rocket. See **Fig. B**. The poster should include information that will tell a viewer how a rocket works, and it should show the various components.

Fig. B

Science Link

Plop, Plop, Fizz, Fizz. Are you familiar with Alka-Seltzer? This product has been around since 1931 and is promoted as treatment for "fast relief from acid indigestion, heartburn, and pain."

Do some research and try to figure out the following:

1. Name the ingredients in Alka-Seltzer.
2. Why does it fizz in water?
3. Why is it sold as a solid, foil-wrapped tablet?

Exploring Careers

Commercial Pilot

ENTRY LEVEL   **TECHNICAL**   PROFESSIONAL

Commercial pilots fly airplanes and helicopters for companies. Most of them transport passengers or cargo, but some dust crops, test airplanes, track criminals, or rescue and evacuate injured people.

Commercial pilots usually work as part of a team that includes a copilot, a flight engineer, and the ground crew. Pilots start the plane's engine, operate controls, monitor instruments, and fly the airplane. Pilots must follow a flight plan as well as Federal Aviation Administration (FAA) regulations and procedures.

Qualifications

A commercial pilot must have a commercial pilot's license with an instrument rating issued by the FAA. Pilots who want to fly helicopters must have a commercial pilot's certificate with a helicopter rating. To qualify for these licenses, an applicant must be at least 18 years old and have at least 250 hours flying experience.

Part of the training may include using a flight simulator, a system that mimics the conditions of flying an airplane. Using the simulator, trainees learn to take off, pilot, and land a plane.

Applicants must pass a strict physical exam. They must have good hearing and 20/20 vision. They may wear glasses, but they cannot have any physical disabilities that might impair them while flying. They also have to pass a written test that covers safe flying, navigation techniques, and FAA regulations. The final test is to demonstrate their flying ability.

Every employer has different guidelines for hiring pilots. Many larger companies require commercial pilots to have a bachelor's degree. Test pilots often need to have an engineering degree.

Outlook for the Future

In a good economy, more people fly, so there is more of a demand for pilots. There is a great deal of competition for the better, more desirable jobs.

Handling Pressure

Good communication skills, the ability to work well on a team, and staying calm under pressure are very important. Pilots must know how to focus on priorities. They must stay alert and be quick to react if something goes wrong.

Researching Careers

Search the Internet to find out about three jobs for commercial pilots. What education and experience are needed? What kind of aircraft will the pilots fly? Write a report about what you find.

More activities on Student CD

Key Points

- Aerospace engineers apply their knowledge of science and technology to developing new aircraft.
- Many forces affect an object in flight.
- Lift and thrust help aircraft overcome forces like gravity.
- Airplanes use airfoils, ailerons, and rudders to fly.
- The flight of a helicopter is controlled by the angle or pitch of the blade.
- Jet planes use engines that have an action and a reaction.
- Rockets are powered by liquid or solid fuel.

Read & Respond

1. Define *aerospace*.
2. Explain the Bernoulli effect.
3. Describe two forces that oppose flight.
4. Name two forces that help overcome gravity.
5. Identify the parts of an aircraft that help control flight. Briefly describe how they work.
6. What force acting on a plane causes it to slow down?
7. Describe how thrust is used to create lift.
8. How does Newton's third law explain how a jet engine works?
9. How is a rocket similar to a jet?
10. What were key developments in air and space technology?

Think & Apply

1. **Contrast.** Which would be more difficult to stop: a rolling bowling ball or a rolling tennis ball? Explain your answer.
2. **Debate.** What are pros and cons to giving the space program a major overhaul?

3. **Compare.** Describe how an airplane propeller works. Does a boat propeller work the same way?
4. **Examine.** Make a sketch of an airfoil. Describe the forces that act on the airfoil as it moves through the air.
5. **Research.** How did being bicycle mechanics help the Wright brothers design their airplane?

TechByte

Going Up. NASA scientists are investigating the possibility of building a "space elevator"—a cable stretching from a point on the equator to 22,187 miles above Earth. Like a satellite, the upper end of the cable would hover continuously over the same spot in orbit above Earth. Scientists propose building the cable with strong molecules of carbon called nanotubes and using magnetically powered "gondolas" to glide up and down it.

CHAPTER 8

Energy & Power Technologies

Objectives

- Explain the relationship between energy and power.
- Identify the five main forms of energy.
- Describe our traditional sources of energy.
- Explain how alternative energy sources can be used.
- List three ways to conserve energy.

Vocabulary

- **energy**
- **work**
- **potential energy**
- **kinetic energy**
- **power**
- **fossil fuel**
- **nonrenewable resource**
- **renewable resource**
- **hydropower**
- **geothermal energy**
- **biomass energy**

This offshore platform in the stormy North Sea is producing a valuable energy source—oil.

Activities

- Get Ready, Get Set, Get Solar
- Making an Energy Web
- Creating an Energy Conversion Machine

What Is Energy?

Imagine you're attending a concert of your favorite band. The amplifiers and sound system blare the voices and instruments of the musicians throughout the auditorium. Colored lights are flashing to the beat of the music, and the audience is moving back and forth to the rhythm. The floor vibrates with the sounds. See **Fig. 8-1**. You are surrounded by energy.

Energy is the ability to do work. Have you ever felt really tired after a hard day's work? You might have run out of energy. Energy is needed to perform work. **Work** is the application of force to make an object move in the direction of the force. Before a concert, heavy speakers need to be set up on the stage. People use energy to put force on the speakers to get them in place. (You can learn more about work in Chapter 6, "Applied Physics.")

There is a direct connection between work and energy. A plane uses energy to carry passengers. Electrons use energy to travel down a wire. Water uses energy when it changes into steam. In all these examples, energy is being used to perform work.

Reading Link

SQ3R. The study method called SQ3R stands for Survey, Question, Read, Recite, and Review. Here's how you can use SQ3R to study this chapter:

Survey: Read the chapter's title and headings. What do these tell you about the chapter? Look at the pictures, tables, and captions to see how they relate to the chapter's headings. Read the Objectives and Vocabulary. These will give you more clues about the chapter's topics. Look at the chapter's Review page. Do you know the answers to any of the questions?

Question: For each heading, write a question that you would expect the textbook to answer. Leave space below each question for an answer.

Read: As you read, look for the answers to your questions and write them down. Reread any parts that aren't clear to you.

Recite: Cover the answers you wrote. Answer the questions from memory. To find an answer you cannot recall, reread the appropriate part of the chapter.

Review: Review all the questions again and try to answer them from memory.

Fig. 8-1. Just imagine all the energy generated at this music concert. Can you name the different types of energy that would be in a concert like this?

Fig. 8-2. A bowling ball can have potential or kinetic energy depending on whether the ball is sitting in the rack or moving down the alley.

Energy can be potential or kinetic. **Potential energy** is energy that is stored. For example, a bowling ball at rest has potential energy. It has the potential to move, but it is currently at rest. When you roll a bowling ball down the alley, you are giving some of your energy to the ball. This is **kinetic energy**, or energy in motion. See **Fig. 8-2**.

Energy powers our world. Communication, transportation, construction, manufacturing, and bio-related systems all depend on energy. Our culture and lifestyle are built around machines and products that consume energy. See **Fig. 8-3**.

 See the interactive labs "The Designed World" and "The Seven Resources of Technology" on the Student CD.

Energy and Power

Energy powers our world, but what is power? **Power** is the rate at which work is done. It can be stated as a mathematical formula:

$$\text{Power} = \frac{\text{Work}}{\text{Time}}$$

From this formula, you can see there are two ways to increase power. You can increase the amount of work or decrease the amount of time.

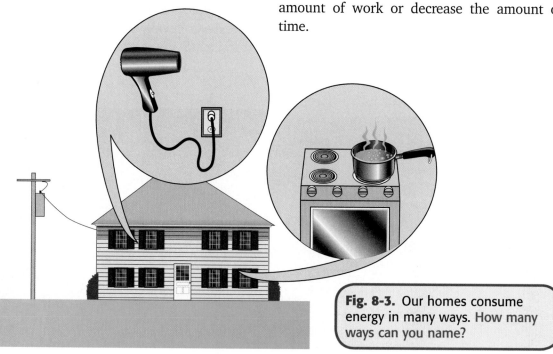

Fig. 8-3. Our homes consume energy in many ways. How many ways can you name?

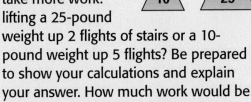

Math Link

Is It Work? In everyday language, "work" is what people do to earn a living. In physics, work is thought of a bit differently. It is the product of a force times the amount of weight being lifted (or displaced) by that force. The formula usually is written as $W = F \times d$ (work = force times distance).

So, which would take more work: lifting a 25-pound weight up 2 flights of stairs or a 10-pound weight up 5 flights? Be prepared to show your calculations and explain your answer. How much work would be done if you just stood still for an hour after someone placed a 25-pound weight in your hands?

Power Systems

Power technologies involve systems for moving and controlling power. Power systems can be mechanical, electrical, or fluid. All include a source (input); transmission and control (process); output (use); and feedback.

• Mechanical systems use the power of physical force. When you push an object, you are demonstrating mechanical power.

• Electrical systems use electricity to power such items as computers, industrial machines, and even the lights you use to see.

• Fluid power systems use fluids under pressure. The fluid may be air, water, or oil under pressure. For example, you might use an air pump when your bicycle tire is flat. (You can learn more about fluid power in Chapter 16, "Hydraulics & Pneumatics," and in the "Fluid Power" lab on the Student CD.)

Forms of Energy

The five main forms of energy are mechanical, heat, chemical, electromagnetic, and nuclear energy.

• Mechanical energy is the energy of moving things. Bicycles, cars, and planes have mechanical energy. You have mechanical energy as you walk down the street. Wind, waterfalls, and ocean tides have mechanical energy. See **Fig. 8-4**. Even the sounds you hear are a source of mechanical energy. What is moving when sound energy is produced?

Fig. 8-4. Can you comprehend all the energy expelled every second at the Niagara Falls?

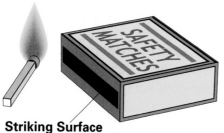

Striking Surface

Fig. 8-5. Heat energy in the match is used to release stored chemical energy. The tungsten in the light bulb is giving off energy and causing the filament to glow red hot.

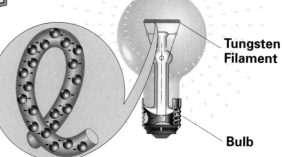

Tungsten Filament

Bulb

- Heat, or thermal, energy is created by the motion of atoms. The faster atoms move, the more heat is generated. Have you ever warmed your hands by a fire? If so, you were soaking up heat energy. See **Fig. 8-5**.
- Chemical energy results when the bonds between atoms are broken. These strong connections store lots of energy. When the connections are broken, energy is released. Wood, propane gas, and gasoline are good examples. When these materials are burned, their chemical bonds are broken and energy is released. See **Fig. 8-6**.

- Electromagnetic energy is the energy of moving electric charges. Power lines bring electromagnetic energy into your home from the power plant. The energy turns motors in your blow dryer, dishwasher, and ceiling fan. Light energy is another form of electromagnetic energy.
- The center of the atom, or nucleus, is the source of nuclear energy. Nuclear energy is a powerful form of energy that is released when the nucleus of an atom is split open or combined with another nucleus. Huge amounts of heat energy, light energy, and radiation are produced from a nuclear reaction. See **Fig. 8-7**.

 For a demonstration, see the "Forms of Energy" lab on the Student CD.

Fig. 8-6. The propane flame in this grill is changing chemical energy into heat energy.

Fig. 8-7. Shown here are the cooling towers of a large nuclear power plant.

Converting Energy

What happens to the heat energy created by a match when the match burns out? Does it just disappear? Is it destroyed? Where does it go? The law of energy conservation answers these questions. This law states that energy can be neither created nor destroyed by ordinary means. However, energy can be converted from one form to another. The heat from the match is transferred to the air in the room. The heat energy will eventually be transferred to the outside air and the atmosphere. It is never lost.

Changes in energy forms are called energy conversions. Energy can be converted from one form to another or transferred from one object to another. It can take many conversions to produce a single action. For example, when a car moves down the road, the mechanical energy is what you see in action. In reality, the mechanical energy of the car comes from burning chemical energy in the fuel. As the fuel burns, heat energy creates expanding gases that move the pistons up and down in the engine. This mechanical energy is eventually transferred to the wheels.

All forms of energy can be converted to other forms. A battery changes chemical energy into electromagnetic energy. A microphone changes mechanical energy in the form of sound waves into electromagnetic energy. See **Fig. 8-8.** What types of energy transformation must take place in order to power your hair dryer?

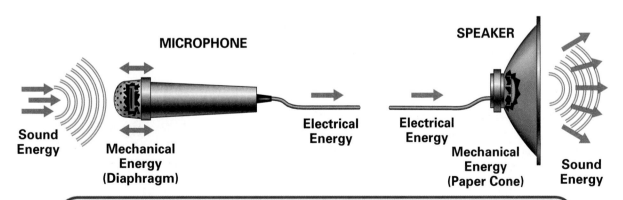

MICROPHONE

Sound Energy

Mechanical Energy (Diaphragm)

Electrical Energy

SPEAKER

Electrical Energy

Mechanical Energy (Paper Cone)

Sound Energy

Fig. 8-8. A speaker and microphone convert energy in opposite ways. The speaker converts electrical energy into sound energy; the microphone converts sound energy into electrical energy.

Traditional Sources of Energy

Most of today's energy comes from fossil fuels. **Fossil fuels** are formed from the remains of plant and animal life over millions of years. Crude oil, coal, and natural gas are fossil fuels. See **Fig. 8-9**. Another traditional source of energy, uranium, is a rocklike material dug from the ground.

Crude Oil

Oil is the major source of energy in the United States. Oil is a fossil fuel. Layers of mud and sand covered the remains of plants and animals over a very long period. The earth's moving crust applied heat and pressure to the layers and transformed the materials into crude oil.

Engineers and scientists locate the oil by studying rock samples in an area. If the site seems promising, drilling begins.

After the crude oil is removed from the ground it is sent to a refinery and separated into petroleum products such as gasoline, jet fuel, fuel oil, diesel fuel, and propane. See **Fig. 8-10**. How many things can you think of that use these products?

Coal and Natural Gas

Like oil, coal and natural gas are fossil fuels. Coal is taken out of the ground. When a deposit of coal is located close to the earth's surface, large holes called strip mines are dug to retrieve the valuable resource. When coal is found deep in the earth's crust, miners have to dig tunnels, or shafts, to get to the coal and transport it up to the surface for processing.

Natural gas can be found in the earth by itself or near crude oil deposits. The United States produces more natural gas than any other country. The gas is mostly used for home heating and cooking.

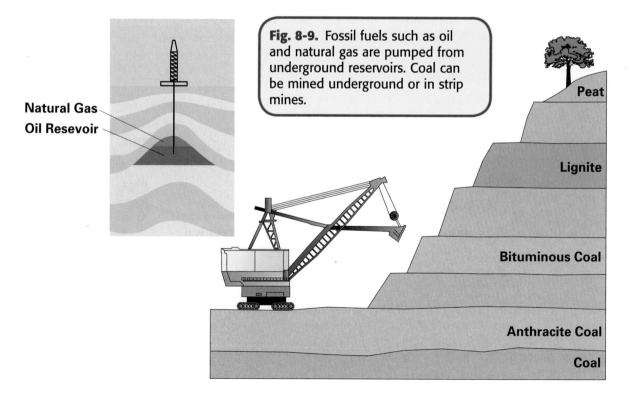

Natural Gas

Oil Resevoir

Fig. 8-9. Fossil fuels such as oil and natural gas are pumped from underground reservoirs. Coal can be mined underground or in strip mines.

Peat

Lignite

Bituminous Coal

Anthracite Coal

Coal

42 gal.	
Gasoline	46.7%
Fuel Oil	28.6%
Jet Fuel	9.1%
Plastics	3.8%
Asphalt	3.1%
Liquidfied Gases	2.9%
Lubricants	1.3%
Kerosene	.9%
Waxes Others	.1%

Fig. 8-10. A barrel of oil can be made into many different products when it is refined at an oil refinery like the one shown here.

Uranium

Uranium is a radioactive material dug from the ground. It is used as a fuel in nuclear reactors.

Unlike fossil fuels, uranium does not need to be burned to release its energy. Nuclei of atoms are split in a controlled reaction. The resulting heat energy is used to generate electricity. See **Fig. 8-11.** One pound of uranium can produce as much electricity as 3 million pounds of coal.

Fig. 8-11. Unlike earlier diesel-powered submarines that could stay submerged for only 48 hours, this nuclear-powered submarine can stay submerged for several months.

Why Do We Need Alternatives?

Unfortunately, the world's supply of traditional energy sources is rapidly dwindling. This is because crude oil, coal, natural gas, and uranium are nonrenewable resources.

Nonrenewable resources are those that cannot be replaced once they are gone. Remember, it takes millions of years for nature to create fossil fuels. We consume these energy sources faster than nature can produce them. See **Fig. 8-12**.

- Consumers' need for products has resulted in industries' need for more power.
- The need for transportation to move faster and over longer distances has resulted in a need for more fuel.
- Growing cities and populations have created a need for more heat and electricity.

Studies have also shown that pollution caused by burning these fuels presents risk for people and the environment. Let's examine these issues.

Health Issues

The burning of fossil fuels by motor vehicles and industrial plants has released many pollutants into the environment. Some, such as carbon monoxide and sulfur dioxide, are gases. Others, such as soot, are fine particles.

Some of these pollutants can be seen as a haze in the sky. They can make your eyes itch and even burn. See **Fig. 8-13**. People with

Fig. 8-13. Gasoline for these cars not only uses up nonrenewable resources but also produces a great deal of air pollution.

asthma and other lung diseases may find it difficult to breathe in areas with high levels of air pollution.

Particles released by fossil fuels can remain in the air for a long time. When people breathe these particles, lung tissues become irritated and damaged. Toxic chemicals in the air can also find their way into the bloodstream.

What will happen if we keep using these fuels? The World Resource Institute reports that many lives will be in danger if we do not change our energy resource usage.

What We Need	How Much We Use*
Manufacturing products	33% of energy produced
Powering homes and businesses	40% of energy produced
Fueling transportation	27% of energy produced

* Percentages are approximate.

Fig. 8-12. Energy Consumption in the United States.

Environmental Impacts

When fossil fuels are burned, they produce sulfur dioxide that is released as a gas. The sulfur dioxide combines with water vapor in the air to produce sulfuric and nitric acid. These two substances are the main ingredients in acid rain. Acid rain can kill fish, trees, and crops. It can also damage buildings, monuments, and statues.

When fossil fuels are burned, they also create carbon dioxide. Carbon dioxide is suspected of being one of the greenhouse gases associated with global warming. See **Fig. 8-14**.

Nuclear waste is the solid material left over from power plants. Besides being a health risk, it also pollutes our planet if not disposed of correctly. Nuclear waste can remain radioactive for thousands of years. Engineers are currently trying to find better ways to deal with this waste.

What Are the Alternatives?

If our traditional forms of energy are running out, then how are we going to power our world? We need to take advantage of our renewable resources. **Renewable resources** are those that are plentiful and/or easy to replace. The sun's energy as well as wind and moving water are considered renewable energy resources. (Strictly speaking, the sun's energy is not renewable. It will someday be used up. However, that day won't arrive for billions of years.)

Solving our energy problems will require new energy technologies. Alternative energy resources will have to be renewable, environmentally friendly, and safe for people. They will also have to satisfy our massive energy needs. While renewable resources provide energy, they are currently not efficient enough to supply all of our energy needs.

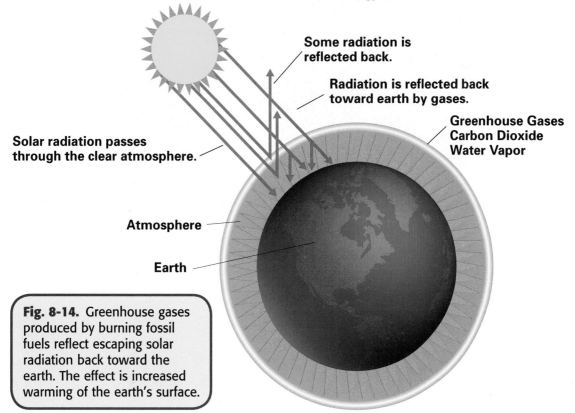

Some radiation is reflected back.

Radiation is reflected back toward earth by gases.

Greenhouse Gases
Carbon Dioxide
Water Vapor

Solar radiation passes through the clear atmosphere.

Atmosphere

Earth

Fig. 8-14. Greenhouse gases produced by burning fossil fuels reflect escaping solar radiation back toward the earth. The effect is increased warming of the earth's surface.

Solar Energy

Our greatest source of energy is the sun. Each day the sun showers the earth with electromagnetic energy. Electromagnetic energy is the same energy found in electricity, light, X rays, and radio waves. The sun produces this energy through a nuclear reaction called fusion. It is estimated that the temperature of the sun is 25 million degrees and that less than one millionth of its energy reaches Earth. However, even this small amount is enough to warm our planet. Engineers have developed devices to capture this energy and convert it to electrical and heat energy.

Photovoltaic Cells. Photovoltaic cells are devices that directly change the sun's rays to electrical energy. The cells are made of silicon, a semiconductive material. When light strikes the cells, electrons begin to flow, producing electricity.

Photovoltaic cells are arranged in panels called arrays. Arrays may be placed on buildings or in fields to capture the sun's energy. Some arrays may even track the sun as it moves in order to increase efficiency. See **Fig. 8-15**. Photovoltaic systems may also include heavy-duty batteries to store the electrical energy.

Photovoltaic cells convert only about 11% of the sun's rays striking the cells into electrical energy. Engineers are working on ways to make the cells more efficient.

Solar-Thermal Converters. Solar-thermal converters use solar collectors to focus the sun's energy and convert it into heat energy. Curved mirrors focus the sun's light onto a receiver. The receiver can be a tank of water on a tower or a pipe with water flowing through it. Temperatures at the receiver can range from 200 to 2,000°F. Water is boiled to create steam. A steam turbine generator then produces electricity.

On a smaller scale, solar-powered water and space heaters are available. Consumers can use them to heat water or their living space. See **Fig. 8-16**.

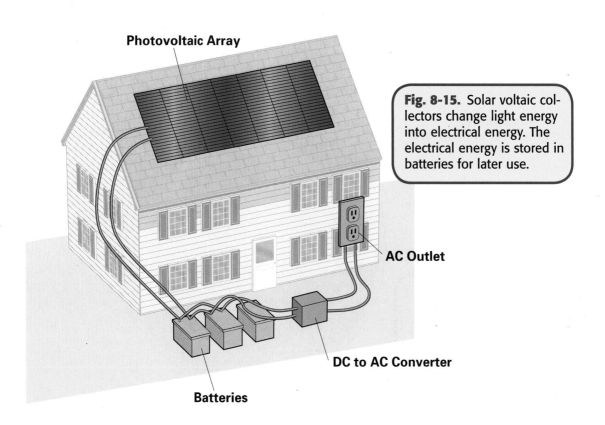

Photovoltaic Array

AC Outlet

DC to AC Converter

Batteries

Fig. 8-15. Solar voltaic collectors change light energy into electrical energy. The electrical energy is stored in batteries for later use.

Impact of Technology
Energy Trade-Offs

Energy production is of concern for many people throughout the world. Along with the fear of running out of nonrenewable fuels, there is also worry over the trade-offs that must be made when producing energy. Making a trade-off means giving up one thing to gain another. For example, hydropower is a renewable energy source, and it causes less pollution than fossil fuels. However, the dams and the lakes created behind them may force people to leave their homes or may harm native plants and wildlife. Even wind farms can have some negative effects on the environment.

Investigating the Impact

Research trade-offs in energy and power technologies.

1. Select one of the energy sources described in this chapter. Find out about the benefits and drawbacks of using that energy source. Do you think the positive effects outweigh the negative? Why or why not?
2. What energy sources are used to produce the electricity used by your community? What trade-offs are involved?

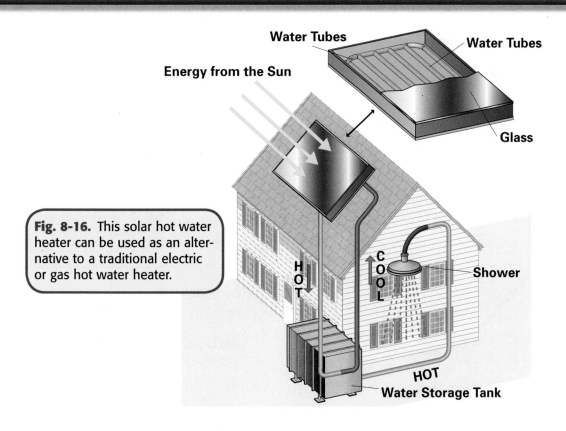

Fig. 8-16. This solar hot water heater can be used as an alternative to a traditional electric or gas hot water heater.

Wind Energy

People have been harnessing the energy of the wind for hundreds of years. Wind has been used to grind grain into flour, pump water, drive factory machines, and power vessels.

Today, wind is used to produce electrical energy. The rotors, or blades, of the wind machine are spun by the mechanical energy of moving air. A shaft connected to the rotors turns a generator, changing the mechanical energy into electrical energy. A gearbox is used to increase the speed of the turning shaft, spinning the generator even faster.

Many individual wind turbine systems can store electrical energy in large batteries and feed off them when wind is not available. See **Fig. 8-17**.

Some stand-alone systems are connected directly to a power company's electric lines. When the wind is not blowing, the consumer taps into the power company's resources. When the wind is blowing, extra power not used by the home is fed back to the power company. The consumer is then given a financial credit.

Wind farms are large fields dotted with hundreds of wind turbine generators that can supply power for many consumers. See **Fig. 8-18**. It is estimated that there are over 15,000 wind farms currently in the United States. Many of these farms are located in California. Wind provides California with one percent of all its electrical needs.

Offshore wind farm projects are being considered by many state and local power authorities. On Long Island, New York, 40 offshore wind turbines are being planned to supply 44,000 homes with electrical power.

Water Energy

Water has been used for thousands of years as a source of energy. **Hydropower** (*hydro* means water) harnesses the mechanical energy of moving water and converts it into electrical energy.

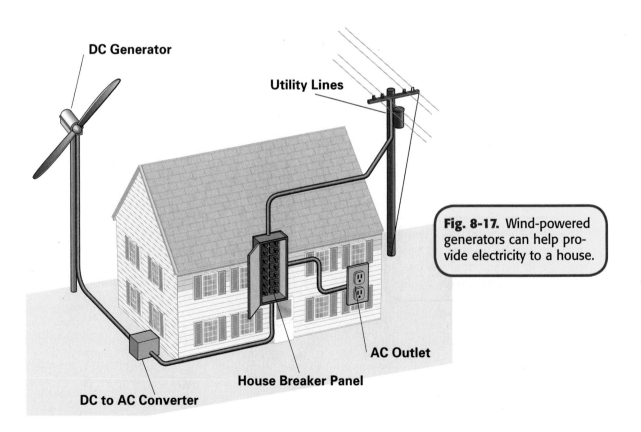

DC Generator

Utility Lines

Fig. 8-17. Wind-powered generators can help provide electricity to a house.

AC Outlet

House Breaker Panel

DC to AC Converter

Hydroelectric power plants use the mechanical energy of the water to spin the blades of a turbine. The turbine is connected to a shaft that turns a generator, producing electrical energy.

Dams are constructed to control rivers and generate electricity. The dam raises the water level, increasing its potential energy. Waterways direct the water from the top of the dam to the powerhouse where turbine generators produce electricity. See **Fig. 8-19**.

Many hydroelectric power plants do not utilize dams. Runoff river systems rely on the natural flow of the water to generate electricity.

The power of ocean waves can also be harnessed to create electricity. The power of incoming waves is trapped and funneled past a turbine generator on the way back to the ocean.

Hydroelectricity provides 12% of the nation's electrical supply. More than 28 million people get their electricity from flowing water.

Geothermal Energy

The earth itself is a massive ball of heat energy. Have you ever seen the Old Faithful geyser at Yellowstone Park? The hot air escaping from the ground is an example of geothermal energy. **Geothermal energy** is energy produced under the earth's crust. Twelve to thirty miles below the earth's crust is the mantle. The rocks in the mantle layer are superheated by a layer of rock below them called the outer core. Temperatures in the outer core reach 9,000°F.

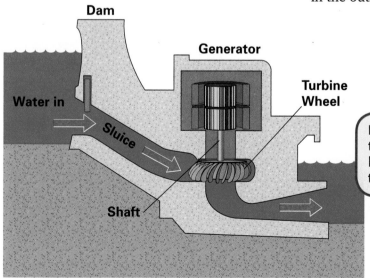

Fig. 8-19. The water falling over the dam turns a turbine. The turbine is connected to a generator that produces electricity.

Hydrothermal systems capture steam from underground reservoirs of water that are super-heated from mantle rocks. The steam is used to heat homes and buildings. In Reykjavik, the capital of Iceland, more than 100,000 homes are heated by hydrothermal systems.

Hot-rock geoenergy systems pump water deep into rock beds that have been heated by the mantle. Temperatures over 400ºF change the water into steam. The steam is then channeled back up to the surface to power steam turbine generators. See **Fig. 8-20**.

Science Link

How Hot Is It? You probably know that water boils at 212 degrees Fahrenheit. Do you think that is hot? Boiling water can certainly give you quite a bad burn. Do you know the temperatures required to melt metals? Tin melts at 450°F, copper at about 2,000°F, and carbon steel at 2,700°F. How hot can it get? Scientists have calculated that the sun may contain temperatures as high as 25 million degrees! Find out how the sun's temperature compares with that of the earth's core.

Fig. 8-20. Deep below the earth, molten rock turns pools of water into steam. Geothermal wells use this natural heat energy to provide steam to turn turbine generators.

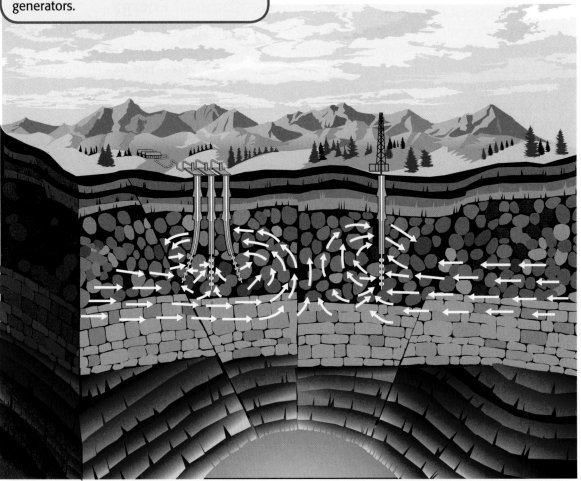

Biomass Energy

Biomass energy is produced from organic materials, such as trees, animal waste, and plants. What is biomass used for?

When you think of corn, you probably think of food. Thanks to biomass technologies, you can run your car on it! Ethanol, a gasoline additive used to extend gasoline supplies, is a product of biomass technology. Ethanol is created from corn or other grains through a fermentation process. Many gasoline manufacturers mix ethanol with their gasoline. See **Fig. 8-21**. In some vehicles, ethanol can also be used as a fuel by itself.

Biomass digesters are used to change animal waste into gases, such as methane. This particular gas can be used in the home for cooking and heating.

Fig. 8-21. Corn and other grains are used to make ethanol, a gasoline additive. This helps extend the gasoline supplies.

Conserving Energy

While engineers are working on ways to make alternative energy technologies more efficient, we need to conserve the energy we use today. How can you help?

- Turn off the lights when you're not using them. Many people leave lights on all over the house. That's a lot of wasted electricity!

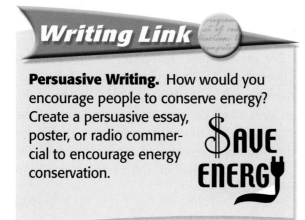

Writing Link

Persuasive Writing. How would you encourage people to conserve energy? Create a persuasive essay, poster, or radio commercial to encourage energy conservation.

- Hold a ribbon up to your home's windows and doors. If the ribbon moves, you have an air leak which affects your heating and cooling.
- In the summer, close blinds and drapes to avoid direct sunlight. In the winter, close blinds and drapes to avoid extra heat loss.
- In an air-conditioned home, set the thermostat at or above 78°F in the summer.
- In the winter, set the thermostat at 60°F before going to sleep or when no one is going to be home.
- When possible, walk or ride a bike instead of using a car. Using buses, trains, or subways also reduces the amount of pollution caused by automobiles.

Our world is full of energy and power. With engineers designing new energy alternatives, and people conserving when possible, energy and power will continue to move us forward.

Get Ready, Get Set, Get Solar

Identify a Need/Define the Problem

The vast majority of automobiles run on fossil fuels, creating a large amount of harmful pollutants. See **Fig. A**. In order to help protect the environment and save on energy costs, scientists and engineers have been developing many alternative technologies, such as solar-powered automobiles. See **Fig. B**.

Design and create a solar-powered car. You will race your car with classmates to see which car is the fastest.

Gather Information

Research the various types of solar-powered automobiles and take notes on how they work and how they are designed.

Materials and Equipment
Select from this list or use your own ideas.

- photovoltaic cell
- DC motor
- wire
- structural material (e.g., cardboard, wood)
- adhesives
- wheels
- steel rods
- wooden dowels

Safety Alert

Look up "Safety Data Sheets" on the Student CD and prepare a data sheet for this activity. As you work on the activity, be sure to follow all safety rules.

Develop Possible Solutions

Prepare several sketches for your solar-powered car. What configuration will the photovoltaic cell be in? What will the design of the car look like? Be sure to include some of the characteristics of solar-powered car designs that you have researched. The more sketches and ideas you develop, the better your design is likely to be.

Fig. A

Model a Solution

1. Choose your best solution.
2. Use the available materials to create a working model of your solution.
3. Attach the wires from the photovoltaic cell to a DC motor. Be sure the wires are connected to the correct terminals unless you want your car to run backwards!

Test and Evaluate the Solution

- Create a simple track where you and a classmate can race each other. (A strip of tape down the center of a hallway will do.)
- Race your car along with a classmate.
- Record the results of the race as well as notes on your car's performance.
- Did your solar-powered car perform as planned? Was your car the fastest?

Refine the Solution

- Could you modify your car to perform better?
- How would your car perform if you had two photovoltaic cells?
- If needed, redesign your solar-powered car and race again. Did your revised car perform better?

Communicate Your Ideas

Create a display for your solar-powered car. Include background information about solar energy, photovoltaic cells, and solar-powered automobiles. Be sure to also include the steps of the design process and the various designs that you came up with. Present your car to the class.

Fig. B

Finding Your Speed. Calculating the average speed is an essential task when it comes to races, but you don't need a radar gun to do it. In order to calculate speed you need to know two things: the distance covered and the time taken to cover that distance. Use this simple formula to calculate the speed of your solar car.

$$\text{Average speed} = \frac{\text{distance (feet)}}{\text{time (seconds)}}$$

Calculate the speed of your solar-powered car after racing it. How would you convert the speed to miles per hour?

Making an Energy Web

Identify a Need/Define the Problem

Do you know how much energy you use in a day? Most people don't. That's because almost everything you do requires some form of energy.

In this activity, you will identify the various forms of energy you use on a daily basis. You will then create an "energy web" out of your data.

Gather Information

Research the different kinds of energy. This chapter is a good place to start. For example, what form of energy do you use when you turn on a stove? See **Fig. A**. For further information, research the Energy Information Administration, which is part of the U.S. Department of Energy. This organization has an informative Web site that further explores the science of energy, including forms, calculations, and sources.

Materials and Equipment
Select from this list or use your own ideas.

• computer
• word processing or graphics program

Fig. A

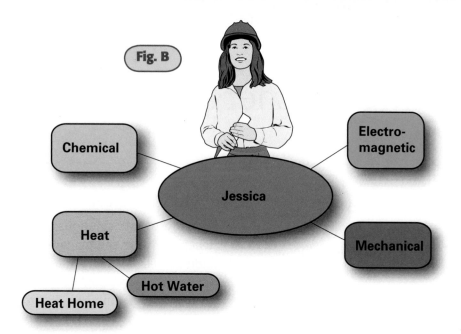

Fig. B

Chemical

Electro-magnetic

Jessica

Heat

Mechanical

Hot Water

Heat Home

Develop Possible Solutions

Develop different formats in which to record your information. Your data must be organized so that it can be easily understood by your classmates.

Model a Solution

1. Brainstorm a list of all the activities you do in a typical day that require energy.
2. Identify the various forms of energy that are required for these activities (e.g., your hot water uses heat energy, which is provided either by natural gas or electricity from the local utility).
3. Identify the various sources for these forms of energy.
4. Record your daily activities, the related energy forms, and the energy sources.
5. Create an energy web similar to the one shown in **Fig. B**. Place yourself at the center of the web. Around you should be the various forms of energy that you use on a daily basis. Located around those forms of energy should be the various activities that you do that utilize those sources.

Test and Evaluate the Solution

• Are there any activities that come up multiple times?
• Which activities are a "need" and which ones are a "want"? Can you cut down on "want" activities in order to have energy to complete your "need" activities?

Refine the Solution

• As you have probably noticed, you use quite a bit of energy throughout the day. Some of this energy may be wasted. List the activities that you do in order of importance.
• What activities can you still do if you had no modern energy sources available to you?
• Try to cut down on "want" activities for several days. Do you find yourself having more energy?

Communicate Your Ideas

Compare your energy web with energy webs created by your classmates. Do you all do some of the same activities? Do you do more activities than other students? Explain how you felt after refining your routine.

Creating an Energy Conversion Machine

Identify a Need/Define the Problem

In the early 1900s, a cartoonist by the name of Rube Goldberg began to draw elaborate ways of performing simple tasks. Many of these designs converted multiple forms of energy. Goldberg's designs served as inspiration for a common game you may have played called Mouse Trap®. See **Fig. A**.

In this activity, you will design and build an energy conversion machine. The machine must meet the following specifications:

- The machine must include at least three different forms of energy (e.g., heat, chemical, and electromagnetic).
- The machine must have at least five places where energy conversion takes place.
- The machine must fit within a given area (e.g., 48 square inches) as assigned by your instructor.
- The machine must accomplish the task specified by your instructor (e.g., pop a balloon).

Gather Information

Research Rube Goldberg and take notes on some of the many ways he accomplished common daily tasks and the various forms of energy that were used in his machines.

Develop Possible Solutions

Prepare several solutions, in the form of sketches, for your energy conversion machine. Include details as to what kind of energy forms you have used and how you will change one energy form into another. Try to develop as many designs as possible so that you have a good final solution. Be sure that your machine is able to accomplish the task specified by the instructor.

Fig. A

Materials and Equipment

Select from this list or use your own ideas.

- assorted materials (student provided)
- adhesives
- fasteners
- possible energy sources (photovoltaic panels, batteries)
- structural materials (e.g., cardboard, wood)

Safety Alert

Look up "Safety Data Sheets" on the Student CD and prepare a data sheet for this activity. As you work on the activity, be sure to follow all safety rules.

Model a Solution

1. Select the design you think will be most effective.
2. Construct the various components of your machine.

Test and Evaluate the Solution

- Set up your machine within the testing area.
- Start your machine.
- Did your machine accomplish the task that is specified?
- Where you able to convert energy from one form to another?
- Was there any loss in energy within the machine?

Refine the Solution

- Were there any points in the machine where your design did not work? If so, what can you change to make the conversion smoother?
- Can you add an additional form of energy?
- Modify your design and test again. Make notes on the redesigned machine.

Communicate Your Ideas

Create a cartoonlike poster for your energy conversion machine. Make sure you include all the various parts and components. Be sure to list the types of energy used, the points where energy conversion occurred, and how the energy was converted.

Science Link

Saving Energy in Reverse. None of Rube Goldberg's humorous concepts have ever evolved into everyday use because they are not practical.

Let's reverse the process. Take a system that is used widely today and see if you can devise a way to make it more efficient (the opposite of a Rube Goldberg concept). Provide scientific evidence of how your improvement will save energy. Write up your idea and include a sketch to share with your classmates.

Exploring Careers

Nuclear Operations Technician

ENTRY LEVEL **TECHNICAL** PROFESSIONAL

Nuclear operations technicians help scientists with laboratory and production activities. The technicians operate the equipment that releases, controls, and uses nuclear energy. For example, they set control panel switches that start nuclear reactors, particle accelerators, and gamma radiation machines. Technicians measure radiation dosage, temperature, and pressure. They monitor instruments, gauges, and recording devices. They also test, inspect, and repair equipment.

Specific job titles for nuclear equipment operators are accelerator operator and reactor operator. An accelerator operator helps scientists study the nucleus of atoms by running the particle accelerators. A reactor operator oversees the operations of nuclear reactors. Nuclear operations technicians work for the federal government, utility companies, or engineering firms.

Qualifications

A nuclear operations technician should have mechanical aptitude and a good understanding of how things are produced and processed. A nuclear operations technician would most likely take courses in chemistry, physics, engineering, and math. Technicians may also need to have government security clearance because they may work with important information and procedures.

Outlook for the Future

The job outlook for nuclear operations technicians is fair when compared with other jobs.

New jobs depend on whether nuclear energy continues to be used on a large scale as an energy source.

Being Responsible

Technicians in this field must be accountable for their actions. They must follow strict procedures for safety, quality assurance, and security. Ways to show accountability include volunteering for more responsibility, helping others when you are done with your own work, and looking for ways to do your job better.

Researching Careers

Find out about jobs for nuclear operations technicians. What are five skills needed for this job? Why is mechanical aptitude needed? Name five activities performed in this job. Make a poster using the information you find.

**More activities
on Student CD**

Key Points

● Energy is the ability to do work. Energy can be potential or kinetic.

● Power systems are mechanical, electrical, or fluid.

● There are five main forms of energy. One form may be converted into another.

● Our traditional sources of energy are nonrenewable.

● Engineers are trying to develop new ways to use renewable resources as alternative sources of energy.

● Conserving energy can help ensure more energy for the future.

Read & Respond

1. What is the relationship between energy and power?
2. Define *work*.
3. Explain the difference between potential and kinetic energy.
4. What are the five main forms of energy?
5. Name the alternative energy sources.
6. Identify the four traditional sources of energy.
7. Why do we need alternatives to nonrenewable resources of energy?
8. Explain how photovoltaic cells work.
9. What is hydropower?
10. Describe three ways to conserve energy.

Think & Apply

1. **Research.** Explore careers in the field of energy conservation.
2. **Formulate.** Ask your parents for copies of their energy bills for up to one year. Compare how much energy was used for each month. Explain why more energy was used some months than others.

3. **Compare and Contrast.** Research how efficient each alternative source of energy is and make a chart comparing each kind.
4. **Propose.** Suggest ways your school can save energy other than the ways listed in this chapter.
5. **Design.** Design a device that harnesses water energy.

TechByte

Garbage Energy. Researchers at Pennsylvania State University in 2004 built the first microbial fuel cell (MFC). The fuel cell uses sewage and bacteria to produce small amounts of electricity. In 2005, researchers in England built a robot powered by an MFC. The robot "eats" flies to replenish the sewage content.

CHAPTER 9

Land & Water Transportation

Objectives

- Identify power sources for each form of land and water transportation.
- Describe the operation of a four-stroke cycle gasoline engine.
- Explain the difference between a diesel engine and a gasoline engine.
- Discuss how vessels are designed to float.

Vocabulary

- **transportation**
- **mode**
- **intermodal transportation**
- **locomotive**
- **maglev**
- **internal combustion engine**
- **diesel engine**
- **vessel**
- **buoyancy**

Water transportation can be on a large or small scale. Which do you prefer?

Activities

- Modeling a Maglev System
- Creating a Cam Operating System
- Sailing Away with a Catamaran

Society on the Move

Is your family mobile? Being mobile means that you are capable of moving around. We live in a society where families are more mobile than at any other time in history.

Your mobility is tested every weekday morning. How do you get to school? Do you walk or ride a bicycle? Maybe you take a train or ride a bus. How do your parents get to work? Do they drive to their workplace? What if they had to go on a business trip far away? They would most likely fly.

Why are we more mobile now than at any other time? As technology evolves, new transportation methods are created. This chapter examines transportation technologies used on land and in the water. (For more information about air transportation, see Chapter 7, "Air & Space Technologies.")

Modes of Transportation

Transportation is the process by which people, animals, products, and materials are moved from one place to another. There are many different modes of transportation. A **mode** is a method of doing something. Modes of transportation can be organized according to the pathways, or "ways," used by transportation systems.

Writing Link

Outlining. Outlining a chapter will show you how it is organized and help you understand what the main topics and subtopics are. Make an outline of this chapter. Here is a start:
  I. Society on the Move
  II. Modes of Transportation
  III. Rail Transportation
       A. Transporting People
          1. Commuter Trains

See **Fig. 9-1.** Land vehicles travel on highways and railways. Water (marine) vessels use seaways and canals, while airplanes travel through airways.

Sometimes it is necessary to move people and products only a short distance. The use of common land vehicles is not always practical in such cases. On-site transport systems are designed for this purpose. For example, malls have elevators and escalators to transport

Fig. 9-1. This cargo ship carries containers on its deck and below the deck. At the destination port, the containers will be loaded onto trucks or railroad cars.

Fig. 9-2. This moving walkway saves a lot of time and energy for the passengers moving from one terminal to another. It also helps control the flow of people through the terminals.

people between floors. Large airports rely on people movers to transport passengers to terminals and gates. These moving walkways help travelers move from one part of the terminal to another quickly and easily. See **Fig. 9-2**.

Combining several modes of transportation to move people or products is called **intermodal transportation**. See **Fig. 9-3**. Let's trace the path of an MP3 player from a factory in Japan to your home.

After being manufactured, the MP3 player likely moved down a factory's conveyor belt to the shipping department. There, the product was packaged and stacked on pallets. The stacked pallets were loaded by forklifts into a large container.

The container was mounted on sets of wheels to become the trailer portion of a tractor-trailer, or semi. The trailer was pulled by the tractor to a ship docked at a harbor. A large crane removed the container from the wheels and placed it on board a large ship.

Fig. 9-3. Intermodal transportation is used to get products to you. What other modes of transportation might have been used to move the cargo shown here?

The ship traveled across the ocean on sea-lanes and docked at a port in the United States. Again, a crane removed the container from the ship. This time the container was placed on a railroad flatcar.

A train pulled by a diesel-electric locomotive moved the products along railways to a warehouse for storage. Trucks carried the product from the warehouse to your local store. Finally, you might have traveled on a bicycle or in a car to the store to purchase your MP3 player and bring it home.

Tracing Transportation. You've just read a description of an MP3 player being transported from a factory to your home. How many different forms of transportation were involved in getting the MP3 player to your home?

Fig. 9-4. Commuter trains can move many people from place to place within a large city. Without these trains, traffic congestion might be unmanageable.

Rail Transportation

Rail transportation can be divided into two basic groups. Some trains move people, while other trains move freight.

Transporting People

Have you traveled by train? If so, were you going someplace near or someplace far? Trains can carry people over a short distance or across the country.

Commuter Trains. Commuter trains operate in large cities, moving passengers short distances. Often referred to as mass transit, these trains provide transportation within the city and to its suburbs. See **Fig. 9-4**. Commuter trains help to relieve the congestion of city automobile traffic and related air and noise pollution.

For example, the Boston Metro Rail System (the "T") brings rail service to 78 cities and local towns. This system's 125 miles of track and 84 railroad stations serve over 200 million passengers each year.

Subways are another form of mass transit. They are underground. Because subway systems aren't on the surface, you never have to cross their train tracks or wait while a train passes.

Long-Distance Trains. These trains can cover long distances much faster than automobiles. They are an alternative for people who don't like to fly.

How fast can long-distance passenger trains go? Japan is where high-speed rail travel was created. Traveling at a maximum speed of 186 mph, the *Shinkansen* (bullet train) carries passengers from Tokyo to destinations west and north.

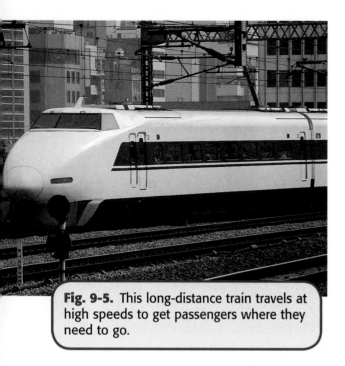

Fig. 9-5. This long-distance train travels at high speeds to get passengers where they need to go.

Rivaled only by the Japanese, the French high-speed rail service, *TGV*, carries passengers from Paris to Belgium, Germany, Switzerland, the Netherlands, the United Kingdom, and Italy. It cruises at a speed of 173 mph. See **Fig. 9-5**.

Transporting Freight

Freight trains carry a large percentage of the materials and products transported in the United States. Freight cars come in many shapes. Boxcars, flatcars, hoppers, and dump cars move solid materials. Tank cars move liquids such as oil, gasoline, and paints. Refrigerated and heated cars carry perishable goods. See **Fig. 9-6**.

Powering Rail Systems

Steam, diesel engines, electricity, and even magnetism have all been used to power rail transportation.

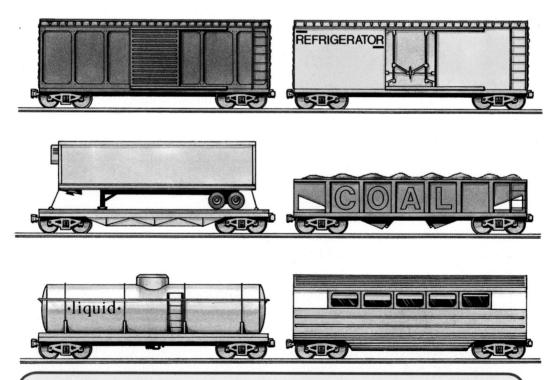

Fig. 9-6. The design of a railroad car is determined by its use. These are six of the more common railroad cars. Would you be able to identify the primary use for each car if some of the cars had not been labeled? Can you tell which is a passenger car?

Fig. 9-7. Because they burn wood and coal, steam-powered locomotives release pollutants into the air. There are very few in operation today.

A **locomotive** is a self-propelled vehicle used to pull or push trains of rolling stock (railroad cars). The first locomotive used the energy of compressed steam to power rail traffic. See **Fig. 9-7**.

Coal or wood fires heated water. The boiling water turned to steam, and the pressure of the steam was used to move pistons. These pistons, in turn, caused the wheels to turn. Steam locomotives used to pull most of the trains in the United States.

Most trains now use diesel engines to turn electric generators. (Diesel engines are discussed later in this chapter.) The generator produces electric current. The current turns an electric motor, which powers the wheels. See **Fig. 9-8**.

An all-electric locomotive also uses an electric motor for power. However, it obtains its electricity from overhead power lines or electrified rails. Some electric locomotives use their powerful motors as generators during coasting periods. This allows them to feed electricity back into the power source.

Fig. 9-8. The diesel locomotive is actually powered by electricity. The diesel engine runs generators for an electric motor that powers the wheels.

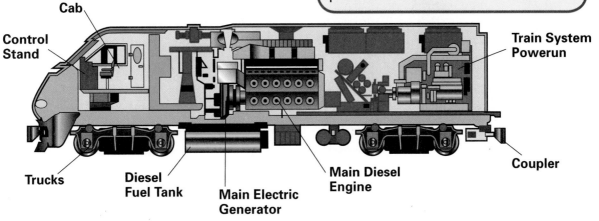

Cab

Control Stand

Train System Powerun

Trucks

Diesel Fuel Tank

Main Electric Generator

Main Diesel Engine

Coupler

Maglev (short for magnetically levitated) trains rise above a guideway and are propelled by magnetic fields. The magnetic fields are created by large electromagnets. See **Fig. 9-9**.

Have you ever experienced the force generated when you try to push together the like poles of two magnets? Have you felt the force of attraction between two unlike poles? It is this push-and-pull force that levitates the train above the guideway and propels it down the track.

Maglev trains have several advantages over trains commonly used today. One difference is that maglev trains do not touch the guideway. With very little friction to rob power, maglev trains can rapidly accelerate to over 300 mph. Without tracks that creep out of alignment, the maglev train's ride is smooth and quiet. Less maintenance is required. Snow, ice, and heavy rain have little effect on the operation of maglev trains. Perhaps their greatest advantage is a reduction in pollution. Maglev trains do not burn gasoline or diesel fuel.

Unfortunately, maglev guideway construction costs millions of dollars per mile. Because travel routes between large cities usually have many stops along the way, maglev trains are not able to travel these routes at top speed. Also, the power for electromagnets must come from electric power plants. Most of them burn polluting fossil fuels.

Science Link

Magnets. Maglev trains work on the principle of magnetism. Opposing magnetic poles can force the train up away from the rails, and the attraction of opposite poles can draw the train forward.

Most people are familiar only with the simple bar and horseshoe magnets seen in schools and in hobby and hardware stores. What is the strongest magnet you have ever seen? Research what the strongest possible magnet (human-made or natural) on Earth could be.

Fig. 9-9. In the maglev system, the magnet in the rails works with the electromagnetic coils in the train car. The result is that the train levitates just above the rails. Note how the lower part of the car wraps around the rail.

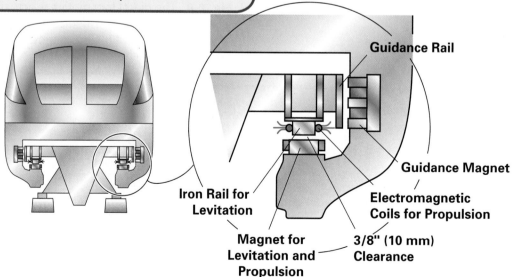

Guidance Rail

Iron Rail for Levitation

Guidance Magnet

Electromagnetic Coils for Propulsion

Magnet for Levitation and Propulsion

3/8" (10 mm) Clearance

Motor Vehicles

Motor vehicles are the most widely used mode of land transportation. Automobiles, vans, buses, motorcycles, and trucks are all motor vehicles. We rely on these vehicles to transport people, products, and materials across town and across the country.

Gasoline Engine

The heart of the motor vehicle is its engine. The vast majority of vehicles today use a gasoline-powered piston engine to achieve motion. The engine creates power by burning, or combusting, fuel inside the engine. Fuel and oxygen are ignited in a controlled explosion. The energy of the explosion is transformed into mechanical energy that powers the vehicle.

Because the fuel is burned inside the engine, it is called an **internal combustion engine**. Most automobiles have an internal combustion engine that burns gasoline. The steam locomotive mentioned earlier had an external combustion engine. Water was boiled outside the engine to create steam. The steam passed through valves to the inside of the engine. The expanding gas pushed on pistons inside the engine, creating mechanical energy and motion.

In both the internal and external combustion engine, power comes from expanding gases within a cylinder. The cylinder is a hollow, closed container, usually made of steel. A piston is a disk or short cylinder that moves up and down within the hollow cylinder. It must fit tightly enough in the cylinder so that hot gases do not escape, but it must be loose enough to move up and down.

Most gasoline engines operate on a four-stroke cycle. This means that four separate operations occur to convert heat energy into mechanical energy. All stages use a piston that slides up and down the cylinder. Each movement of the piston is a different stroke. See **Fig. 9-10**. After a cycle is completed, it begins again.

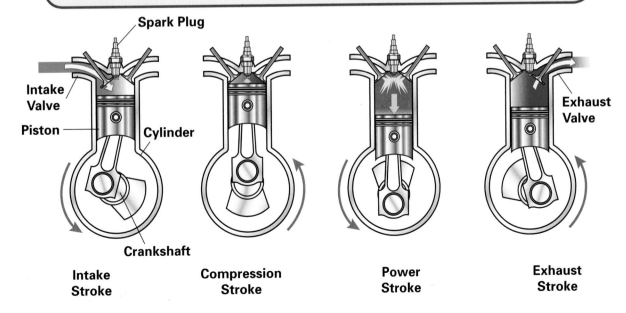

Fig. 9-10. This is how a four-stroke-cycle engine works. Note where in the cycle the intake and exhaust valves operate. The crankshaft is what transfers the power to the wheels of a vehicle.

Spark Plug

Intake Valve

Piston

Cylinder

Crankshaft

Exhaust Valve

Intake Stroke

Compression Stroke

Power Stroke

Exhaust Stroke

- **Intake stroke.** A mixture of gasoline and air is drawn into the cylinder as the piston moves down.
- **Compression stroke.** The second stroke compresses the fuel mixture by moving the piston up in the cylinder.
- **Power stroke.** A spark plug produces an electric spark that ignites the compressed fuel. The burning fuel explodes creating expanding hot gases. These gases push the piston back down the cylinder. A series of shafts and gears connected to the piston transfers energy from the moving piston to the wheels.
- **Exhaust stroke.** The piston moves back up to force unburned gasoline and waste gases out of the cylinder.

 For a demonstration of the four-stroke cycle, see the "Internal Combustion Engines" lab on the Student CD.

Most automobile engines have four, six, or eight cylinders. Each cylinder operates through the four-stroke cycle. When your family car stops at a traffic light, the pistons are moving up and down ten times per second. As your car accelerates to 50 miles per hour, the pistons will move up and down over 80 times a second!

Diesel Engine

A **diesel engine** is an internal combustion engine that burns fuel by using heat produced by compressing air. The diesel engine operates in a different way than the gasoline engine. You

 can see a demonstration of this process in the "Internal Combustion Engines" lab on the Student CD.

During the intake stroke, the diesel engine injects only air into the cylinder. See **Fig. 9-11**. The compression stroke squeezes the air. This causes the air to increase in temperature. It becomes much hotter than the air in a gasoline engine. At this point, diesel fuel is sprayed into the cylinder.

The hot air ignites the fuel. An explosion takes place. The hot, expanding gases drive the cylinder down, beginning the power stroke. As in the gasoline engine, the power stroke provides the mechanical energy needed to turn the wheels of the vehicle. Why doesn't a diesel engine require a spark plug?

Both gasoline and diesel engines have advantages. For example, while the gasoline engine has better acceleration, the diesel engine is more powerful. Each engine also has disadvantages. For example, both engines pollute the air and diesel engines are very noisy.

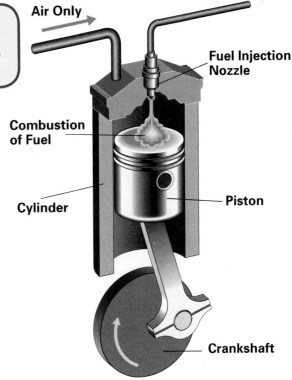

Air Only

Fuel Injection Nozzle

Combustion of Fuel

Cylinder

Piston

Crankshaft

Electric Vehicles

Electric vehicles use electricity to turn the wheels of the vehicle. As fuel costs rise and our nation's dependence on foreign countries for oil increases, auto manufacturers are turning to new sources of energy to power motor vehicles.

There are many different designs of electric vehicles currently available. Each of the major auto manufacturers has committed time and money to developing electric vehicles.

Battery-electric vehicles rely on the power stored in rechargeable batteries to turn electric motors in the vehicle. When not in use, these vehicles have to be recharged by plugging them into outlets. Battery-electric vehicles are usually low-speed vehicles. Many airports use this mode of transportation to move people and cargo throughout the terminals. See **Fig. 9-12**. Electric vehicles can also be found in factories and neighborhoods as alternative transportation.

Fig. 9-12. One reason battery-electric vehicles are used in airports is because they don't give off any pollutants. Where else are battery-operated vehicles used?

Impact of Technology

Jobs in Transportation

Throughout history, advances in the modes of transportation have made earlier forms obsolete or less desirable. As transportation changed, jobs related to transportation also changed.

For example, horses used to play a major role in moving people and goods. Horse-drawn wagons delivered farm products to town. Fire engines were pulled by horses. Families hitched up the horse and buggy for a Sunday drive. To help maintain the huge number of animals, every town had one or more blacksmiths to make and fit horseshoes. As the number of horses dwindled, so did the number of blacksmiths.

Investigating the Impact

Technology leads to changes in jobs. Some changes are negative; some are positive.

1. What new jobs were created as the automobile displaced the horse?
2. Name other examples of how a change in technology caused changes in the job market. What kinds of jobs were eliminated? What new jobs were created?
3. Technology also changes the way jobs are done. Ask a parent or other adult how his or her job has changed as new technology came into the workplace.

Hybrid Vehicles

Chances are you won't see a battery-electric vehicle fly by on the highway. However, you might see a hybrid-electric vehicle pass you. Hybrid-electric vehicles use a gasoline engine and electric motor to propel the vehicle. The addition of the electric motor helps the vehicle achieve tremendous fuel efficiency.

There are two kinds of hybrid vehicles. A parallel hybrid vehicle can use either the electric motor or gasoline engine to propel the vehicle. See **Fig. 9-13**. It may also use a combination of both power sources. The electric motor is perfect for traveling around town and the gasoline engine is great for high-speed highway travel. A generator connected to the gasoline engine helps to charge the batteries for the electric motor.

A series hybrid vehicle uses the electric motor to turn the wheels of the vehicle. A generator connected to the gasoline engine charges the batteries that power the electric motor. In a series hybrid vehicle, the gasoline engine does not power the vehicle directly. See **Fig. 9-14**.

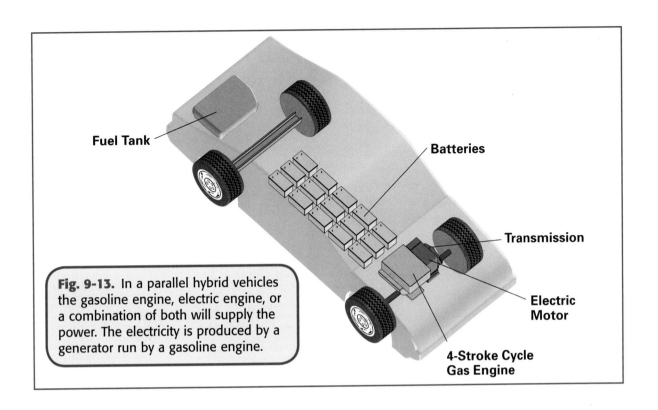

Fuel Tank

Batteries

Transmission

Electric Motor

4-Stroke Cycle Gas Engine

Fig. 9-13. In a parallel hybrid vehicles the gasoline engine, electric engine, or a combination of both will supply the power. The electricity is produced by a generator run by a gasoline engine.

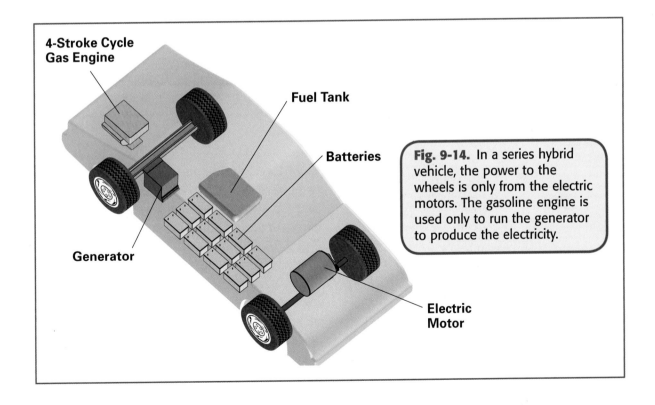

4-Stroke Cycle Gas Engine

Fuel Tank

Batteries

Generator

Electric Motor

Fig. 9-14. In a series hybrid vehicle, the power to the wheels is only from the electric motors. The gasoline engine is used only to run the generator to produce the electricity.

Water Transportation

Have you ever traveled by boat? If so, you've used water transportation. This transportation system uses vessels. **Vessels** transport people, products, and materials over waterways.

Types of Vessels

Vessels are designed to perform specific jobs. The design depends upon what the vessel will transport. Vessels are usually classified as commercial, recreational, or utility.

Commercial vessels transport people and products to make a profit. Ocean liners and ferries are commercial ships that transport people. Cruise ships even have restaurants, theaters, and recreational activities to meet the passengers' wants and needs.

Various kinds of commercial vessels transport products. Cargo ships move many of the products you see in stores. Barges carry a variety of solid materials, such as coal, iron ore, and even garbage. See **Fig. 9-15**. Container ships

Fig. 9-16. Tankers are used for the transoceanic shipment of goods. They are especially useful in transporting oil and, as shown in this photo, natural gas.

Fig. 9-15. Barges provide economical delivery of low-cost products such as coal and grain. What factors would limit the use of barges on rivers? What other types of cargo would be suitable for shipment by barge?

carry large containers packed with products. Freighters are used to transport large items such as automobiles. Tankers are vessels that carry cargoes of liquid products such as oil. See **Fig. 9-16**.

Utility vessels do work on water. Tugboats and towboats perform the delicate task of moving ocean liners and large freighters in and out of harbors. Such ships are too big to maneuver unaided in shallow water and tight spaces. Commercial fishing boats are utility vessels. Many of the large North Sea fishing boats are set up not only to catch the fish, but also to process them onboard the vessel.

Recreational vessels are used for fun. Sailboats, speedboats, and cabin cruisers are just a few kinds of recreational vessels.

How Do Vessels Float?

Have you ever experienced the feeling of being weightless while under water? You were experiencing the force of buoyancy. **Buoyancy** (BOY-ann-see) is the upward force a fluid places on an object. This force is referred to as upthrust. Objects float because of upthrust. For something to float, the upthrust has to be equal to or greater than the weight of the fluid an object displaces. This is called Archimedes' principle, named after a Greek inventor who discovered how buoyant forces work.

How does Archimedes' principle tie into boats? A vessel's hull is what floats on water. See **Fig. 9-17**. The main design feature of the hull is the shape of the bottom. Boats that sit deep in the water have a displacement-style hull. Displacement-style hulls are very buoyant. This means they float very easily. They ride smoothly in the water. These hulls can hold a great amount of weight. Tankers, barges, ocean liners, and freighters have displacement style hulls.

Boats requiring speed usually have a planing-style hull. Planing-style hulls are designed to rise out of the water at high speeds. Water moving across a hull creates drag and friction. Friction robs energy from the boat. With the hull partly

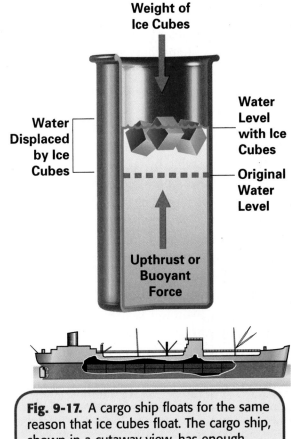

Fig. 9-17. A cargo ship floats for the same reason that ice cubes float. The cargo ship, shown in a cutaway view, has enough buoyancy to float both itself and the cargo of containers it carries.

out of the water, less surface area is in contact with the water. This reduces friction and drag. Speedboats and many other recreational vessels have planing-style hulls. See **Fig. 9-18**.

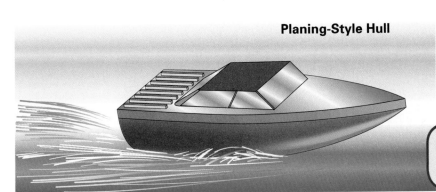

Planing-Style Hull

Fig. 9-18. A planing-style hull is used on boats designed for speed.

Powering Vessels

Like land vehicles, water vessels have different powering systems that have been developed over time. Sailboats use wind energy for power. The size, shape, and arrangement of the sails allow the wind to push or even pull the vessel through the water. See **Fig. 9-19**. A sail called a lateen sail allows ships to move across the direction of the wind by taking a zigzag path.

However, if we only used wind as a source of power, boats might sit for days waiting for a strong wind. Today, many vessels used for long-distance transportation use steam-turbine engines and gas-turbine engines to power their massive propellers.

Some commercial ferries use hydrofoils to travel fast. A hydrofoil is like an underwater wing. At certain speeds, its winglike structure creates lift, much like an airplane wing. (See Chapter 7, "Air & Space Technologies" for more on wings and lift.) The hull of the ship rises out of the water, reducing drag. See **Fig. 9-20**. Some hydrofoils are powered by waterjets. These are called jetfoils. Water is taken in at the front of the vessel and forced out the back using large pumps. The high-speed waterjet propels the vessel. Propellers with specially designed shafts are also used to propel hydrofoils.

Air-cushion vehicles float on a cushion of air. Large fans create a pocket of air under the vehicle. The vehicle rides on this cushion. These vehicles work equally well on land or water. They are sometimes called Hovercraft®, or ground-effect vehicles. See **Fig. 9-21**.

Fig. 9-19. The wind moves between the sails and creates a powerful suction, which pulls the vessel through the water. This is how a sailboat can sail against the wind. The sailboat follows a zigzag course to keep the sails at the correct angle to the wind. The keel adds stability.

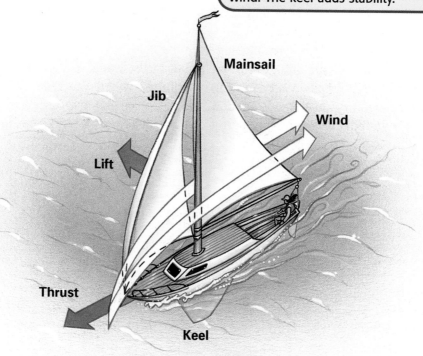

Fig. 9-20. The hydrofoil takes its name from the use of underwater foils. The rapid flow of water over the foil creates lift. Forward propulsion is made possible either by waterjets or by propellers. This craft is powered by waterjets.

Fig. 9-21. Hovercraft provide transportation for short trips over water. A large fleet of hovercraft is used in the English Channel between England and France.

Transportation Engineering

Do you know what land and water transportation systems have in common? Designing these systems requires the work of many engineering professionals.

Vehicle Design

Vehicles are composed of many subsystems. Many connections have to be made to make a car work properly. Mechanical engineers make these connections. They develop engines and engine systems that work with other parts of a vehicle. Take a look underneath a car on a lift in a repair shop. How many connections can you see?

Materials engineers develop new materials that will bring down a vehicle's manufacturing cost and improve efficiency. (You can learn more about materials engineers in Chapter 4, "Materials Science.") The materials will determine how well the vehicle works, how long it will last, and how safe it will be. For example, many cars are now engineered to absorb most of the impact in an accident. Since the car takes most of the damage, the people inside the car are safer.

Electrical and chemical engineers working for car manufacturers are busy developing new long-lasting batteries for electric and hybrid vehicles. The easier it is to use electric and hybrid vehicles, the less we may rely on other vehicles that pollute the air.

Vessel Design

Naval architects design ship hulls, steering systems, internal support structures, and floating off-shore structures. They use their knowledge of math and science to calculate the buoyancy of large vessels and to determine how these vessels will react in rough wave conditions. The safety and efficiency of vessels relies on the skills of naval architects. See **Fig. 9-22**.

Some of the biggest changes in vessel design have more to do with the design process than actual changes in the vessel itself.

Fig. 9-22. When designing the overall structure of a ship, naval architects must consider the safety of the ship and crewmembers in times of rough seas.

Fig. 9-23. This perfectly scaled model of a proposed design for an America's Cup sailboat is being tested in a wind tunnel. If the tests are favorable, the actual boat will be built.

The computer has become an integral tool in marine design and architecture. Using CAD software, engineers can create plans in hours rather than weeks. The detailed drawings of hull designs and even sail designs can be changed as fast as data about their efficiency is collected from testing.

Velocity prediction programs (VPP) assist designers in calculating the speed of a vessel under different conditions. This software takes into account the friction on the hull from water, wind speeds, and hull length and shape. VPP even predicts how fast the ship will move.

Towing tanks or water tunnels are used to test the hydrodynamics and drag of model vessels. Some tanks are over 900 feet long. The models are towed down the tank while forces acting on the hull are measured.

The use of wind tunnel testing on racing yachts has become a necessary tool in sail design. Yacht makers create one-ninth scale models of their boat design and test the sail shape and size in wind tunnels that create race wind conditions. The simulation is so accurate that full-size sail testing is usually not needed. See **Fig. 9-23.**

Modeling a Maglev System

Identify a Need/Define the Problem

Maglev trains have advantages over traditional trains that run on tracks. For this activity, your challenge is to design and build a maglev vehicle, or one car of the train, and a guideway for it to ride on. The project should demonstrate the advantages of a maglev train.

Gather Information

Gather resources that show the basic design of a maglev system. How does the system work? Does it matter what types of magnets are used? Does it matter what direction the magnetic poles are facing?

Develop Possible Solutions

Sketch designs for the ideas that you have. These should show the guideway and the placement of magnets on the vehicle and guideway. See **Figs. A** and **B**. Using graph paper, draw a pattern of the side and top view of your maglev vehicle.

Safety Alert

Look up "Safety Data Sheets" on the Student CD and prepare a data sheet for this activity. As you work on the activity, be sure to follow all safety rules.

Materials and Equipment

Select from this list or use your own ideas.

Guideway
- magnets 1" × ¾" × ³⁄₁₆"
- wood (for base)
- acrylic plastic (for sides)
- ¼" dowel rod

Body
- foam insulation board

- cardboard
- foam plastic

Chassis
- masonite
- magnets (same kind as for guideway)
- standard material processing tools and equipment

Model a Solution

1. Choose the solution that you think will be most effective.
2. Construct the guideway.
3. Create the chassis (the supporting frame of the structure).
4. Shape the vehicle body and attach it to the chassis.
5. Attach magnets to the chassis.

Test and Evaluate the Solution

- Test the model on the guideway.
- Does the vehicle levitate on the guideway?
- Can the vehicle travel in either direction?

Refine the Solution

- Are there design changes that would allow better operation?
- Create any necessary changes and test the vehicle again.

Communicate Your Ideas

Demonstrate the operation of your maglev system to the class. Be sure to point out any modifications that you made after testing. Share what you have learned about the advantages of maglev trains.

GUIDEWAY

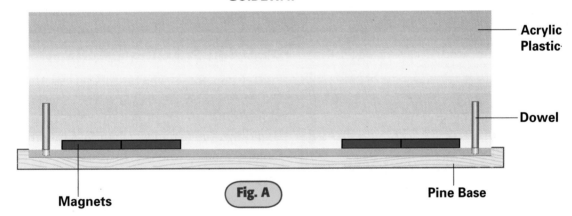

Acrylic Plastic

Dowel

Magnets

Pine Base

Fig. A

MAGLEV CAR

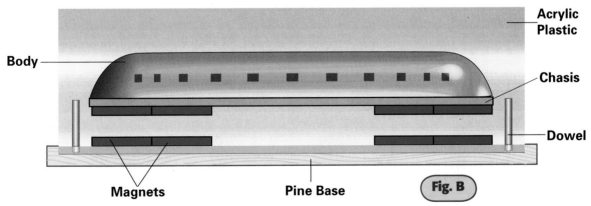

Acrylic Plastic

Body

Chasis

Dowel

Magnets

Pine Base

Fig. B

Creating a Cam Operating System

Engines used in transportation change motion from one direction to another. The direction of motion is changed by using a cam. Cams are arranged along the camshaft in the car's engine.

A cam is an offset wheel. Cams are connected to rods. See **Fig. A**. A second rod, or follower, rests on the top of the cam. As the cam turns,

the top rod moves up-and-down. To see how a cam works, go to "Mechanisms" on the Student CD.

Safety Alert

Look up "Safety Data Sheets" on the Student CD and prepare a data sheet for this activity. As you work on the activity, be sure to follow all safety rules.

Identify a Need/Define the Problem

Design, build, and test a cam system that operates a device by changing rotary motion into linear motion. At least one part of the device must move as a result of the cam's action.

Gather Information

What items do you know of that use a cam? Toys and tools may be a good place to look for ideas. The Internet may also be used to research how a cam operates.

Develop Possible Solutions

Develop three sketches of devices that can be operated by a cam system. Consider the complexity of the design and the amount of building time you have when selecting the best solution. Select the design you feel best meets the specifications.

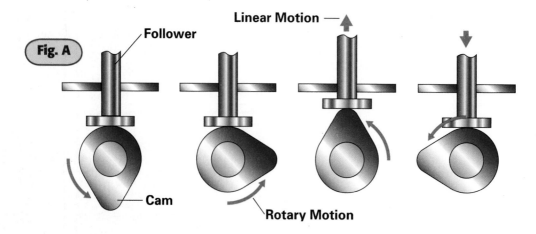

Fig. A

Follower

Linear Motion

Cam

Rotary Motion

Model a Solution

1. Prepare any patterns you might need for the construction of the device.
2. Construct the cam system. A suggested cam design is shown in **Fig. B**.
3. Place the cam system within the device.
4. Attach the device to the shaft of the follower.

Test and Evaluate the Solution

- Test the system.
- Does the device change rotary motion into linear motion?
- Does your device move in the expected way?

Refine the Solution

- Are there design changes that would allow better operation?
- Create any necessary changes and test the system again.

Materials and Equipment
Select from this list or use your own ideas.

- ¼" dowel rod
- container for cam operating system (orange juice can or milk carton)
- plastic tubing (must fit snugly over the dowel rod)
- cardboard
- foamcore board
- construction paper
- brass paper fasteners
- pipe cleaner
- glue
- scissors
- pliers

For information about hand tools or power tools used in this activity, see the Student CD.

Communicate Your Ideas

Demonstrate the operation of your cam system to the class. Be sure to point out any modifications that you made after testing. Share what you have learned about how cam systems operate.

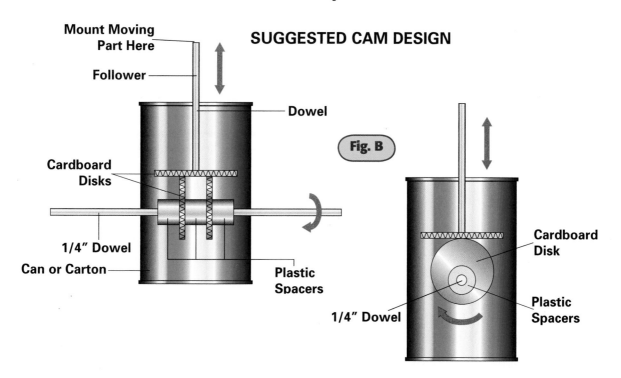

SUGGESTED CAM DESIGN

Mount Moving Part Here
Follower
Dowel
Cardboard Disks
1/4" Dowel
Can or Carton
Plastic Spacers

Fig. B

Cardboard Disk
Plastic Spacers
1/4" Dowel

Sailing Away with a Catamaran

Identify a Need/Define the Problem

A catamaran (CAT-uh-ma-ran) is a boat with two hulls. Design and build a wind-powered catamaran. You will need to meet the following specifications:

- The vessel will sail down a 12' trough.
- The vessel must be stable in the water.
- The trough will be made of PVC pipe and filled with water.
- The vessel will be powered by the wind generated from a 20" box fan.

Gather Information

Collect photos and drawings showing catamaran designs. Study the shape of the hull and sails.

Develop Possible Solutions

Prepare sketches of possible hull designs. Remember that the catamaran must easily fit within the trough. Select the design that you feel will be most effective and make a detailed drawing. Three views of a possible catamaran design are shown in **Fig. A**.

Model a Solution

1. Create patterns for the hulls and sails. Be sure both hulls are identical.
2. Assemble the hulls.
3. Design a system to support the sail.
4. Install the sail system.
5. Complete construction of the catamaran.
6. Set up your trough and fan. See **Fig. B**.

Materials and Equipment

Select from this list or use your own ideas.

- high-density construction foam
- insulation board
- sail material (fabric or plastic)
- modeling materials (dowels, balsa strips, etc.)
- 20" box fan
- material processing tools
- glue gun, cool melt
- trough 8" wide and 12' long

For information about hand tools or power tools used in this activity, see the Student CD.

Safety Alert

Look up "Safety Data Sheets" on the Student CD and prepare a data sheet for this activity. As you work on the activity, be sure to follow all safety rules.

Test and Evaluate the Solution

- Place the catamaran at one end of the water-filled trough.
- Is the vessel stable? Does it sit level in the water? (Weight may have to be added to the hull to level it in the water or to have it sit deeper in the water.)
- Test the sail by using the fan as a wind source. Does your catamaran complete its path down the trough?

CATAMARAN DESIGN

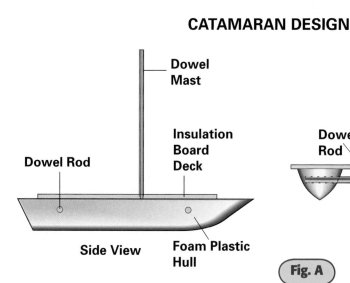

Side View

Front View

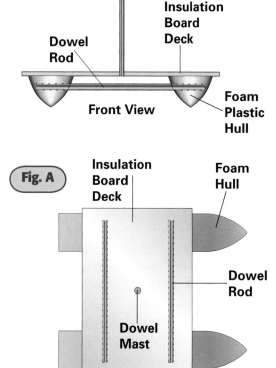

Fig. A

Top View

Refine the Solution

• A keel may be needed if your vessel fell over or sailed in an unstable way. Research the function of a keel. Prepare keel designs. Add the keel to your catamaran.

• Based on your initial testing, make additional modifications to your catamaran until it sails smoothly down the trough.

Communicate Your Ideas

Create a presentation of your catamaran. Include your initial designs, your catamaran, and your trough. Be sure to point out any modifications you may have made after testing.

Math Link

Calculating Wind Speed. The speed of vessels is usually measured in knots, which stands for the number of nautical miles traveled per hour. One knot equals about 1.15 miles per hour. If a catamaran was traveling 8.2 knots, how many miles per hour would that be?

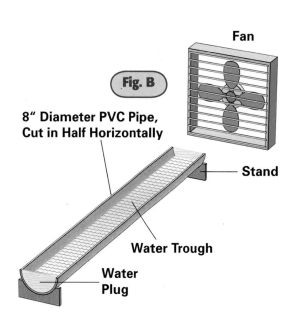

Fig. B

Fan

8" Diameter PVC Pipe, Cut in Half Horizontally

Stand

Water Trough

Water Plug

Exploring Careers

Truck Driver

ENTRY LEVEL TECHNICAL PROFESSIONAL

Truck drivers drive large trucks to move goods from one place to another. These trucks can carry several tons of cargo such as automobiles, food, furniture, and livestock.

Before they leave the base of operations, truck drivers make sure the truck is in good operating order. They check the truck's fuel and oil levels, inspect the truck's brakes and lights, and make sure they have safety equipment on board. They also check that their cargo is secure before leaving. When driving, truck drivers must stay alert for traffic accidents. They use a radio or telephone to get instructions from a dispatcher at the base of operations. Drivers keep safety records and may help load and unload the truck.

More than 10% of truck drivers are self-employed and own their own trucks. These drivers lease their trucks and services to trucking companies.

Qualifications

A truck driver must have good coordination, vision, and hearing. He or she must have a quick reaction time and be in excellent health. Truck drivers must have a CDL (commercial driver's license). To get this license, they must pass a written test and show that they can safely drive a truck. They also must pass a test issued by the Department of Transportation.

Drivers who drive trucks from one state to another must be at least 21 years old and pass a physical exam every two years. Also, they must not have a criminal record involving motor vehicles or drugs.

Outlook for the Future

The job market for truck drivers is expected to grow as the need for products increases. In addition, experienced truck drivers will retire or leave for other jobs, making more openings available.

Following Directions

In this career, if you follow directions, you will be a better worker, save time, and be more efficient. When you listen carefully, you will learn new information. To better remember directions, take notes and imagine the directions in your mind. Be sure to ask questions when you need to.

Researching Careers

Find out about jobs for truck drivers by looking at want ads. How many years of experience are needed? Is any special experience or education required? Is there an age requirement? Write a brief summary of what you find.

More activities on Student CD

Key Points

- Transportation is the process by which people, animals, products, and materials are moved from one place to another.
- Some trains transport people while other trains transport freight.
- A gasoline piston engine creates power by burning fuel inside the engine.
- Some vehicles are powered by electricity or a combination of electricity and gasoline.
- Vessels transport people, products, and materials across waterways.
- Many different kinds of engineers work to design vehicles and vessels.

Read & Respond

1. Name four sources of power for rail transportation.
2. What is intermodal transportation?
3. What are three examples of on-site transportation?
4. How is a diesel engine different from a gasoline engine?
5. How is a hybrid car more fuel-efficient than a regular car?
6. Describe the cycles of a four-stroke cycle gasoline engine.
7. Identify three sources of power for vessels.
8. What part of a vessel is responsible for making it float?
9. Why isn't wind the primary source of power for vessels?
10. Name three modes of transportation.

Think & Apply

1. **Assess.** Describe how motor vehicles have changed the way people live and work. Include positive and negative effects.
2. **Appraise.** Recreational boating is a billion-dollar industry in the United States. Discuss the impacts recreational boating has had on shore communities.
3. **Relate.** Explain how buoyancy is used to raise and lower a submarine.
4. **Categorize.** Create a table of the different power sources for rail transportation, showing the advantages and disadvantages of each.
5. **Extend.** You've learned about various engineers in the transportation industry. Interview someone in the industry to find out more information.

TechByte

Plastic Power. Internal combustion and diesel engines require large amounts of lubricating oil, which, like the fuel they burn, is derived from petroleum (oil). Chemists have recently found a way to make lubricating oil from recycled plastic. Each year, Americans throw away 25 million tons of waste plastic but recycle only 1 million tons. Recycling most of the plastic waste—and saving oil—would strike a double blow for the environment.

CHAPTER 10

Weather

Objectives

- Explain the difference between weather and climate.
- Identify the conditions that create weather.
- Describe three ways to gather forecasting data.
- Explain how computers are used to help forecast weather.

Vocabulary

- **atmosphere**
- **weather**
- **meteorologist**
- **climate**
- **radiant energy**
- **evaporation**
- **humidity**
- **precipitation**
- **air pressure**
- **front**
- **forecasting**

A hurricane's eye is the calm point in the middle of the storm.

Activities

- Measuring Pressure with Barometers
- Measuring Temperature with Thermometers
- Measuring Humidity with Hygrometers

What Is Weather?

What do a farmer, an airline pilot, a cruise ship captain, and a school bus driver all have in common? They all depend on weather information to help them do their job.

The farmer's crops need rain and a surprise freeze can destroy an entire season's profit. The airline pilot seeks to avoid thunderstorms that create air turbulence so the passengers will have a smoother flight. The cruise captain must sail around giant waves caused by storms at sea, and your school bus driver must carefully maneuver the bus along icy roads and through flooded intersections.

Current and accurate weather forecasting is vital to these and millions of other people whose daily lives are affected by weather. Daily weather forecasting helps you decide what to wear to school tomorrow or when to plan an outdoor event. Much of what we do is affected by weather. What exactly is weather?

The big blue marble we call Earth is surrounded by a blanket of mixed gases called **atmosphere**. Oxygen (21%) and nitrogen (78%) make up most (99%) of this gaseous mixture, which extends upward from Earth's surface for several hundred miles.

The atmosphere is divided into several layers, which are separated by air density and temperature. All of our weather and climate are created in the bottom layer of the atmosphere known as the troposphere, which extends just five to nine miles above our heads. See **Fig. 10-1**.

Weather is the condition of the atmosphere in regard to temperature, moisture, wind, and clouds. Conditions in the atmosphere can change fast and frequently. Air temperature and pressure can rise and fall while wind speeds increase and decrease. The air within the atmosphere may be dry or contain a lot of moisture. Weather is about the changes that take place at a particular area within the atmosphere for a short period of time.

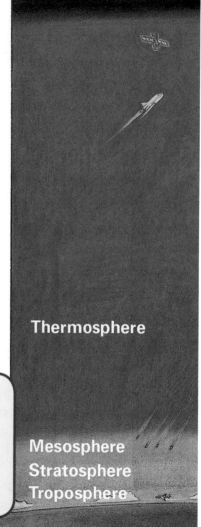

Fig. 10-1. The air surrounding Earth is about 600 km (372 miles) thick. This atmosphere is divided into layers. All weather takes place in the first 15 km of the atmosphere, which is called the troposphere. The stratosphere (15–50 km) contains the ozone layer. Most meteors that approach Earth will burn up in the mesosphere (50–85 km). The space shuttle orbits within the layer called the thermosphere. Beyond the thermosphere is the exosphere, which merges with outer space.

Thermosphere

Mesosphere
Stratosphere
Troposphere

Fig. 10-2. This meteorologist is studying information from weather satellites and weather stations. The information will be used in making weather predictions.

Weather scientists, or **meteorologists**, observe, record, and forecast changes in the weather, based on changes within the atmosphere. See **Fig. 10-2**.

What Is Climate?

What is the weather like where you live? How cold is it in winter? How much rainfall do you get during the spring? Over a long period of time, you could keep a record of changing conditions like temperature, rainfall, snowfall, and wind speed. You would then be able to describe the average weather conditions for your hometown. **Climate** is the word we use to describe the average weather conditions for an area over a period of many years.

Weather may change daily, but the climate in the area where you live stays the same year after year. For example, the climate in the southwest United States is, on average, hot and dry all

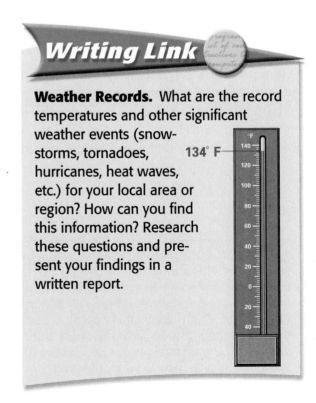

Writing Link

Weather Records. What are the record temperatures and other significant weather events (snowstorms, tornadoes, hurricanes, heat waves, etc.) for your local area or region? How can you find this information? Research these questions and present your findings in a written report.

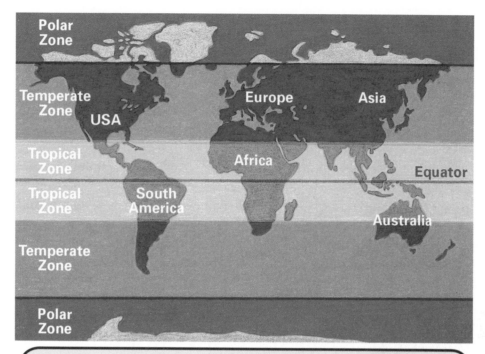

Fig. 10-3. Earth's climate can be divided into three types of zones, based on the temperature and distance from the equator. What climate zone is Australia located in?

year. However, the weather forecast for Taos, New Mexico, might be rainy and cool for a few days. Remember, weather affects a small area for a short period of time, while climate reflects the average weather conditions for a large area over many years.

Climate Zones

Earth's climate can be divided into three areas, or zones, that have similar weather conditions. The major factors that determine climate zone are the area's temperature, moisture, and latitude. (Latitude refers to a region's distance from the equator.) The three major climate zones are the polar, tropical, and temperate zones. See **Fig. 10-3**.

If you live in a polar zone, you had better dress warmly for that walk to school. The average temperature in the polar zone rarely moves above freezing. Greenland, as well as parts of Canada and Alaska, are all in the northern polar zone.

Walking to school in a tropical zone would definitely make you sweat. Tropical climates have the warmest weather, rarely falling below 64°F, with very high humidity. There's no need for snow skis in tropical climates unless you go high into the mountains (where the high altitude brings cooler temperatures).

Most of the United States is within the northern temperate zone. Areas in the temperate zones have the greatest range in both temperature and rainfall. Regions of the temperate zones that are closer to the equator tend to have higher average temperatures and milder winters than areas that are farther away. As you move farther from the equator, snow is more common in the winter.

What Creates Weather?

What makes the clouds go away or the rain come down? The interaction of heat energy, moisture, air pressure, and wind within the atmosphere causes weather.

Temperature

What specifically causes air temperature outside to change? Why is it usually warmer at midday and cooler by late afternoon or early evening? What makes a tropical climate sweltering hot and a polar climate zone freezing? The answer to these questions is sunlight. The sun provides the atmosphere with all of its natural heat energy.

Energy from the sun moves toward the earth in waves. When energy moves in waves, it is called **radiant energy**. Radiant energy waves stream through the atmosphere as sunlight and strike the earth's surface. The temperature of the earth is determined by the amount and angle of radiant energy striking it.

Not all of the radiant energy actually hits the earth. Some of it is reflected back into the atmosphere as it collides with dust and water droplets in the air, heating up the atmosphere. See **Fig. 10-4**.

The radiant energy that does strike the earth's surface is absorbed by the ground, water, and buildings and is changed into heat. The air above these surfaces is warmed by this heat and rises. Cool air sinks below the rising warm air, and the process continues. See **Fig. 10-5**. The transfer of heat from the earth's surface to the air causes changes in temperature.

The angle at which the sunlight strikes the earth's surface also helps to determine the air temperature. The greatest heating occurs when the sun's rays are most direct, the period during which the rays strike the surface at an angle near 90°. As you can see in **Fig. 10-6A**, the sun is directly overhead (forming a 90° angle) around noon, making midday the time of day that receives the most heat. Now look at **Fig. 10-6B**. Notice that areas on and around the equator receive most of the sun's radiant energy. Observe how close the tropical zones are to the equator. How does this explain the temperature in tropical climates?

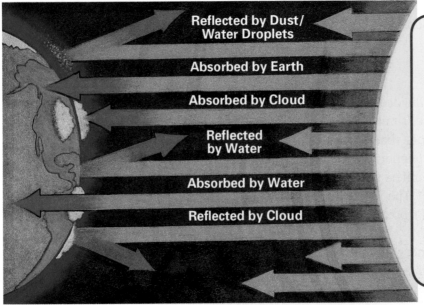

Reflected by Dust/ Water Droplets

Absorbed by Earth

Absorbed by Cloud

Reflected by Water

Absorbed by Water

Reflected by Cloud

Fig. 10-4. Radiant energy travels a vast distance from the sun to the earth's surface. Most of that energy is absorbed by the atmosphere or reflected back into the atmosphere by particles of dust or water. The energy that does make it to the earth's surface is absorbed and radiated back out as heat energy. Less than one millionth of the sun's radiation reaches the earth.

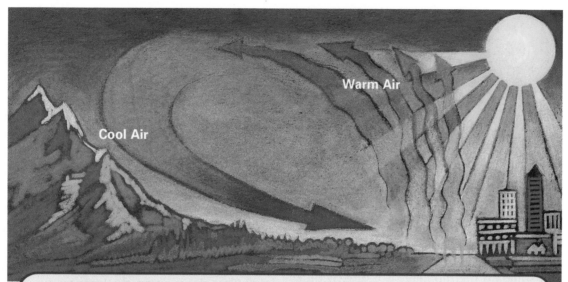

Fig. 10-5. Have you ever seen waves of heat rising from a roadway or sidewalk on a hot summer day? Radiant energy from the sun is absorbed by the ground, water, buildings, and all surfaces on the earth. As these surfaces heat up, they heat the air above them. If you place your hand above the hot sidewalk, you can feel the heat rising. At some point, this warm air cools and begins to drop; then the process begins again.

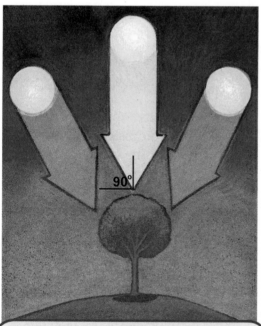

Fig. 10-6A. As the sun moves across the sky, radiant energy strikes the earth at different angles. The sunlight is more intense at noon and less intense at sunrise and sunset.

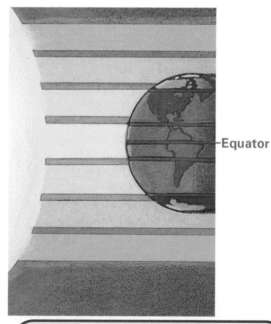

Fig. 10-6B. Notice that the sun's rays are most intense in the areas along and around the equator. What does this suggest to you about the weather in tropical climates?

Fig. 10-7. Digital thermometers are easy to read.

Measuring Temperature. Most frequently, thermometers are used to measure air temperature. A standard thermometer is a tube with a bulb on the end. The bulb is filled with a fluid. When air temperature rises, the fluid expands and moves up the tube. What do you think happens as the temperature cools?

Air temperature can also be measured digitally by computers. Sensors attached to the computer detect changes in air temperature. Those changes are converted into electrical energy that the computer can read and store. Since the sensors are so accurate, and the results are so easy to read, many thermometers today are digital. See **Fig. 10-7**.

Fig. 10-8. When the ground cools at night after a hot day, dew will collect on the grass. If the temperature gets close to freezing, the dew will become frost.

Humidity

Moisture is constantly circulating throughout the atmosphere in a water cycle. This cycle involves evaporation, condensation, and precipitation.

Moisture gets into the air through the process of evaporation. **Evaporation** takes place when liquid water is turned into gas by the sun's radiant energy. This gas is called water vapor. **Humidity** is the amount of water vapor in the air. As the molecules of water vapor in the air increase, the damp and sticky feeling associated with high humidity increases.

Just how much water vapor can the air contain? That depends on the temperature of the air. The warmer the air is, the more water vapor it can hold. As the land cools during the night, it is hard for the air to hold its humidity. The water vapor condenses and forms water droplets, or dew, that covers the grass in the morning. See **Fig. 10-8**. If the nighttime temperature falls below freezing, these droplets become frost.

Science Link

Relative Humidity. You have probably heard weather forecasters talk about relative humidity. The relative humidity is a ratio. It is the actual amount of water vapor in the air compared to the amount of water vapor the air could hold at that temperature. For example, a relative humidity of 50% means there is half as much water vapor in the air as there could be. What would 25% relative humidity mean? If the relative humidity is 100%, then the air cannot hold any more water vapor. In what two ways might that condition be changed so that more evaporation could take place?

Fig. 10-9. Fog forms when warm moist air passes over cool ground. This often happens in low-lying areas where the ground is usually cooler.

When warm, moist air passes over a cool surface, the air itself may cool, causing the water vapor in the air to condense and stay suspended in the air. This results in fog. See **Fig. 10-9**. Fog can also occur when cold air passes over warm water.

Clouds

Clouds are formed when warm, moist air rises into the atmosphere. As the molecules of water vapor move up, they slowly cool down in temperature. The cooling forces the water vapor to change back into a liquid. When these small droplets of liquid stick to dust particles in the atmosphere, a cloud is formed. A cloud is really a mixture of water droplets and dust particles suspended in air. See **Fig. 10-10**.

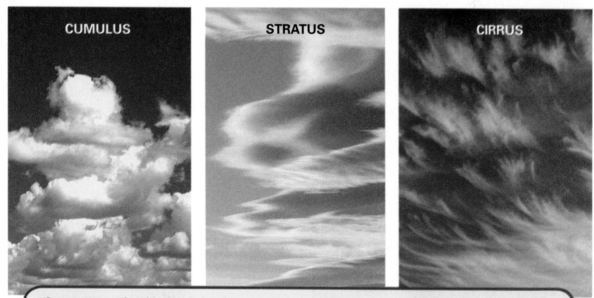

Fig. 10-10. What kinds of clouds are in the sky? Clouds come in many different shapes. Those that look like cotton balls are known as cumulus clouds. Smooth, gray clouds that cover the sky are called stratus clouds. Feathery clouds are known as cirrus clouds.

Precipitation

Sometimes the droplets of water suspended in the cloud bump into each other and stick together. The droplets get bigger and bigger as they combine. When the drops become too large to stay suspended in the cloud they fall to Earth as **precipitation**. Rain, snow, sleet, and hail are all forms of precipitation.

When the air is warm enough, the droplets fall in liquid form, or rain. However, when water vapor comes in contact with cooler air within the atmosphere, it sometimes changes into a solid form. This solid form is snow. See **Fig. 10-11**. When raindrops pass through an extremely cold layer of air, they freeze and become ice pellets called sleet.

Fig. 10-12. When raindrops fall through colder air, they can form small pellets called sleet. When ice pellets are caught in updrafts, they increase in size and form hailstones.

Fig. 10-11. As the air gets colder, water vapor can change directly into snow.

Sometimes, as it falls through the atmosphere, the rain may strike ice pellets in a cloud. The raindrops stick to the pellets and become part of the ice mass. Strong winds or updrafts may keep these pellets suspended in the cloud for a long time, allowing layer after layer of raindrops to stick to the pellets. The pellets get bigger and bigger until they fall from the sky as hail. See **Fig. 10-12**. Hailstones the size of softballs have been recorded in parts of the United States.

Air Pressure

Have you ever given your friend a ride on your shoulders? Remember the feeling of the weight pushing down on your body with a force from above? The weight of your friend was placing pressure in your body from your shoulders to your toes. The atmosphere above our heads also places a force on our body as well as everything else on the earth's surface.

Air pressure, also known as atmospheric pressure, is a measure of the force of air pressing down on top of us and the earth's surface. The air pressure at any spot on the earth is equal to all the weight of air directly above that point.

The amount of air pressure is affected by three factors: air temperature, water vapor, and elevation. Warm air exerts less pressure than cool air. Moist air exerts less pressure than dry air. As elevation, or altitude, increases, the atmosphere becomes thinner and less dense, exerting lower pressure on the earth's surface.

Later in this chapter, you will see how air pressure can move weather systems in and out of an area.

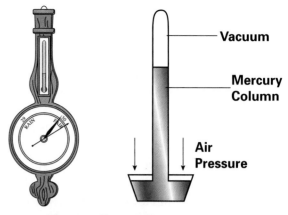

Mercury Barometer

Math Link

Elevation and Temperature. Have you ever had the opportunity to hike up a relatively tall mountain? Did you experience cooler temperatures near the top than at the bottom? As it turns out, as the elevation on Earth increases, the resulting temperature decreases. Can you uncover why this is so?

Shown below are the altitude and temperature along one mountain during a spring day. Make a graph to show this data.

Altitude (feet)	Temperature (F)
0	59.0
1,000	55.4
2,000	51.9
3,000	48.3
4,000	44.7
5,000	41.2
6,000	37.6
7,000	34.0
8,000	30.5
9,000	26.9
10,000	23.3

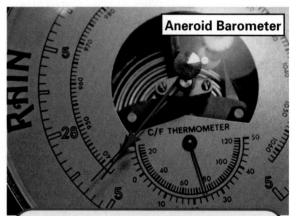

Fig. 10-13. In the mercury barometer, the mercury rises as air pressure increases. In the aneroid barometer, a flexible piece of metal connected to a pointer bends as air pressure changes.

Measuring Air Pressure. Air pressure is measured with a barometer. The two basic types of barometers are mercury and aneroid.

A mercury barometer uses a tube filled with mercury placed in a dish of mercury. As air pressure places a force on the dish of mercury, it makes the column of mercury in the tube rise. When air pressure drops, the mercury falls.

Aneroid barometers work differently from mercury barometers. An aneroid barometer uses a flexible piece of metal that bends as the air pressure changes. A pointer connected to this metal indicates the air pressure. See **Fig. 10-13.**

Wind

What causes a breeze? Where does wind come from? When air is heated, it becomes less dense and rises, causing a low-pressure area. Cooler, denser air drops below the warm, rising air, which creates a high-pressure area. Winds are formed by the movement of air from high-pressure areas to low-pressure areas.

Believe it or not, there are different types of winds. Local winds cool off your neighborhood and can blow from any direction. Global winds blow from a specific direction and travel very long distances. Both types of winds are caused by differences in air pressure due to the unequal heating of the earth's atmosphere.

Local Winds. What's the best thing about the beach on a hot summer's day? It's the breeze. During the day, air above the land is much warmer than air above the water. As the warm air above the land rises, cooler air from above the sea moves inland and slides under the rising warmer air. This flow of air from sea to land is a sea breeze, which is one type of local wind. See **Fig. 10-14**. What do you think a land breeze is?

Global Winds. Unequal heating of the earth's surface also creates a large global wind system. Global winds are created when warm air from over the equator rises and moves toward the polar regions. Cooler air from the polar zones drops down and heads toward the equator, sinking below the rising warm air. Like local winds, differences in atmospheric temperatures cause the air to move, creating wind. The patterns of global winds usually remain the same and are very predictable. Local winds are less predictable.

Measuring Wind. Meteorologists need to know the direction and speed of the wind to help make weather predictions. A wind vane is used to show the direction of the wind. An anemometer is used to measure the speed of the wind. See **Fig. 10-15**.

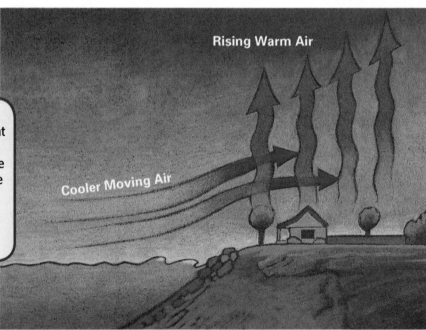

Fig. 10-14. Land and water gain and lose heat at different rates. During the day, the sun's rays heat the land more quickly than the water. As the warm land air rises, the cool sea air moves in, creating a sea breeze.

Fig. 10-15. When your local weatherperson forecasts weather conditions, wind is usually described by its direction and speed. The point of the arrow on the weather vane indicates the wind's direction. The cups on the anemometer spin and show the wind's speed on a meter.

Fronts

Air sometimes moves in large blocks known as air masses. What happens when a mass of cold air contacts a mass of warm air? Well, it's like oil and water—two different types of air masses just don't mix.

When two air masses of different properties (temperature, pressure, or moisture) collide with each other, a front is formed. A **front** is a line where two air masses actually touch.

Fronts often produce unstable and potentially violent weather. When a cold air mass pushes under a warm air mass, the resulting cold front may produce thunderstorms. The weather that follows the thunderstorm is usually cold or cool and fair. See **Fig. 10-16.**

Fig. 10-16. During a cold front, a cold air mass rapidly pushes a rising mass of warm air away, resulting in rain and thunderstorms. During a warm front, a warm mass of air runs into a mass of cold air and slides over it. The result of this front is often steady rain.

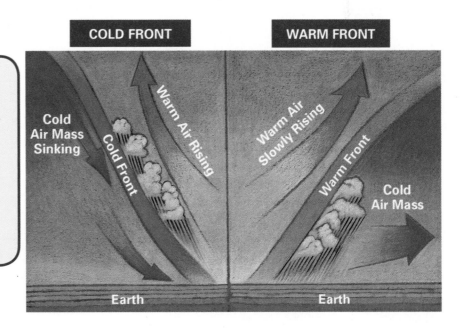

COLD FRONT   WARM FRONT

Cold Air Mass Sinking

Warm Air Rising

Cold Front

Warm Air Slowly Rising

Warm Front

Cold Air Mass

Earth   Earth

A warm front occurs when a warm mass of air slides over a cold air mass. This type of weather condition usually produces lots of steady rain. The weather that follows the rain is usually warmer or hot and more humid.

Sometimes different types of air masses meet and stall and no air movement takes place. This type of front is known as a stationary front. Stationary fronts can cause rain to fall or the sun to shine where you live for days and days.

Storms

A storm is a violent disturbance in the atmosphere caused by the collision of different types of fronts. Rainstorms and snowstorms are caused by warm air masses meeting cold air masses. In the summer, this condition can cause a steady rain. In the winter months, this condition can cause a snowstorm. If the temperature is low enough and the wind is fast enough, a snowstorm can turn into a blizzard.

Thunderstorms

Thunderstorms take place when a cold air mass moves in on a warm air mass. Thunderstorms produce heavy rain. They can be very dangerous because they also produce lightning.

Lightning is a discharge of electrical energy from a cloud to another cloud or to the ground. See **Fig. 10-17**. What causes lightning? As molecules within a cloud bump into each other, they develop an electrical charge. It's similar to the charge you often pick up when walking on a carpet. When you touch a doorknob, the charge is suddenly released.

Where does the thunder come from? As the lightning discharges, it heats up the air around it. The air expands so rapidly that it creates a crackling thunder sound, much like the air expanding from a balloon when you pop it with a pin.

Cyclones and Hurricanes

A cyclone is created in low-pressure areas when rising warm air is replaced by cooler air. As the air moves up, it begins to spin. The spinning air carries with it rain and high winds.

A hurricane is a very powerful cyclone that forms over tropical waters. The winds created by a hurricane can reach speeds of over 150 miles per hour, causing great damage when they come ashore. In 2004, Hurricane Charley caused Florida $14 billion in property damage, leaving many people homeless. Hurricane Katrina, which struck the Gulf Coast in 2005, was the costliest natural disaster in United States history.

Fig. 10-17. Lightning results when molecules within a cloud rub against each other, building up a static charge. The cloud gets so filled with static electricity that the electrons jump between it and the ground or between it and another cloud.

Impact of Technology

Global Warming

Have you heard the expression "global warming"? It sounds as if a globe is warming up. In fact, the globe to which the term refers is our earth. Over the last 100 years or so, the average temperature around the globe has risen just a bit. If the rise in temperature continues, there would eventually be catastrophic results, such as flooding from the melting of the polar ice caps.

Investigating the Impact

1. To what do scientists attribute the phenomenon of global warming?
2. What actions can be taken to limit or possibly even reverse the trend toward warmer climates?

Tornadoes

Tornadoes form over land rather than water. A tornado is a whirling, funnel-shaped cloud that can contain winds traveling up to 220 miles per hour. A tornado can cause massive destruction. The bottom of the tornado acts like a giant vacuum cleaner when it touches ground, sucking up soil, trees, cars, and even houses. See **Fig. 10-18**. Each year, tornadoes cause millions of dollars in property damage and can even kill people in their paths. One tornado has a record of causing over one billion dollars in property damage!

Fig. 10-18. Tornadoes can leave a path of destruction several hundred yards wide. As the tornado twists and turns through an area, it is not uncommon for one building to be destroyed while a nearby building goes untouched.

Forecasting the Weather

What will the weather be like next weekend for the family picnic? The newspaper says sunny with clear skies and temperatures in the 80s. How can anyone predict the weather so far in advance?

Making predictions is called **forecasting**. Meteorologists use science and technology to forecast the weather, often weeks in advance.

There are many techniques used to forecast weather. Some are simple and can be used by anyone. Other techniques are very complex and require expensive, complicated equipment and a lot of training. These more advanced methods produce the most accurate forecasts.

Some forecasts are very simple in practice. For example, in some areas of the country, weather conditions change very little from day to day. Persistence forecasting is often used in these areas. This type of forecasting assumes at the time of the forecast that weather conditions will not change. In areas like southern California where weather patterns may stay the same over large blocks of time, the weather is usually predicted to be persistent and not change over the next few days.

Doesn't sound too reliable, you say? Reliable methods of forecasting start with collecting data about weather.

Gathering Data

Temperature, moisture, air pressure, and wind direction and speed are all clues used by meteorologists to forecast weather. The first step in forecasting weather is to gather data about these conditions nearby as well as in other locations.

Why is it important to know the weather in other locations? Weather in the United States

Fig. 10-19. Weather satellites take photographs of atmospheric conditions from space and relay these images back to Earth. This technology allows meteorologists to obtain pictures of weather conditions anywhere in the world.

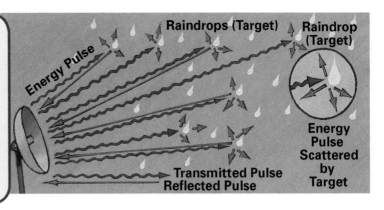

Fig. 10-20. Radar can locate precipitation by sending pulses of energy into the atmosphere. The pulses are scattered when they come in contact with targets, such as raindrops or snowflakes. Some of this scattered energy is reflected back to radar antennas on Earth. The size and quantity of targets determine the amount of reflected energy.

Energy Pulse

Raindrops (Target)

Raindrop (Target)

Energy Pulse Scattered by Target

Transmitted Pulse
Reflected Pulse

usually moves in a pattern from the west coast to the east coast. If you can determine the speed of a weather front moving in a certain direction, you can forecast when it will hit an area. If a storm front is moving 250 miles a day and is 1,000 miles away from your town, then you can predict that the effects of that front will be within your area in four days. That's why conditions that develop on the west coast may affect weather in the Midwest. If we want to forecast weather in California, we must study atmospheric conditions over the Pacific Ocean.

Using information from a network of weather stations located all over the world, the National Weather Service provides meteorologists with important atmospheric data. Remote weather stations are equipped with instruments such as thermometers, barometers, wind vanes, and anemometers. These instruments record the air temperature, air pressure, wind direction and speed, and humidity in their local area.

To obtain a "big picture" view of the atmosphere, meteorologists can rely on satellites, radar, and other specialized equipment to help gather data from high in the atmosphere.

Satellites. Designed to send pictures back to Earth, satellites photograph cloud formations and reveal the world's weather in pictures. Weather satellites are positioned all around the planet and are constantly taking photographs that can be used in forecasting. By studying sat-

ellite pictures, meteorologists can record cloud development and other patterns that are valu-

able for weather prediction. See **Fig. 10-19**. For more information, see the "Weather Satellites" lab on the Student CD.

Radar. Television weather reports often show radar images of storms as they move across the globe. These images are created when radar antennas transmit (send) electromagnetic energy pulses into the atmosphere. The energy pulses are scattered in different directions when they collide with objects in their path, such as snowflakes or raindrops. Some of the scattered energy is reflected back to Earth and detected by the radar antennas, which act as receivers as well as transmitters. See **Fig. 10-20**. The larger the object, or target, the stronger the return signal. Computers then convert the reflected images into photographs that can be analyzed by meteorologists.

Doppler radar is an advanced form of radar. Doppler radar images, which appear in different colors on the screen, can show the direction and the speed of a snowstorm or rainstorm as well as the amount of precipitation that is falling. One of Doppler's main uses has been to warn people of tornadoes before they hit. Thanks to Doppler, many people have been able to take shelter ahead of time and avoid harm during tornadoes.

Weather Flights. Fasten your seat belt if you happen to board a weather plane. Your destination could be the center of a fierce hurricane with winds up to 125 miles per hour.

In an attempt to gather important information about storms, scientists and meteorologists fly into hurricanes in special planes designed to gather data on wind speed, precipitation, air temperature, and pressure.

While in the storm, the aircraft lowers instruments into the moving air. After a short time, the plane moves into the calm of the hurricane's eye (center). From this position, the people in the aircraft can see the ocean and ground below and can study the movement of the storm. See **Fig. 10-21**.

Making Forecasts

Meteorologists study the data provided by weather technology. They apply their knowledge of weather science to interpret the data and to predict what the weather might do next.

The data that is collected is often used to create weather maps. These maps provide meteorologists with a graphic picture of weather

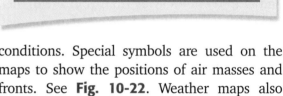
conditions. Special symbols are used on the maps to show the positions of air masses and fronts. See **Fig. 10-22**. Weather maps also identify the direction, or track, that fronts are moving in, and they show air temperature, atmospheric readings, and wind speed data.

Fig. 10-21. Weather planes use sophisticated technology to gather data about weather conditions. These planes have the ability to fly directly into a storm to collect information that would otherwise be impossible to obtain.

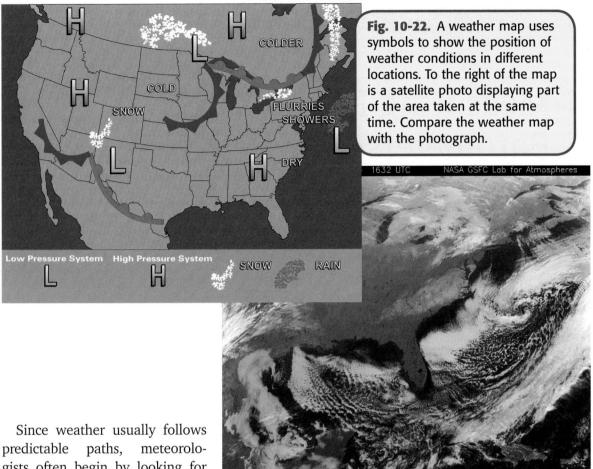

Fig. 10-22. A weather map uses symbols to show the position of weather conditions in different locations. To the right of the map is a satellite photo displaying part of the area taken at the same time. Compare the weather map with the photograph.

Since weather usually follows predictable paths, meteorologists often begin by looking for weather patterns. What are these patterns? Cold fronts moving into warm air masses usually cause thunderstorms. Also, remember that weather usually moves from west to east in the United States. This is also a weather pattern. So, if the weather map shows a cold front moving across Pennsylvania in an easterly direction and a warm front is stalled in New Jersey, it might be a good prediction that thunderstorms may develop in the New Jersey area within a day or so.

Of course, weather patterns are only one factor that must be analyzed by meteorologists when they forecast the weather. As you have learned, all the collected data must be analyzed. One tool that meteorologists can use to process weather information gathered from all over the globe is a supercomputer.

Using supercomputers for weather analysis is sometimes called numerical weather prediction. This method requires gathering lots of information about atmospheric conditions, such as temperature, pressure, precipitation, wind speed, and direction. This data is fed into computers, which then create weather models based on the information.

The computers search through a database of previous weather models that have been accumulated over many years, looking for matches in weather conditions. The assumption is that similar conditions will create similar weather. A meteorologist's final forecast is based on his or her interpretation of the models and knowledge of local weather patterns.

Measuring Pressure with Barometers

Identify a Need/Define the Problem

Working in teams, design, build, and evaluate barometers using different liquids to see which liquid provides the most accurate measurement of air pressure. See **Fig. A**.

Gather Information

The liquid in a barometer moves as atmospheric pressure pushes on it. Will any liquid work? Will barometers using different liquids give different pressure readings? Gather common household liquids such as water, alcohol, hydrogen peroxide, vinegar, and motor or cooking oil.

Develop Possible Solutions

Observe the viscosity (the measure of a material's ability to flow) of each liquid. Will thicker, more viscous materials work better in a barometer than a thin, less viscous material? Each team will be testing water and two other liquids to determine which provides the most accurate measurement. Choose the two liquids you wish to test against water in your homemade barometers.

Materials and Equipment
Select from this list or use your own ideas.

- 3 pieces ⅛" diameter glass tubing, 12"–14" long
- colored water
- cooking oil, vinegar, or other test liquids
- safety goggles
- leather welding gloves
- propane torch or Bunsen burner
- tongs (to hold glass tubing)
- mounting board
- drill
- spade or Forstner bit
- epoxy glue
- small triangular file (to score glass tubing so it can be snapped)

 For information about the tools listed here, see "Hand Tools" and "Power Tools and Machines" on the Student CD.

Safety Alert

Look up "Safety Data Sheets" on the Student CD and prepare a data sheet for this activity. As you work on the activity, be sure to follow all safety rules.

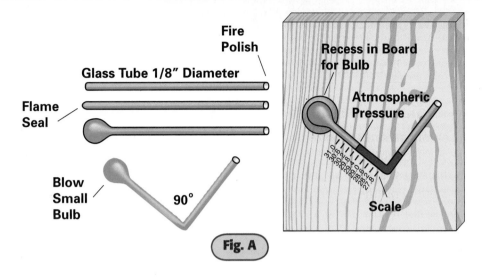

Fire Polish

Glass Tube 1/8" Diameter

Flame Seal

Blow Small Bulb

90°

Recess in Board for Bulb

Atmospheric Pressure

Scale

Fig. A

Model a Solution

1. Follow all safety rules listed in the "Safety Alert" boxes. Also, your instructor may need to complete certain parts of this activity that are dangerous.
2. Cut three pieces of glass tubing to length (12"–14"). Follow steps 3–8 for each tube.
3. Using a propane torch, fire polish one end of the tube by rolling it in the flame. Be sure you *do not seal the end.*

4. Allow the tube to cool for five minutes.
5. Seal the opposite end of the tube by rolling that end in the flame until it closes and seals. Allow the sealed end to cool for five minutes.
6. Create a bulb on this sealed end by placing that end of the tube in the flame. Allow it to get red-hot. Then *gently* blow into the polished end of the tube until a small bulb forms on the sealed end. (This is called glass-blowing, and it is how some shapes are made.)
7. Allow the tube to cool.
8. Heat the center of the tube until it glows. Bend the tube slowly until it forms a 90° angle.
9. Prepare three mounting boards by drilling a recess into the wood using a spade or Forstner bit. Be sure the recess is a little larger in diameter than the bulb you created, and deep enough so the tube can lay flat on the board.
10. Attach each tube to a board with epoxy glue.
11. Using an eyedropper, partially fill one tube with colored water. Fill each of the remaining tubes with your other test liquids.

Measuring Pressure with Barometers (continued)

Test and Evaluate the Solution

- Allow the barometers to sit for 24 hours.
- Mark the level of the liquid in each tube. Check an Internet site, such as one for a local TV news station or the National Weather Service, to find out the current barometric reading in your area. Write this measurement next to the level line you marked on each of your barometers.
- Hang your barometers on the wall or sit them up vertically and observe them over the next few days. Mark off a measurement on the barometers each day.
- What, if any, differences did you observe in your three barometers? Did one barometer change more quickly or more noticeably than the other barometers?

Refine the Solution

- How could the barometer be improved to be more accurate, based on the knowledge that you have from testing it?
- What would happen if glass tubes of a different width were used to make the barometer?
- If you decide to make barometers with a different size of tube, check with your instructor first.

Communicate Your Ideas

Compare your findings with those of the other teams in your class. Overall, what type(s) of liquids seemed to most accurately indicate changes in air pressure?

Math Link

Units of Pressure. You learned in this chapter that air pressure is a measure of the force of air pressing down on top of us and the earth's surface. Air pressure has been measured at sea level (the natural level of a large body of water) at about 14.7 pounds per square inch.

Knowing air pressure at sea level, see if you can figure out these challenges.

1. How much air pressure is being exerted per square foot?
2. The United States is the only country that commonly uses psi to measure air pressure. What is the metric conversion for psi?

Measuring Temperature with Thermometers

Identify a Need/Define the Problem

Design and build a working thermometer by testing a variety of fluids that react to changes in surrounding temperature. A commercial thermometer is shown in **Fig. A**.

Fig. A

Gather Information

As the temperature around a thermometer changes, the fluid inside the thermometer rises and lowers. As the surrounding temperature gets hotter, the fluid inside the glass tube expands and rises. As the temperature cools, the fluid contracts and lowers. You read the

Materials and Equipment

Select from this list or use your own ideas.

- 3 half- (.5-) liter bottles
- 3 corks
- drill
- glass tubing
- water and at least two other chosen liquids
- food coloring
- commercial thermometer

 For information about the tools listed here, see "Hand Tools" on the Student CD.

thermometer by locating the level of the fluid and reading the temperature marked next to that level. Research any additional information about thermometers that you may need.

Develop Possible Solutions

Mercury is usually the material used inside a thermometer, but any material that reacts to changes in temperature can be used. Test three fluids—water as well as two other everyday household liquids, such as vinegar, saltwater, alcohol, or cooking oil—to indicate the temperature in a homemade thermometer. Research the properties of all these fluids, such as their freezing point and boiling point.

Measuring Temperature with Thermometers (continued)

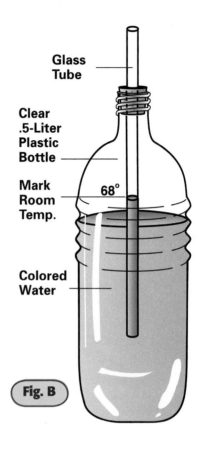

Glass Tube

Clear .5-Liter Plastic Bottle

Mark Room Temp. 68°

Colored Water

Fig. B

Model a Solution

1. Select the design that you think will be the most effective. An example is shown in **Fig. B**.
2. Locate three clear plastic .5-liter water or beverage bottles.
3. Find three corks to fit snugly into the top of each bottle. Drill a hole in the center of each cork. The hole should be large enough that a glass tube should slide through it. Make sure the hole is not too tight, since the glass tube may break if you push too hard.
4. Remove the cork and tube components from each bottle.
5. Fill one bottle ¾ full with colored water. Fill the remaining bottles ¾ full with the two other liquids that you wish to test. If needed, add coloring so that the fluid level in the tube can be easily observed.
6. Replace the cork-and-tube components so that they fit snugly in the bottle opening.
7. Let the thermometers you have made, as well as a commercial thermometer, sit for a while at room temperature.

8. Mark each plastic bottle with a line even with the level of the fluid inside the glass tube. Label this with the current room temperature shown on the commercial thermometer. Make a chart on which to record the temperatures on the four different thermometers as well as the time at which the temperatures were read.

Test and Evaluate the Solution

- Place the bottles and the commercial thermometer in a sunny window. After the liquids in the glass tube rise, mark the bottle with a line even with the level of the fluid in the tube. Then label this with the current temperature shown on the commercial thermometer. Record this data and the time on your chart.
- Experiment by placing the thermometers in different places and marking and recording the data.
- Were there any differences in how the various fluids reacted to the changes in temperature? For example, did one fluid seem to rise or fall more quickly?
- Which fluid seemed to reflect the most accurate changes when compared to the commercial thermometer?
- How can knowing the freezing or boiling point of a fluid help to select the best fluid for your thermometer?

Refine the Solution

How could the thermometer be improved to be more accurate, based on the knowledge that you have from testing it? What would happen if narrower or wider glass tubes were used to make the thermometer?

Communicate Your Ideas

Share your chart showing the results of your experiment with the class. Demonstrate what you have learned about measuring temperature.

Science Link

Expanding Your Knowledge. Not all materials expand when heated and contract when cooled. Research why some materials react to heat one way while others react to heat another way. Be prepared to explain the differences to your classmates. What uses could there be for knowing whether a material expands or contracts or for knowing the specific amount a material expands or contracts?

Measuring Humidity with Hygrometers

Identify a Need/Define the Problem

Design and build a working hygrometer by testing a variety of materials or objects that react to changes in surrounding humidity. A commercial hygrometer is shown in **Fig. A**.

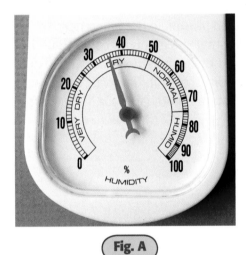

Fig. A

Gather Information

As the humidity in the air changes, material inside the hygrometer expands and contracts. The material expands when humidity increases, and it contracts when humidity decreases. The hygrometer is read by locating the level of the material, or the needle that it is connected to.

Materials and Equipment
Select from this list or use your own ideas.

- pinecone
- modeling clay
- commercial hygrometer
- one foot long human hair
- thin strips of wood
- materials students identify

Pinecones can serve as a natural hygrometer. The pinecones "open" in dry air, and "close" in dampness or rain. Find other objects that can serve as natural hygrometers.

Develop Possible Solutions

Any material that reacts to changes in humidity can be used to create a hygrometer. Test three materials or objects, a pinecone as well as two other materials, to indicate the humidity. An example is shown in **Fig. B**.

Model a Solution

1. Select the design that you think will be the most effective.
2. Construct the hygrometer according to your plans.

Test and Evaluate the Solution

- Let the hygrometers you have made, as well as a commercial hygrometer, sit for a while at room temperature.
- Take notes on the appearance of the hygrometers and mark them in some way if you can. Also take note of the current room humidity shown on the commercial hygrometer. Make a chart on which to record the humidity on the four different hygrometers as well as the time at which the readings were taken.
- Observe the hygrometers once a day for a week and record your findings in a chart.
- Experiment by placing the hygrometers in different places and marking and recording the data.
- Were there any differences in how the various materials reacted to the changes in humidity? For example, did one seem to rise or fall more quickly?
- Which material seemed to reflect the most accurate changes when compared to the commercial hygrometer?

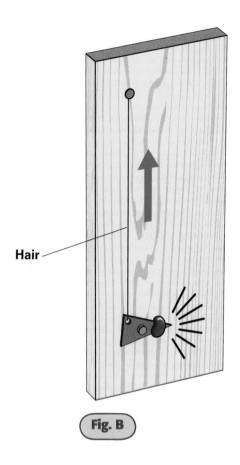

Hair

Fig. B

Math Link

Pies, Bars, and Lines. What kind of chart did you make to record your data? Did you make a standard column and row chart? There are many ways to record data. Research how pie charts, bar charts, and line charts are used to record data. Take whatever kind of chart you made in this activity and convert the data into different chart forms.

Refine the Solution

- How could the hygrometer be improved to be more accurate, based on the knowledge that you have gained from testing it?
- If necessary, revise your hygrometer design. You may even decide to use alternative materials than the three you originally selected.

Communicate Your Ideas

Share your chart showing the results of your experiment with the class. Was your chart similar to other classmates' charts? Why or why not?

Exploring Careers

Meteorologist

ENTRY LEVEL | TECHNICAL | PROFESSIONAL

Meteorologists study the atmosphere—the air covering the earth—and how it affects land and water. They observe temperature, air pressure, humidity, and wind velocity. They use this information to predict the weather. They also learn about past and present climate trends.

Meteorologists research the weather to learn how it affects air and sea transportation, forestry, air pollution, and agriculture. They study global warming, ozone depletion, and droughts.

Meteorologists gather data using equipment such as weather balloons, satellites, and radar.

Because most weather stations operate 24 hours a day, someone has to be there at all times. Meteorologists work rotating shifts. Also, meteorologists work more hours during weather emergencies to stay current with developments.

About 50% of meteorologists work for the federal government, and the other 50% work for private weather consulting companies, radio and television stations, air carriers, or state governments.

Qualifications

Meteorologists must have strong knowledge of math and science. A bachelor's degree in meteorology or atmospheric science is required, and many employers want a master's degree. A meteorologist takes courses in the analysis and prediction of weather systems, atmospheric dynamics and thermodynamics, physical meteorology, and instrumentation. Students also learn physics, ordinary differential equations, and physical sciences.

Outlook for the Future

The job outlook for meteorologists is fair. More jobs with private companies will be available in the future than with the government.

Enthusiasm

Enthusiasm is a good quality to bring to this or any other career. It will help you to do a better job, and other people will share your enthusiasm. To make work seem more like fun, you should be able to link what you enjoy doing to your job tasks.

Researching Careers

Find out about jobs for meteorologists. Contact a college and ask about its meteorology graduates. What classes did they take? In what jobs do the graduates work? Write a report describing what you learned.

**More activities
on Student CD**

Key Points

- Weather is the condition of the atmosphere in regard to temperature, moisture, wind, and clouds.
- Different areas of the world have different climate zones.
- Weather is caused by many different changes in the earth's atmosphere.
- Fronts, storms, and wind are caused by air and temperature changes.
- Forecasting weather involves gathering and interpreting weather data.

Read & Respond

1. Describe how climate is different than weather.
2. When do thunderstorms take place?
3. How does moisture get in the air?
4. Name the six conditions that cause weather.
5. What is a front?
6. Describe three ways to gather data for forecasting weather.
7. Describe a Doppler radar image.
8. How are computers used in helping forecast weather?
9. What are the two kinds of wind?
10. Identify the three climate zones.

Think & Apply

1. **Connect.** How might weather forecasting technology impact food production and the economy?
2. **Design.** Design a model of a home in a tropical zone. Then design a model of a home in a polar zone. Present your models and explain your design decisions.

3. **Hypothesize.** Do you think we will ever be able to control the weather? What would be the positive impacts? What would be some negative impacts?
4. **Summarize.** How does radiant energy determine temperature?
5. **Assess.** How has technology changed the process of weather forecasting over the last 100 years?

TechByte

Chaotic Weather. Chaos theory has given new life to proposals for weakening or diverting a hurricane. According to the theory, tiny changes made early in the life of a chaotic system can become hugely magnified over time. Meteorologist Ross N. Hoffman suggested that previously attempted interventions, such as dropping silver iodide in a storm's clouds to trigger rainfall, could, at the right time and place, have such a snowballing effect.

CHAPTER 11

Graphic Communication

Objectives

- Define *graphic communication*.
- Describe hardware and software used in graphic communication.
- Explain how type, graphics, and color are used in publishing.
- Identify five major printing processes.

Vocabulary

- **graphic communication**
- **scanner**
- **resolution**
- **page layout program**
- **leading**
- **typeface**
- **relief printing**
- **offset lithography**
- **screen printing**
- **gravure printing**

This large web press operator is checking the magazine that is being printed.

Activities

- Creating a Document
- Communicating Through Symbols
- Making the News

What Is Graphic Communication?

Graphic communication is the process of using words and images to send messages. It involves designing, preparing, and reproducing materials electronically or on paper for publishing.

In this chapter, you will learn how graphic communication technologies are used to prepare printed documents. You will also learn some of the important things that need to be considered when designing for publication and about the major printing processes. In Chapter 12, "Digital Multimedia," you will find that many of the same principles described here apply to electronic documents such as those designed for publication on the Web.

Today, most printed products are designed using computers. Text and graphics can be easily combined on the same page. The documents can be small, such as a business card or single-page newsletter, or as large as catalogs or textbooks.

Hardware

What's needed to design documents electronically? First, you need a computer. Computers used in professional publishing usually have a lot of memory and are very fast.

Large-screen monitors allow the user to view an entire page at a large size. These are ideal for publishing work. Designers often need to view two facing pages at the same time on a computer. Two facing pages are often called a spread. See **Fig. 11-1**.

Graphics are an important part of publishing. To place a photograph into a document, a scanner can be used. **Scanners** are devices that can change images such as photographic prints into an electronic form that computers can use.

If the photos were taken with a digital camera, a scanner won't be needed. The photos can be electronically transferred to the computer for editing. You will learn more about digital cameras in Chapter 12.

Color printers are useful for previewing color pages and for printing small quantities of documents. They are not usually used to print many copies because they are slow and the ink or toner is expensive.

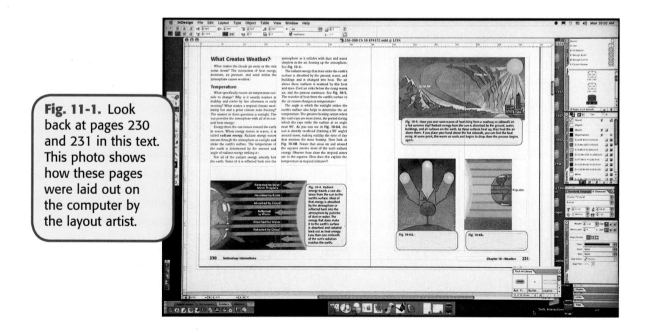

Fig. 11-1. Look back at pages 230 and 231 in this text. This photo shows how these pages were laid out on the computer by the layout artist.

Fig. 11-2. Can you see the differences in the resolution in these two photos? The photo on the left is 50 dpi and the one on the right is 600 dpi.

The quality of the documents depends on the printer's resolution. In printing, **resolution** refers to the number of dots per inch (dpi) of ink on printed images. As the number of dots per inch increases, images become clearer. See **Fig. 11-2**. Clearer images can show more detail. Home and business office laser printers typically print at 300 or 600 dpi. Commercial publishers use printers that have a much higher resolution.

Software

In addition to the right hardware, publishers need the right software. **Page layout programs** are software programs that are used to combine text and graphics in a document. For example, users can place text in columns and add headlines. They can also add photos, drawings, charts, and graphs.

Fig. 11-3. Cropping a photo allows you to eliminate unwanted parts of the photo. The inset photo shown here has been cropped from the larger photo.

Impact of Technology

Forgery and Plagiarism

Have you heard news stories about forgery or plagiarism? Forgery is the process of making or adapting objects or documents with the intention to deceive (such as writing someone's signature on a contract). Plagiarism is taking someone's ideas or works (such as writing or music) and presenting them as one's own. Both practices are illegal and unethical.

Computers, color printers and copiers, and digital cameras have brought many benefits, but they also have made forgery and plagiarism easier. The ability to forge or plagiarize is limited only by a person's integrity. The mechanics are no longer a significant obstacle.

Investigating the Impact

Consider the ethical, social, and economic impacts of plagiarism and forgery.

1. Why is it wrong to claim someone else's work as your own? What can or should be done about plagiarism?
2. If a news photo is altered to remove something gruesome or embarrassing, is that forgery?
3. Is it realistic to blame forgery and plagiarism on the technology that has made them easier to do? Ask your classmates what they think of the issue.

Publishing projects usually begin with text prepared on a personal computer (PC) using a word processing program. You have probably used word processing in school or at home to create school reports or to write letters. Some page layout programs can be used for word processing, but most layout programs work best with text created from a word processor. The text file is saved and imported (brought into) the page layout program.

Several other types of programs are used with page layout programs. Drawing programs are used to create art. In addition, artists use them to create logos, modify typestyles, and add special effects such as shadows.

Image-editing programs modify drawings and photographs. Photo editors can change the size, shape, color, and brightness of images. They can also crop (remove unwanted portions of) photographs. See **Fig. 11-3**.

Clip-art software contains photographs and drawings that can be imported into a document. Hundreds of clip-art programs are available. Some contain thousands of images. Some copyright-free clip art is available on the Internet.

When people buy licenses to use these programs, the price usually includes permission to use the artwork for certain personal purposes.

However, some clip art is designed for commercial use only and can be expensive. See **Fig. 11-4**. To use the photos, publishers have to get permission from the artist or photographer. They also sometimes need to pay a fee.

Designing Documents

When designing a document, publishers must keep in mind the document's purpose and audience. A well-designed publication will attract the reader's attention. This allows the publication to communicate the desired message. See **Fig. 11-5**. Three important things to consider when designing a document are
• Type
• Graphics
• Color

Type

Printed letters are measured by their height, and their size is specified in points. There are 12 points in a pica and 6 picas in an inch. See **Fig. 11-6**.

Perhaps some of the computer programs you have used allowed you to choose type size by specifying points. The text used in books and newspapers is usually between 8 and 12 points. Large type is used where there is a need to capture attention. Such type is used in headlines, callouts, and advertising. See **Fig. 11-7**.

Fig. 11-5. What design elements make you want to pick up a magazine and look through it?

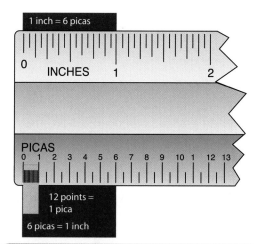

1 inch = 6 picas

0 INCHES 1 2

PICAS
0 1 2 3 4 5 6 7 8 9 10 1 12 13

12 points = 1 pica

6 picas = 1 inch

Fig. 11-6. In publishing, measurements are in points and picas. The pica is used to measure the length of a line of type. The point is used to measure typefaces.

Math Link

Points and Picas. There are many systems of measurement in the world. Some are industry specific. The printing industry, for example, uses points and picas. There are 12 points in 1 pica. One inch equals 72 points, or 6 picas. See if you can fill in the missing data in the table.

Inches	Points
1/16	
	9
1/4	
	36
	54

In addition to type size, designers must decide how much leading to use. **Leading** (LED-ing) is the space between lines of type. Choosing the right leading is important. Lines of type that are set too close together or too far apart are difficult to read.

Line length must also be considered. Long lines of small type and short lines of large type are difficult to read. Look at several different publications to see how line length affects readability.

Have you noticed that the same letters can often look different? A **typeface** is a set of letters, numbers, and symbols that have the same design. In computer programs, typefaces are called fonts.

Fig. 11-7. The larger the point size, the larger the type size.

This type is 6 points.

This type is 12 points.

This type is 18 points.

This type is 24 points.

This type is 36 points.

The three main categories of typefaces are serif, sans serif (pronounced SAN serif), and decorative. The difference between a serif and a sans serif typeface is shown in **Fig. 11-8**. Decorative typefaces are designed to capture attention. Dozens of decorative typefaces are available. However, some are difficult to read when they are used for more than a few lines.

Typefaces are often available in type styles such as **bold** and *italic*. Sometimes **bold** and *italic* styles are combined to make ***bold italic***. These type styles are used to call attention to new terms and important ideas. Why do you think certain words are in bold style in this chapter?

Graphics

The use of graphics should be planned carefully to add interest to the publication. You may have heard of the phrase, "show, don't tell." A good graphic can present a great deal of information in a small space.

Printed graphics fall into two categories: drawings and photographs. Maps, charts, and cartoons are examples of drawings. Drawings with lines but no shading are referred to as line art. Drawings and photographs can be in black and white or color.

Reading Link

Evaluating Design. Bring to class examples of well-designed and poorly designed graphic communication. They could be magazine articles, advertisements, or any other printed product. Tell what makes the design a good one or a poor one. Suggest ways to improve the poorly designed products.

Graphics may come from a variety of sources. You've already learned about scanners, digital cameras, and clip-art programs. Many companies have an online database of graphics they own. Photo editors scan these databases to find the graphics they need and then purchase permission to use them in a publication.

Original illustrations can be created in drawing programs. They can be used to draw any shape or provide any background that can be imagined. Some drawing programs are designed for specific purposes such as making charts. CAD (computer-aided design) programs are used to create technical drawings. For more on CAD, see Chapter 2, "Computer-Aided Design."

Below is one example of a serif typeface. The red arrows point to the small horizontal lines that extend from the ends of most characters. These lines, called serifs (SER-ifs), help guide your eye across the page.

Technology

Below is one example of a sans serif typeface. Sans (SAHNZ) is a French term that means "without." Note the strokes that form the letters are simple and direct.

Technology

Fig. 11-8. Shown here is a serif typeface compared with a sans serif typeface.

Cyan

Magenta

Yellow

Black

Writing Link

Designing Your Message. Good design can make the written word more effective. Suppose you are working for the owner of a small restaurant. The owner wants to increase business by advertising in the local newspaper. How would you rewrite and redesign the following ad to gain attention and make people want to come to the restaurant?

Enjoy home-style cooking at The Full Plate restaurant. Every Tuesday we have a lunch special, which is a chicken dinner. You get two pieces of chicken, salad, and a roll for only $4.99. Our pies are the best in town. We are open from 6:00 a.m. until 2:00 p.m. Monday through Saturday. Breakfast is served all day. Our address is 2022 W. River Road.

Color

Color encourages people to read a document. For example, color makes pie charts and graphs more effective. However, it must be used carefully. Don't print blocks of text in color. Such text can be difficult to read. Color also increases the cost of production.

Did you know that printouts are made with only four colors? Magenta, cyan, yellow, and black can be combined to produce any color found in nature. See **Fig. 11-9**. For more on four-color printing, see the Student CD.

Process colors were used in printing this book. Printing with process colors requires a page layout program that is more advanced than those used in most technology classes.

Sometimes, a user may use only two colors, such as black and another color. For example, a document could have black type with an occasional word highlighted in red for importance. The red would be called the document's spot color.

Printing

Before a document is printed, certain decisions need to be made. For example:
• How many copies are needed?
• When are they needed?
• What is the budget?
• What printing process should be used?
• Where should the item be printed?

More than 2 million people are employed in the field of graphic communication. Many of these people work in the commercial printing industry.

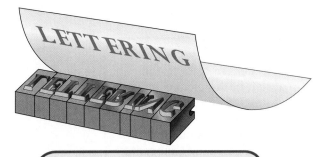

Fig. 11-10. Relief printing is when the letter is raised above the rest of the printing plate.

Relief Printing

Relief printing uses a raised surface to put ink on paper. Letterpress printing is a relief printing process that was popular for about 500 years. In letterpress printing, a raised metal surface containing the image is coated with ink. The inked surface is pressed against the paper to transfer the image. See **Fig. 11-10**.

Flexography is a relief printing process that uses a raised surface made of rubber or plastic. See **Fig. 11-11**. It is similar to using a rubber stamp except that it uses printing plates that are mounted on cylinders. Flexography is often used in the packaging industry because it works well for printing on a wide range of materials, such as paper, foil, and plastic.

Offset Printing

Offset lithography uses a flat surface and is based on the principle that oil and water don't mix. A flat printing plate with an image on it is wrapped around a cylinder. The plate is dampened by a water roller. The water does not stick to the image and remains only on the non-image (non-printing) areas. Ink from a second roller is attracted to the dry image area and repelled by the wet non-image area. The ink is then transferred (or offset) to another surface called a blanket. Next, the image is transferred from the blanket to the paper. See **Fig. 11-12**.

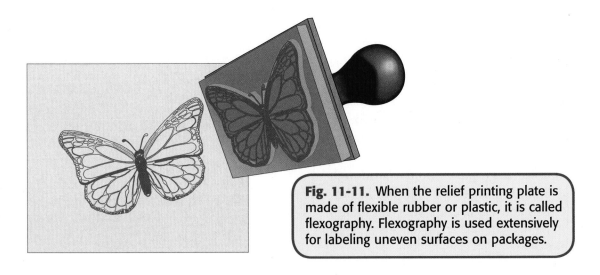

Fig. 11-11. When the relief printing plate is made of flexible rubber or plastic, it is called flexography. Flexography is used extensively for labeling uneven surfaces on packages.

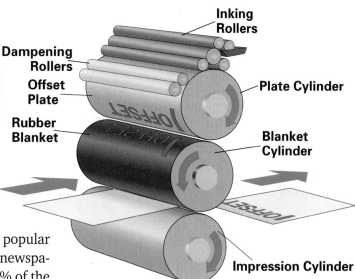

Inking Rollers

Dampening Rollers

Offset Plate

Rubber Blanket

Plate Cylinder

Blanket Cylinder

Impression Cylinder

Fig. 11-12. Offset lithography is the most popular type of printing. It prints by using the principle that ink (oil) and water don't mix.

Offset lithography is one of the most popular printing techniques. It is used to print newspapers, magazines, and books. Almost 50% of the printing in the United States is done using offset lithography.

Screen Printing

Screen printing uses a stencil attached to a mesh (screen) stretched in a frame. The frame is placed on top of the paper or other item being printed. Ink is placed in the frame and pulled across the screen with a squeegee, forcing ink through the screen and onto the paper.

Screen printing is often used to print items such as T-shirts and glassware because they cannot be run through most other printing presses. See **Fig. 11-13**.

Fig. 11-13. Screen printing forces ink through openings in the screen and onto the printed surface. Most T-shirt printing is done using this method.

Gravure Printing

Gravure printing uses a recessed surface. The image is etched in a plate, which is then wrapped around a cylinder. The plate is inked and a blade is used to wipe off the excess ink. The ink remaining in the etched areas is transferred to paper as it passes between the plate cylinder and another cylinder called the impression cylinder. Gravure is an expensive process that produces high quality results. It is used to print some magazines and to produce all U.S. paper money. See **Fig. 11-14**.

Digital Printing

Digital printing has changed the way documents are printed. Many documents are now being produced in homes and offices. Computer files are sent directly to a printer. There is no need to make printing plates and use a press.

Most traditional printing is expensive. The cost per piece goes down as the quantity goes up, but printing small numbers of books, magazines, etc., is costly. High-speed digital printers can produce small quantities of high-quality publications for a low price. The process, sometimes called printing on demand, is being used to print some books. Instead of printing thousands of extra books, a few hundred copies are printed instead. If the book sells well, additional copies can be printed on short notice. See **Fig. 11-15**.

Planning a Newsletter

Sometimes, individuals or small companies handle all aspects of publishing. They do the planning, designing, and printing. They use a personal computer and a printer. This type of small-scale publishing is often called desktop publishing. Newsletters are among the most common products of desktop publishing. Companies publish newsletters to provide information to employees and to promote their services and products. You might publish a newsletter to inform the school or community about activities in your technology class.

Some important decisions should be made before starting a newsletter. Whether creating a newsletter by yourself or with a group, brainstorming and the problem-solving process can be used to answer some important questions.

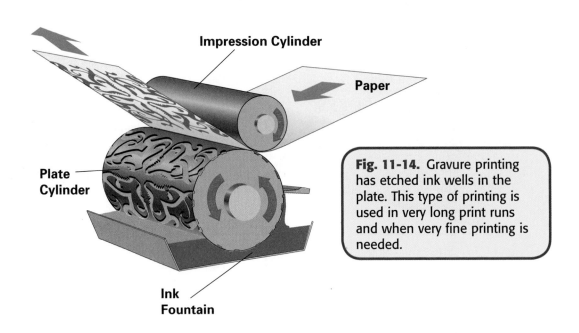

Fig. 11-14. Gravure printing has etched ink wells in the plate. This type of printing is used in very long print runs and when very fine printing is needed.

Impression Cylinder

Paper

Plate Cylinder

Ink Fountain

Fig. 11-15. This high-speed digital printer can print several hundred finished books in just a few minutes.

- Who is the intended audience? Will the newsletter be read by students, teachers, and/or the community?
- What kinds of articles will interest them?
- What do you think would be a good name for the newsletter?
- How often will it be published?
- What will the page size be?
- How many pages will it have?
- How will it be printed?

Once you've answered these questions, you can start designing the document.

Design Guidelines

Thousands of newsletters are produced by individuals and organizations each week. Although they may focus on different topics, most newsletters follow similar guidelines. Here are some helpful design tips:

- Use only two typefaces. Select a serif typeface for the body text. For occasional emphasis, use the same typeface in bold. For headings and headlines, use a sans serif typeface.
- Include white space on every page to avoid a cluttered look.
- Print with black ink. If you want to add color, consider using black ink on paper of an interesting color. Color paper is less expensive than color ink.
- Choose carefully when selecting photos and drawings. Many beginners use too many graphics. Place graphics in several locations before making a final decision.
- Experiment with margin size and column width. On 8½" × 11" paper, most readers prefer two columns.
- Maintain a consistent look throughout the document. This means that the design of page two and the design of page six should be similar.
- Seek feedback from readers.

Remember that newsletters are a great way to convey information. Graphic communication technologies will continue to provide a means for people to express their ideas in an inexpensive and attractive format.

Creating a Document

Identify a Need/Define the Problem

Use a computer and any desktop publishing software to compose a single-page document that incorporates the following items:
- Music, film, book, or product review.
- Headline(s), by-line, and body text.
- Picture or other graphic surrounded by the body that conforms to the shape of the graphic (often described as "tight" text wrap).

The document should be printed on an 8½" × 11" sheet of paper. See **Fig. A**.

Gather Information

Using the "Help" section of your desktop publishing software, explore the various features that it contains. You will want to explore such features as "text box" and "columns." These features will help you lay out and compose your document.

Develop Possible Solutions

Develop possible designs for your document. You will need to prepare several design sketches. In your rough sketch, you should document the locations of your graphic elements and the type and style of font that you want to use.

Materials and Equipment

Select from this list or use your own ideas.

- computer
- desktop publishing software
- advanced word processing program

The interactive lab "Principles of Design" on the Student CD will teach you about proportion, balance, and other characteristics of design. Understanding these principles can help you create a good design.

Model a Solution

1. Create your document using desktop publishing software or an advanced word processing program.
2. Print the document and review it.
3. Seek comments from classmates regarding the appearance of your document.

Test and Evaluate the Solution

- Did the text wrap the way you had planned?
- Did you use the software appropriately?
- Is the text in your document easy to read?
- Did your classmates easily understand your message?

Refine the Solution

- Based on the feedback that you received from your classmates, you may want to revise your document.
- Show your document to your classmates and get new feedback. Is your new document better looking or easier to read?

Communicate Your Ideas

Mount your document on a piece of larger paper. Identify the various components of your document, such as the type and style of font used. On the back, write a narrative of the step-by-step process you used in creating your document.

Headline or Title

By-Line

Body Text

Fig. A

T Communicating Through Symbols

Identify a Need/Define the Problem

As the world's population continues to expand, public areas, such as airports, shopping malls and libraries, require a universal method to communicate for people who use different languages. These institutions must use international symbols in order to communicate a message. See **Figs. A** and **B**.

Create an international symbol that meets the following specifications:
• It is new and original.
• It can be read from 30' away.
• It must represent a recognizable place, object, action, or idea.

Gather Information

Research and brainstorm some international symbols and list some of the related elements they share: size, shape, style, and color.

Develop Possible Solutions

Develop a possible design for your international symbol. You will need to prepare several design sketches. Make sure the symbol meets the specifications. You may want to develop a sign for an ice cream stand, a movie theater, or perhaps an Internet cafe.

The interactive lab "Elements of Design" on the Student CD can help you learn about the role of color, shape, and other factors in creating a good design for a product.

Math Link

20/20 Vision. Do you know what it means to have 20/20 vision? It refers to the distance you can see compared to a person having perfect vision. For example, if someone with perfect vision could read a sign at 40 feet, and you had to get closer, such as within 20 feet, you would be described as having 20/40 vision. (Each eye has a separate rating; some people could have 20/40 vision in one eye and 20/20 vision in the other.)

Create a chart that has rows of text printed in ¼-inch increments, from ½ to 2½ inches in height. Select several people who think they have 20/20 vision. Record the furthest distance for each height of lettering that a person is able to read. Can you determine which of your classmates have 20/20 vision?

Name	Height of Text (in inches)		
	½	¾	1
Allyson			
Derek			
Mackenzie			

Materials and Equipment

Select from this list or use your own ideas.

- computer
- drawing software

Model a Solution

1. Compare your sketches. Select the one that you think is the best solution for your purpose.
2. Use a drawing program to illustrate your international symbol.
3. Print and analyze the symbol.
4. Seek comments from classmates regarding the appearance and intended meaning of your international symbol.

Test and Evaluate the Solution

- Does your symbol communicate a message?
- Were there any uncertainties as to what you were trying to communicate?
- Does the symbol meet all your specifications?
- Can the use of a different color be used to modify the message?

Fig. B

Refine the Solution

- Were your classmates able to read and understand your sign?
- Take the feedback that you received and redesign your sign so that it is easier to understand. For example, can the use of a different color be used to aid in understanding?
- Print off your final symbol.

Communicate Your Ideas

Create a collage of the step-by-step process you went through when designing your international symbol. (A collage is essentially a series of images or shapes pasted together that tell a story.) In your collage, you will want to have images of your rough sketches, your final product, and the steps you took in creating this international symbol. You may want to find images of things like a computer, a screen shot of a drawing program, and the idea or purpose the symbol represents to communicate to others the resources that you used.

Fig. A

[T] Making the News

Identify a Need/Define the Problem

As you learned in this chapter, newsletters can be used to inform, persuade, or promote. Make your own newsletter for this activity. Your newsletter should include the following:

- Four pages of text and graphics
- At least two fonts
- Bold, italic, and regular styles
- One photograph (from a digital camera or output from a scanner)
- One or more electronic clip-art images
- One or more original art pieces (drawn using the computer, such as with a paint program)
- One or more images surrounded by text
- Two or three columns of text on a page
- Attractive nameplate (title)
- Use of color (if possible)

Your teacher may organize students into publishing teams.

Gather Information

Collect several newsletters either from your school, town, or local library. Identify the various elements that they have in common and the various layout styles. Takes notes on the similar features and the layout ideas that you like.

Develop Possible Solutions

Decide upon the information you plan to present in your newsletter. Estimate how long you would like each article to be. Identify suitable illustrations, such as photographs, clip art, and electronic art. Prepare rough sketches showing the placement of text and art on each page. See **Fig. A**.

Design notes you may want to keep in mind include the following:

- Including white or negative space on every page helps avoid a cluttered look.

Fig. A

- Select drawings and photographs carefully. Many beginners use too many graphics. Place them in several locations before making a final decision.
- Experiment with margin size and column width. On 8½" × 11" paper, most readers prefer more then one column.

 See the interactive lab "Principles of Design" on the Student CD to learn more about what is needed for good design.

Model a Solution

1. Compare your sketches. Select the one that you think is the most effective layout for your newsletter.
2. If you are working in a team, make sure that responsibilities have been assigned for every part of the newsletter. Each person must know exactly what he or she is expected to accomplish.
3. Use any publishing or drawing programs to compose or illustrate the part of the newsletter for which you are responsible.
4. Print the newsletter.

Test and Evaluate the Solution

- Does the newsletter meet all specifications?
- Do the four pages seem to blend in well with each other?
- Is the newsletter easy to read?
- Is the newsletter interesting to read?
- Is the newsletter pleasing to look at?

Refine the Solution

- Give a copy of your newsletter to another publication team. Have that team review your work and give you feedback.
- Make any needed changes. Pay special attention to any typos or other errors. Mistakes on your newsletter will distract your audience.
- Print the newsletter again. Assemble all pages of the newsletter. Submit the newsletter for final evaluation.

Communicate Your Ideas

Create a portfolio of the process you went through to create your newsletter. Be sure to include the various resources you used such as the hardware and software that was used. You will want to communicate your thought process and the various ideas that you developed, so be sure to include your initial rough sketches.

Science Link

Science in the News. What do you think the headlines might have been when Nicolaus Copernicus determined that Earth revolved around the sun or when Marie and Pierre Curie discovered radioactive materials? How about when Louis Pasteur demonstrated that microbes can cause disease?

Select an important scientific development of the past. Research the background of your important moment in science and outline the points that should be noted. Create a newsletter mock-up. How would a newsletter from an earlier time period look different from a newsletter today?

Exploring Careers

Graphic Artist

ENTRY LEVEL | TECHNICAL | PROFESSIONAL

Graphic artists create illustrations and designs. They are key members of a design team. They help an idea come to life by making a visual image using computers or freehand drawing.

Graphic artists work mainly for advertising and publishing companies, such as book and newspaper publishers. Graphic artists may create Web sites, illustrate books, lay out pages of books, draw cartoons, or do animation. They may design logos, packaging, and advertising materials.

More than 60% of graphic artists are freelancers. This means they are self-employed. They work for a client on an as-needed basis until the work is finished. Then they move on to work for another client. The remaining 40% of graphic artists work as staff members for companies.

Qualifications

Graphic artists must have artistic ability, be creative, and know how to use computer software to create graphics products. A high school degree is needed. Although a college degree is not always required, it can lead to better jobs that pay more. Some graphic artists learn their skills on the job.

Outlook for the Future

The job outlook for graphic artists is very good. It is expected that there will be more new jobs available. New people will also be needed to replace workers who become managers, retire, or leave the field to take other jobs.

Creativity

In this career, you should be able to come up with new and fresh ideas and be a good problem solver. Looking at ordinary things in a new way shows creativity. Brainstorming may be a good way to come up with new ways to see something. Good communication skills and the ability to work well with other people on a team are also very important.

Researching Careers

Interview a graphic artist to learn about this career. What is the person's educational background? How many years has she or he worked as a graphic artist? Ask to see some of the artist's projects. Using what you've learned, create a pamphlet about this career.

CHAPTER 11 Review

More activities on Student CD

Key Points

- Graphic communication technologies are used to prepare documents for publishing.
- Both hardware and software are necessary when publishing with computers.
- Publishers use type, graphics, and color to help convey the document's message.
- Commercial printers use different printing processes, depending on what is being printed.
- Newsletters are one of the most common products of desktop publishing.

Read & Respond

1. Define *graphic communication*.
2. What is dpi?
3. List five kinds of hardware that might be used in graphic communication.
4. What software is used to combine text and graphics?
5. Identify the three categories of typeface.
6. Why are type, graphics, and color so important to publishing?
7. What four colors are used in process printing?
8. Identify the five kinds of printing discussed in this chapter.
9. What is desktop publishing?
10. Why is it good to have white space on every page?

Think & Apply

1. **Assess.** Research digital cameras. Determine whether some digital cameras are better than others for use in publishing.
2. **Relate.** *Technology Interactions* uses many of the design elements discussed in this chapter. Write a report identifying the design elements used in this book.
3. **Formulate.** Identify the overall message of this chapter. Do you think it succeeds in getting its message across? Why or why not?

4. **Design.** Using information you learned from this chapter, design a newsletter highlighting the past year's events in your life.
5. **Produce.** Create two documents that have the same message but use different design elements. Give half the members of your class one document, and half the other. Does one half of the class understand your message better than the other half? Determine why or why not.

TechByte

Covert Printing. The U.S. government has long used highest-quality engraving technology to print paper money. But in the digital age, foiling counterfeiters poses special challenges. Since 2000, the government has issued redesigned $10, $20, and $50 bills with added security features. The bills use a wider variety of colors; color-shifting ink, which changes color as the bill is tilted; an embedded security thread that glows yellow in UV light; and watermarks, faint images embedded in the paper.

CHAPTER 12

Digital Multimedia

Objectives

- Define *multimedia*.
- Compare and contrast audio and video signals.
- Describe how digital cameras work.
- Identify the steps in planning and producing a multimedia presentation.

Vocabulary

- **multimedia**
- **amplitude modulation (AM)**
- **frequency modulation (FM)**
- **digital music player**
- **video on demand (VOD)**
- **Internet**

Ask your grand-parents if they ever had access to thousands of songs like on this iPod. What do you think the next step is in taking your music with you?

Activities

Digital

- A Technology Campaign
- Creating a Storyboard
- Multimedia Teaching

Electronic Communication

Imagine starting your day with the sound of a clock radio next to your bed. After listening to the weather forecast, you get out of bed and begin to prepare for another day at school. On the way to school, you listen to several songs on your MP3 player or Apple iPod™.

In your technology class, the teacher introduces a new topic by showing a DVD produced by NASA. You think about what it would be like to be an astronaut on board an international space station. In biology class, you dissect a frog—on a computer screen. After school, you use the Internet to find more music to listen to on your portable player.

Throughout the day, you have used electronic communication. Information was sent and received using electronic devices such as radios, music players, and computers. See **Fig. 12-1**.

Fig. 12-1. The use of electronic devices like these that we use throughout the day seems so commonplace. Can you think of multimedia devices that do not have some type of electronic device incorporated into their design?

Analog vs. Digital

Electronic communication may use analog (ANN-uh-log) or digital signals or a combination of both. Analog signals are continuous and variable. Digital signals are separate, distinct signals in the form of binary code. For an explanation of binary code, see the Student CD's interactive lab.

More and more electronic communication uses digital signals. Digital radio and photography have become very popular. Digital television is going to replace traditional analog TV.

In this chapter, you will learn about multimedia communication. **Multimedia** is the combination of several forms of communication, such as text, video, photographs, spoken words, and music. Much of the multimedia communication uses a digital format.

Radio

Radio is an example of audio communication. "Audio" refers to sound. Sound is a form of energy. Sound waves produced by vibrations act on the ear so that we can hear. When we speak, our vocal cords vibrate to help create the words we want others to hear. These vibrations spread out in the same way that waves spread when you drop a pebble into a pond. When the sound waves reach other people, their eardrums vibrate and they hear the sound.

The distance between the peaks of any two successive (one after the other) waves is called the wavelength. The number of vibrations per second is the frequency of the sound. The size of the vibrations is called amplitude. Loud sounds have greater amplitude than soft sounds. See **Fig. 12-2**.

In radio communication, sound waves are changed to electromagnetic waves. The waves are sent through the atmosphere, a cable, or a combination of both. When the waves reach a receiver, they are changed back into sound.

Electromagnetic waves travel very fast— nearly one billion feet per second. This great speed makes it possible for an electromagnetic signal to reach its destination almost instantly. To learn more, see the "Electromagnetic Spectrum" interactive lab on the Student CD.

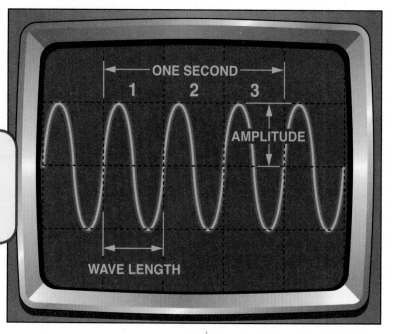

Fig. 12-2. This wave diagram illustrates amplitude and wave length. It also shows frequency, which is three cycles per second. Most waves, even ocean waves, can be measured this way.

Fig. 12-3. The broadcast booth of a radio announcer does not have to be very big, but it must be soundproof and have all the necessary electronic equipment within easy reach.

Radio Broadcasting

All radio stations have a studio and a control room. Usually they are adjacent rooms separated by a large window.

The studio is where the on-air performers work. Radio studios are designed to broadcast live and recorded sounds. Studios are sound-proofed to prevent outside noise from interfering with the broadcast. Disc jockeys, the other announcers, and guests create the live portion of a broadcast. See **Fig. 12-3**. Compact discs (CDs) are usually played for the music. Commercials are typically prepared in advance and stored on tape.

In the studio, a microphone picks up the sound created by the voices of the disc jockey and the announcers. The microphone changes the sound energy into electrical energy. The signal is sent to an audio console in the control room.

The console is operated by an engineer who combines live sound with recorded music and commercials. This process is called mixing. Mixing is a complex task that involves adjusting many controls.

After mixing, the program signals are sent to a transmitter. The transmitter combines the program signals with carrier waves that "carry" the program signals away from the transmitter.

Like sound waves, radio waves have frequency, amplitude, and length. However, sound waves are mechanical energy. They are pressure waves that need a medium such as air or water. Radio waves are electromagnetic energy that can travel through space.

Radio broadcasts can be transmitted by **amplitude modulation (AM)** or **frequency modulation (FM)**. In AM radio transmission, the amplitude (strength) of the carrier wave changes. In FM broadcasting, the frequency of the carrier wave changes.

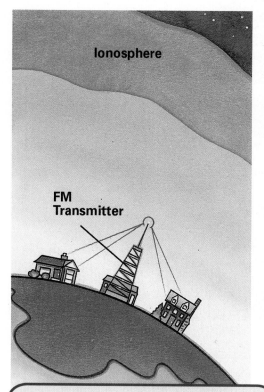

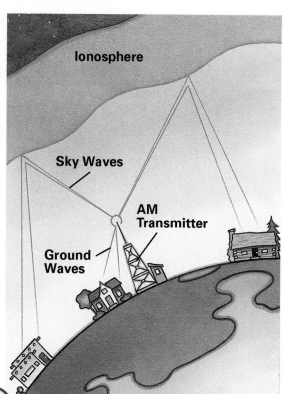

Fig. 12-4. FM broadcasts cannot be received beyond the horizon. AM broadcasts send out ground waves and sky waves. The sky waves bounce off the atmosphere and can travel long distances.

The radio waves from an AM broadcast go a long distance because they travel near the ground and bounce off the ionosphere (a lower level of Earth's atmosphere) and back to Earth. Radio waves from FM broadcasts do not bounce off the ionosphere. This is why FM broadcasts do not travel as far as AM broadcasts. However, FM radio broadcasts are usually of better quality. They are affected less by static than AM broadcasts. See **Fig. 12-4**.

The distance a radio broadcast travels also depends on the power of its transmitter. An AM broadcast of 50,000 watts has a range of several hundred miles. Powerful FM broadcasts of 100,000 watts have an effective range of 50 to 60 miles.

Science Link

AM Radio Waves. Have you ever listened to AM radio at night, especially late at night when some local stations cease broadcasting until the next morning? If you have, there's a chance that you heard broadcasts from hundreds, even thousands of miles away.

Why is it that AM radio stations can reach listeners only a hundred or so miles away in the daytime but can travel much farther at night? See if you can research the answer.

Every station has an assigned frequency. Frequency is measured in hertz, or vibrations per second. The AM frequency band goes from 535 to 1705 kilohertz. (One kilohertz equals 1,000 hertz.) The FM band goes from 88 to 108 megahertz. (One megahertz equals one million hertz.)

Satellite Radio

Satellite radio is a new digital technology that makes it possible for you to listen to the same radio station as you travel long distances, even thousands of miles. Satellite radio stations broadcast a variety of music, news, sports, and talk shows.

To get satellite radio you need a special receiver and a subscription. Many new automobiles come with satellite radio receivers, and models for home and portable use are also available.

There are several satellite radio providers, but they operate in a similar way. Each company owns several satellites that are in orbit about 22,000 miles above the earth. Radio programs are produced in ground stations and transmitted to the satellite, which bounces the signals back to Earth. Satellite radio receivers unscramble the data, which includes dozens of channels. In addition to the sound, the signal includes other data such as the name of the song and the artist. This information is displayed on the radio receiver.

The signal is very clear and produces sound quality similar to that of a CD. In cities where tall buildings may block the satellite signal, ground transmitters repeat the signals so that subscribers receive a clear signal. See **Fig. 12-5**.

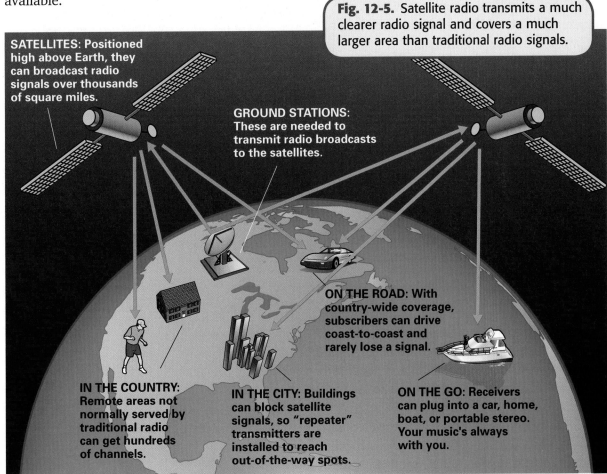

Fig. 12-5. Satellite radio transmits a much clearer radio signal and covers a much larger area than traditional radio signals.

SATELLITES: Positioned high above Earth, they can broadcast radio signals over thousands of square miles.

GROUND STATIONS: These are needed to transmit radio broadcasts to the satellites.

ON THE ROAD: With country-wide coverage, subscribers can drive coast-to-coast and rarely lose a signal.

IN THE COUNTRY: Remote areas not normally served by traditional radio can get hundreds of channels.

IN THE CITY: Buildings can block satellite signals, so "repeater" transmitters are installed to reach out-of-the-way spots.

ON THE GO: Receivers can plug into a car, home, boat, or portable stereo. Your music's always with you.

Abbreviations. An abbreviation is a shortened form of a word or phrase. In this chapter many abbreviations are used—DVD, CD, AM, and FM, to name just a few.

Abbreviations using only one syllable or a portion of the word are usually written in lowercase letters and followed by a period, such as "in." for "inch." Abbreviations using the first letter of each word in a phrase are written in uppercase letters with no period, such as CD for "compact disc." Abbreviations that form a word are called acronyms. SCUBA, for example, stands for self-contained underwater breathing apparatus.

Compact Disc

C D

With your classmates, brainstorm abbreviations used in your technology class and textbook. Prepare a list of correctly written abbreviations followed by their longer forms.

Digital Music Players

Portable music players that use tapes and CDs have been popular for more that 25 years. Now these devices are being replaced by digital music players.

Digital music players work differently than other music players. They have a solid-state memory that requires no moving parts. Digital music players are more reliable and never skip. This makes them ideal for people who enjoy listening to music as they exercise.

Some digital music players are called MP3 players. That's because they use a format called MP3, which is a system that compresses music

and video data so that it requires less storage space. Using MP3 technology, the data required to play one song from a CD is reduced to about one-tenth of the original file size. This makes it possible to download songs from a computer to a digital music player very quickly.

To get music for your digital music player, all you need to do is use the Internet to search for the music you want, pay for it, and then download it. Some music is available for free, but to legally obtain most music, you can expect to pay about a dollar per song. After downloading the files into your computer, you can use software that came with your digital music player to transfer the files to your player.

Now you are ready to put on earphones and listen to your music. Some digital music players, such as the iPod, give you the choice of selecting specific songs from a play list or using a feature called shuffling that plays songs in random order. You can even create library entries that group similar songs together. See **Fig. 12-6**.

Fig. 12-6. Some digital music players will let you listen to special radio stations and news broadcasts as well as store thousands of songs and/or photos. What do you think will be the next new feature?

Television

Television combines audio and video communication. ("Video" refers to a moving image.) Television emerged as an important communication system in the late 1940s. At that time, all television signals were sent through the air. Television broadcasting was similar to radio broadcasting. Electronic signals traveled from a transmitter to an antenna, which sent the signals to a receiver in the TV.

Today most television stations still broadcast over the air, but nearly three of every four homes receive their television signals by cable or satellite. Satellite systems are especially popular in remote areas. Both cable and satellite systems provide better reception and more programs than over-the-air analog broadcasting systems.

Television today is also used in many ways by schools, businesses, and indus-

try. Distance learning allows students in small, rural schools to participate in advanced classes offered in a larger school many miles away. Banks and stores use closed-circuit TV for security. Video teleconferencing has become a popular way of conducting business meetings. Teleconferencing combines telephone and satellite systems. Participants at different locations can see each other on video monitors.

Producing a TV Program

Television production is exciting and challenging. Most of the shows on television are prerecorded. Some programs, such as the news and sporting events, are done live. See **Fig. 12-7**. Although the production of live and prerecorded TV programs is not exactly the same, the processes they require are similar.

Television programs are carefully planned. The producer hires the writers to prepare a script. The producer also hires a director to turn the script into a TV program that viewers will

Fig. 12-7. A nightly news broadcast may be simple in presentation, but it involves many people besides just those you see on the TV screen.

find informative or entertaining. Many directors require their staff to produce storyboards of the program. Storyboards include sketches that show what should happen in a scene and text to summarize the dialogue. They also identify the camera shots needed.

The producer and director usually work together to select the people who will appear in the TV program. The people who appear are referred to as talent. They include announcers, actors, newscasters, reporters, and hosts of game and talk shows. Selecting the right talent is important. A program that starts with a good script but uses the wrong people will not be successful.

Many television programs must be rehearsed (practiced). Before the actual broadcast or taping session, the studio must be prepared. Microphones and lighting are set and tested. Most shows are recorded using two or more cameras. Studio cameras are similar to home video cameras. However, they are much larger and have wheels so that they can easily be moved around the studio.

The control room is a busy place. Monitors show the image captured by each camera. A technician uses a switcher to select the camera specified by the director. See **Fig. 12-8**.

Here is an example of how the switcher is used. Imagine a news broadcast with two anchors. As one anchor finishes a story, the director might say, "Camera two." A minute later, as the other anchor reads another story the director might say, "Camera one." After a few stories, a technician uses the switcher to cut to a commercial. While all of this is happening, engineers make sure that the picture and sound quality are satisfactory.

Television Broadcasting

Over-the-air television transmission is similar to AM and FM radio transmission. The transmissions are electromagnetic waves. Video is sent as AM; sound is sent as FM.

The frequency a station uses for broadcasting is called its channel. Channels 2 through 13 are VHF (very high frequency) channels. VHF signals have a frequency between 54 and 216 mega-

Fig. 12-8. Have you ever noticed all the different camera angles used in a TV broadcast? The angle that is shown on your TV screen is determined by a director and a technician switching to different cameras.

hertz. Channels 14 to 69 are UHF (ultra-high frequency). They range between 470 and 806 megahertz.

Over-the-air television broadcasting has an effective range of up to 150 miles. Cable, microwave, and satellite systems carry TV signals over longer distances. The networks often use cable to send their programs to local stations. Also, most communities have cable companies that install cables to carry the signals into homes.

Microwave towers throughout the country relay signals from one tower to the next. At each tower, the signal is amplified before being sent to the next tower. When the microwave signal reaches the TV station, that station changes the microwave signal into a television signal.

Satellites also relay signals around the globe to dish antennas at television stations, cable companies, and homes. Some satellite TV networks rent out the receiving equipment. Other providers require the user to buy it.

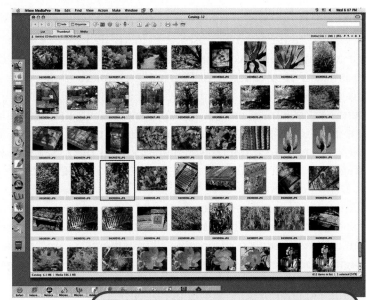

Fig. 12-9. With photo-editing software, you can print the photos you like best. You can even enhance them before you print them.

Video on Demand

Video on demand (VOD) systems use digital technology to make television interactive. With VOD, you can view your favorite programs and movies when you want. Some of these systems also let you play video games.

Cable television networks now provide a variety of on-demand services, such as pay-per-view and on-demand video. Pay-per-view movies are available on digital cable networks at about the same time they are released on DVDs. Fewer people are renting DVDs since there is no need to visit the store.

True on-demand video systems work like a VCR or DVD player. They allow almost complete control, including the ability to start, stop, pause, forward, and reverse a program. Advertisers are very interested in video on demand because it can be used to target people with particular interests. For example, dog owners could be sent commercials promoting a new pet shop that will be opening in the neighborhood.

Digital Photography

Digital photography is rapidly replacing traditional film photography. Digital images are easy to transfer from your camera to your computer. Once they are stored in your computer, they can be viewed on the monitor, sent by e-mail to friends and family, printed using your own printer, or e-mailed to a lab for printing.

Commercially produced digital prints cost about the same as prints made from film. An advantage of digital photography is that instead of printing every picture you have the option of making prints of only your best pictures. See **Fig. 12-9.** You also can use photo-editing software to improve the pictures before printing.

Camera Lenses

Digital cameras have lenses that focus light on a light-sensitive computer chip. Most cameras use a chip called a CCD (charge-coupled device). When light hits the CCD, the device sends an electrical charge to a processor that creates the image in digital form. Usually this data is saved to a removable memory card or stick.

Digital cameras use several different kinds of lenses. Fixed-focus lenses are found on the least expensive cameras. This kind of lens is similar to those used on disposable film cameras. Digital-zoom lenses increase the size of part of the picture but reduce the overall quality of the image. Optical-zoom lenses magnify the actual image and can produce high-quality pictures.

Just like our eyes, digital cameras analyze light intensity. The intensity of the three main colors (red, green, and blue) is also analyzed. The brightness of each color is recorded and stored. During editing, some programs allow you to view and adjust each of the individual colors.

Resolution

Digital images are made up of tiny dots called pixels. Look carefully at a digital photograph on your computer monitor and you should be able to see the individual pixels. The more pixels a camera has, the more detail it can capture. This characteristic is called resolution. See **Fig. 12-10**.

Low-resolution cameras produce low-quality pictures. These pictures may be acceptable for viewing on a computer, but they will not make very good prints. Digital cameras are described by how many pixels they produce. For example, a 3-megapixel camera can produce an image with up to 3 million pixels. Such a camera can produce images that can make good quality 4" × 6" or 5" × 7" prints. A camera with more than four megapixels can produce a good quality 8" × 10" print. Cameras with more than 10 megapixels are available for professional use.

Aperture and Speed

Many digital cameras have settings that allow you to control the resolution of the picture being taken. Higher resolution settings will produce better pictures, but they create larger files that take up more memory space on your camera.

Fig. 12-10. This close-up of the larger photo shows the pixels that make up the photo. The more pixels a camera has, the finer detail it can record.

Impact of Technology

Does the Camera Change How We See?

Has technology such as digital photography changed the way we "see" things? When photography was first introduced (about 150 years ago), it was a source of fascination and amusement. Then came the Civil War, and the camera became a tool for newsgathering. For the first time, noncombatants saw actual scenes of death and carnage. For many people, it changed the image of war from an intangible, idealistic struggle to a flesh-and-blood reality.

Investigating the Impact

Ask a parent or other adult what he or she thinks about the following:

1. How could modern news photography be used to benefit society?
2. In what ways might photography be misused by individuals or organizations (including governments) for their own gain?

Many digital photographers use cameras that automatically adjust the amount of light that reaches the sensor. This is done by controlling the size of the opening that lets light into the camera and controlling the amount of time that the light is let in. The opening is called the aperture, and the time control is called shutter speed. The right combination of aperture and shutter speed is needed to produce good quality photographs. See **Fig. 12-11**.

Fig. 12-11. The aperture and shutter of a camera control the amount of light that is recorded on the film or CCD. The main difference between this digital camera and a film camera is that a CCD is used in place of film.

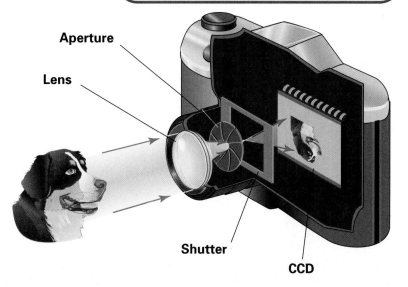

Aperture

Lens

Shutter

CCD

Digital Video

Digital video cameras have a charge-coupled device (CCD) similar to those used in still digital cameras. The CCD captures the image and converts it to digital data that is recorded on magnetic tape, miniature CDs/DVDs, or miniature flash memory cards similar to those used in digital cameras.

A video image is made up of a series of still images that are shown quickly to create the illusion of motion. Each image is a frame, and approximately 30 frames are recorded each second. See **Fig. 12-12**.

For many years film movie cameras were popular. Then video cameras that recorded directly on tape were introduced. The problem with both film movies and early videos was that they were difficult to edit. With film you could use scissors to remove unwanted parts of a movie. To edit video, you could connect two VCRs together and try to press the Record and Stop buttons at the right time to cut out what you didn't want.

Today if you have a digital video camcorder, a computer, and editing software, you can create videos that others will enjoy watching. Using a cable and the software, you can transfer the video to your computer. This process is called capturing. Once the video is captured, you can begin editing.

Each time you push a button to turn the video camera on and off to record a single scene, you are creating a video clip. Using the storyboard feature of the video-editing program, you can electronically arrange the sequence of the clips. Unwanted clips are easy to delete. See **Fig. 12-13**. Even after assembling your video clips in desired order, you will probably want to do additional editing. For example, you may want to shorten a particular clip from fifteen to ten seconds.

You will probably want to add transitions between the clips. Fades are popular transitions. As one scene ends it fades out and a new scene fades in. Special effects can also be added. You will also probably want to edit the audio portion of your video. The editing software will probably allow you to have several audio tracks so that narration and background music can be added to the original audio recorded with the video clip.

When you have finished editing, you can transfer your video to a DVD, share it with friends over the Internet, or import it into a multimedia presentation.

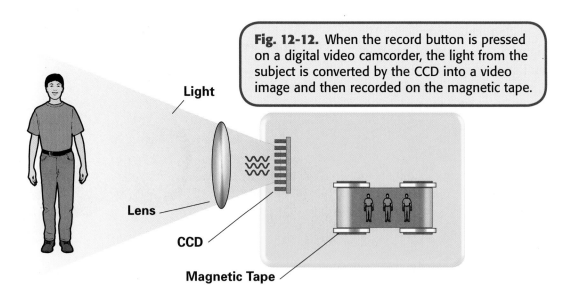

Fig. 12-12. When the record button is pressed on a digital video camcorder, the light from the subject is converted by the CCD into a video image and then recorded on the magnetic tape.

Light

Lens

CCD

Magnetic Tape

The Internet

The **Internet** is a global network of computers. You may have heard it referred to as "the information superhighway." The Internet is also our society's largest source of multimedia. When you visit a Web page, do you see a screen filled up with words only? The answer would almost always be no.

A single Web page can provide information through many forms of media. Text, images, audio, and video can all be used to communicate information online. How can these different forms of media all be used together?

What if you were looking up information about a CD from your favorite band? A Web page about the specific CD might give a text description of what songs are included. The page might also have images of the band or the CD cover. By clicking on the titles of songs listed on the page, you might access an audio file or a video file and hear music samples before deciding to buy the CD.

Multimedia Production

Now that you know about some of the important communication systems, you can create your own multimedia presentation. Most multimedia presentations are developed on a computer. Often they are interactive. This means that the user can control parts of the action or decide how the program is used.

Multimedia can be delivered in a number of different ways, such as CDs, DVDs, the Internet, and PowerPoint presentations. In Chapter 13, you will learn how animations are being used to create exciting interactive video games.

Creating a multimedia presentation is similar to using the design process to create a new product. **Figure 12-14** shows the major steps involved in planning and preparing a multimedia presentation. You can see these steps in action in the "Multimedia Ad" interactive lab on the Student CD.

Planning

To be effective, your multimedia presentation must be carefully planned. First, identify what you are trying to communicate. Brainstorm and list the content that might be included in the presentation. Decide who the audience will be and begin to think about the best way to deliver the presentation. Should it go on the Internet so that it's available to everyone? Do you want it to go on a CD or DVD that will be given to certain people? Perhaps you want to use your multimedia product to support a live presentation.

Designing

During the design phase think about what you will need to capture and maintain the interest of the audience. How long should the presentation be? How can multimedia add to rather that detract from the presentation? Make an outline and develop a storyboard.

MULTIMEDIA PRODUCTION CYCLE

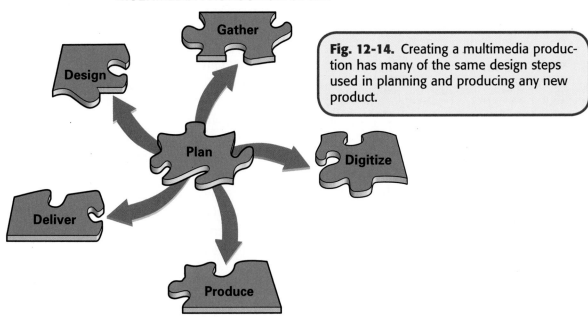

Fig. 12-14. Creating a multimedia production has many of the same design steps used in planning and producing any new product.

Fig. 12-15. With the proper software, creating a Web page for the Internet is not very difficult.

Gathering Material

After you have decided on the content of the presentation, you can begin to gather the specific material that is needed. Keep in mind that you may need to give credit for some of the ideas you plan to include in the presentation. Also remember that some material—including music, video, and still images—cannot be used without permission because of copyright laws.

Digitizing

You will probably need to digitize some material. You will want to capture and organize all the images, sounds, text, and video that will make up the presentation. You may need to prepare text by word processing. Original photographs or drawings may need to be scanned.

Producing

To produce your presentation you will be using a computer and software. For a simple PowerPoint presentation you may need only the Microsoft PowerPoint® software and the files of text and graphics you want to include. If the presentation is more complex, you may also need to use software for editing video and sound. To put the presentation on the Internet, software packages for page design can be purchased. If

you are interested in Web page design, look on the Internet for free tutorials and software to help you get started. See **Fig. 12-15**.

Delivering

When it is complete, you will want to deliver the presentation. If you are planning a live presentation, rehearse it and ask others to provide feedback. If the presentation is going on a CD or DVD, test it on several computers to make sure it works properly. Be sure to test each of the interactive elements. For Web-based multimedia presentations, try to identify possible problems before putting it on the Internet and test it again online.

A Technology Campaign

Identify a Need/Define the Problem

A new campaign to effectively communicate the value of technology education is needed. Your team has been selected to design and implement a multimedia campaign to promote technology education. The campaign must inform the public about technology education and encourage students to sign up for technology classes. Your campaign will be created using multimedia software, such as Microsoft PowerPoint. See **Fig. A**.

The presentation will be evaluated on the following:

- Clear explanation of the goals of technology education.
- Appropriateness and description of the technology classes that you have selected.
- Persuasiveness.
- Transition—smooth flow from one segment or idea to another.
- Correct grammar and spelling.
- Sound and animation.
- Overall appearance.

Gather Information

You may want to interview students and teachers involved in the technology program, as well as look at the course curriculum to get ideas as to what you might want to showcase.

Each team member should also become familiar with multimedia software. While each team member will focus on a different aspect of the campaign, understanding the whole will help your teammates develop the individual parts.

Materials and Equipment

Select from this list or use your own ideas.

- computer with multimedia software
- digital camera
- digital video camera
- digital editing software

Develop Possible Solutions

After you've gathered information, brainstorm to identify what to include in your team's campaign. List things such as benefits of technology education, course titles, class activities, and student opinions. Narrow down the list of ideas and images to be included in your presentation. Develop some time lines or storyboards to help guide your presentation of the information. You will want to develop several alternatives so that you can choose the best sequence of information.

Model a Solution

1. Each individual team member should select one aspect of the program as his/her primary focus. Write your portion of the script and be sure to select any multimedia elements (video, stills, etc.) that you may need.
2. Meet as a team. Each member should describe his or her progress. Devote part of the meeting to an exchange of ideas and helpful feedback.
3. Revise your individual segment.

4. As a team, use the multimedia software to integrate the contribution of each member.
5. Review the evaluation criteria. Determine if changes to the program are necessary.

Test and Evaluate the Solution

- View the presentation as a group.
- Is there a clear focus and direction of your presentation?
- Did you utilize the various features available to communicate your message effectively?
- Are other students able to understand the various benefits of taking technology education classes?

Refine the Solution

Using the information from your testing and evaluation, review and fine-tune your program. You might do the following:

- Proofread your presentation. Mistakes can distract your audience.

Fig. A

- Add a title and credits. Credits can be used to communicate which team member did which part of the campaign.
- Save the final program on a disk or CD. Submit your work to your instructor for final evaluation.

Communicate Your Ideas

Create a portfolio that describes the process you followed in creating your multimedia campaign. Be sure to include information about each step in the design process as well as the contribution that you made. This portfolio should serve as a guide that tells a viewer what you have accomplished and how you did it.

Creating a Storyboard

Identify a Need/Define the Problem

Create a storyboard for a short comedy skit to be played on the local public access channel. The segment may either be a short skit or a joke assigned by the instructor or selected by the team members.

Gather Information

Research storyboards. Storyboards usually contain the following:
- **Script.** This is the text of what will be said. The script also indicates who will be talking.
- **Scene.** Tells where the action will be taking place. This is often sketched to show the placement of the "action."
- **Sequence of events.** Specifies what will be taking place and for how long.
- **Camera direction.** Specifies which camera will be used for each sequence. Specifies type of shot to select: close-up or not; narrow, normal, or wide field of vision; still or pan (moving) technique. Plan for three cameras.

Develop Possible Solutions

Would it be practical to tell everyone involved what is expected of them? Could everyone remember and understand purely spoken instructions? Would it be useful to give all participants written notes describing the actions

Materials and Equipment

Select from this list or use your own ideas.

- paper
- tape
- cardboard or foam core board

to come? What advantages might a storyboard have over these techniques? Prepare several possible storyboard designs with these questions in mind. An example is shown in **Fig. A**.

Model a Solution

1. Decide on a plan of action. What part of the storyboard should be prepared first?
2. Prepare each part of the storyboard's requirements.
3. Compare and contrast all the components of your storyboard. Do they merge smoothly? If not, some parts probably need to be modified.
4. Arrange the individual sketches and components on the cardboard or matboard.

Test and Evaluate the Solution

- Are there any questions from your classmates that would indicate that the storyboard is lacking cohesion?
- Could another group use your storyboard as is and successfully present a program?

Fig. A

Refine the Solution

- Based on the feedback that you have gotten from your classmates, revise your storyboard to reflect any changes or alterations you decide to make to the storyboard.
- Complete a final edit of your storyboard.
- Submit your storyboard for approval.

Communicate Your Ideas

Create a document that will explain the process that you underwent in order to complete the assignment. Be sure to explain the purpose of each storyboard component.

Math Link

Timing Is Everything. When planning a story, time is an important factor. You have to allow a specific amount of time for each scene in order to deliver your message.

Imagine that you are making a three-minute skit. Your storyboard has 15 scenes, and you want each scene to have an equal amount of time. How long should each scene take?

Multimedia Teaching

Identify a Need/Define the Problem

Have you noticed a common element in all of the activities in this textbook? They all follow the engineering design process you learned about in Chapter 1. Understanding the design process is essential for participating in these activities, as well as in designing any product in general.

For this activity, you and your team will create a video that explains the stages of the design process used in your technology education class. You will need to develop a video for students who are either new to the lab or were absent during the explanation of the design process. Your video should include the following:

- Titles and credits.
- Step-by-step demonstration of the design process.
- Examples of activity completed in class.
- Lab safety notices.
- Appropriate voiceovers or soundtracks.
- Transition between scenes.

Gather Information

Many video editing software programs allow users to incorporate a wide variety of special effects, in addition to standard editing features. Examples include sound and animation. Look over and understand your editing software package. Become familiar with the various features available.

Materials and Equipment

Select from this list or use your own ideas.

- computer
- video editing software
- digital video camera
- tripod
- blank recording media (tapes, CDs, DVDs, etc.)

Develop Possible Solution

What are you instructing the user about? What information does the user need to have? How will you introduce each stage to the user? Brainstorm answers to these various topics in order to help identify the content of your video. You will want to develop some rough storyboards to give yourself a guideline for making the video.

Model a Solution

1. Chose the storyboard that most effectively models your chosen solution.
2. Assign roles and responsibilities to individual team members for each design step.
3. Record your raw footage on video cameras. See **Fig. A**.
4. Use digital editing software to capture the footage from your video device.
5. Edit your video.

Fig. A

6. Add effects, such as titles and transitions, to your video.
7. Add any voice over and soundtracks.
8. Add a title and credits to your video.
9. Present the video to your classmates for review and critique.

Test and Evaluate the Solution

- Does your video effectively communicate its core message?
- Did you take advantage of all the features of the editing software?
- What advantage does digital video editing have over traditional editing?

Refine the Solution

Review the feedback and recommendations that you have obtained from your self evaluation and classmates. Refine any footage or effects before transferring your video onto a CD or DVD. Items you might change include the following:

- Transitions to make the video more appealing or captivating.
- Soundtracks to provide entertainment.
- Additional demonstrations for clarity.

Communicate Your Ideas

Keep a daily journal of your work. Include information about your daily progress as well as the steps you have taken in order to complete your activity. You will also want to include any feedback that you received and how you responded to it.

Exploring Careers

Professional Photographer

ENTRY LEVEL | TECHNICAL | PROFESSIONAL

Professional photographers are paid to take pictures of people, places, and events. Many photographers specialize. They may be commercial photographers, scientific photographers, news photographers or photojournalists, or fine arts photographers.

More than 50% of photographers work as freelancers. This means they are self-employed. Newspapers, magazines, television broadcasters, and advertising agencies may hire freelancers or have a staff of full-time photographers.

Qualifications

A photographer must be creative and have a good imagination. He or she must have a good technical understanding of photography. A college degree is not always required, but entry-level photojournalists and commercial photographers generally need a bachelor's degree.

Photographers must be able to work well with others, have artistic ability and computer skills, be detail oriented, and have good hand-eye coordination. If self-employed, photographers need to have good business skills.

Outlook for the Future

The job outlook for photographers is good, but there is competition for the better jobs. Because of the growing number of Internet magazines and newspapers, more digital images are needed. On the other hand, if fewer print newspapers and magazines exist, fewer photos are needed for these publications.

Interpersonal Skills

This career requires good people skills. Good manners, such as saying please and thank you, are important. Learn to listen to what people have to say. Use people's names when talking to them, and do not gossip.

Researching Careers

Interview a professional photographer. What is this person's education and experience? What does he or she like to photograph the best? Make a PowerPoint presentation of what you learn.

CHAPTER 12 Review

More activities on Student CD

Key Points

- Electronic communication may use analog or digital signals or a combination of both.
- Radio is audio communication that travels as electromagnetic waves.
- Television is video communication and is usually received by cable or satellite.
- Digital cameras use a CCD to help create an image.
- The Internet is our society's largest source of multimedia.
- Creating a multimedia presentation is similar to using the design process.

Read & Respond

1. Define *multimedia*.
2. What kind of wave is used for both radio and television?
3. What are the advantages of satellite radio?
4. How do television signals use AM and FM signals?
5. How are digital music players different from other music players?
6. How are video on demand systems similar to VCRs and DVD players?
7. How is a CCD used in a digital camera?
8. What's the difference between a digital zoom and an optical zoom?
9. Digital video is recorded on what kinds of media?
10. What are the major steps in planning and producing a multimedia presentation?

Think & Apply

1. **Hypothesize.** While digital photography is more popular, many people still use film photography. What advantages might film photography have over digital photography?
2. **Extend.** Explain how the activities in radio and television control rooms are similar.
3. **Produce.** Write a one-minute audio script to promote an upcoming event in your school.
4. **Assess.** What forms of multimedia do you use now that you didn't use five years ago? Has this changed your school or home life? If so, how?
5. **Summarize.** Compare and contrast radio and television broadcasting.

TechByte

Wayback When. The Internet may be the most important communications medium today—and yet many Web sites are here today, gone tomorrow. Computing expert Brewster Kahle decided in 1996 to do something about that problem by creating a public digital library called the Internet Archive. Through a front-end called the Wayback Machine, users can access Web sites now "dead." The archive at first used 20 terabytes (20 trillion bytes) of data storage. Now it is well past 100 terabytes.

CHAPTER 13

Animation

Objectives

- Discuss the uses of animation.
- Describe the information included on a storyboard.
- Compare and contrast the four types of animation.
- Describe Web animation techniques.
- Discuss how video games are developed.

Vocabulary

- **animation**
- **persistence of vision**
- **stop-motion animation**
- **motion-capture animation**
- **storyboard**
- **modeling**
- **key frame**
- **GIF**
- **streaming animation**

Many animated movies today are three-dimensional. Two of the most recognizable 3D characters today are Shrek and Donkey.

Activities

- Creating a Flip-Book
- Make Your Own GIF
- Presenting Video Games

What Is Animation?

Animation is the creation of simulated movement by using a series of still images. For many years, animation has been used to produce cartoons. Today, it is also used to produce special effects in commercials, live-action movies, video games, and Web sites.

Animation was first used in the 1800s in toys. One toy, the zoetrope (ZO-uh-trope), used a cylinder and a long strip of paper with a series of images. See **Fig. 13-1**. When the cylinder was rotated, the images could be viewed through slits in the cylinder. The images appeared to move. The zoetrope and similar toys led to the development of motion pictures. To create your own zoetrope, see the interactive lab on the Student CD.

Another simple animation device, also shown in **Fig. 13-1**, is a flip-book. The flip-book is easy to make. It consists of a sequence of drawings placed on top of each other and

Reading Link

Motion Pictures. The zoetrope was just one of several early animation devices. Search the library or Internet to read about the phenakistiscope, praxinoscope, or thaumatrope. Prepare a report describing the device. Tell who invented it and how it worked. You might even try making one as a demonstration model.

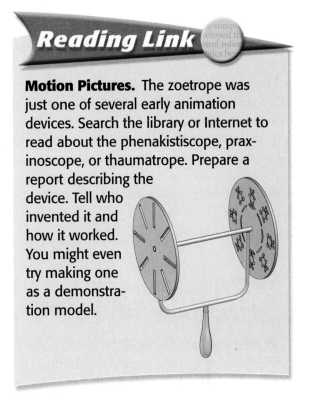

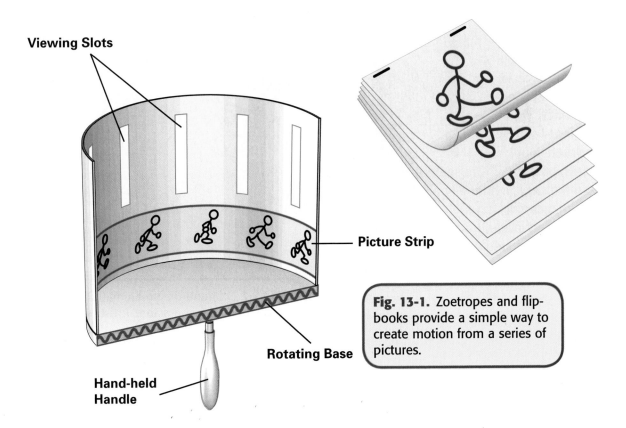

Viewing Slots

Picture Strip

Rotating Base

Hand-held Handle

Fig. 13-1. Zoetropes and flip-books provide a simple way to create motion from a series of pictures.

fastened together along one edge. Each drawing is slightly different from the drawing before. When the pages are flipped rapidly, the image appears to move.

The image appears to move because the eye sends signals to the brain faster than the brain can process them. Before the brain has finished processing one image, it receives another. This blending of individual images into one image that seems to move is called **persistence of vision**. Television programs and movies create the illusion of movement by showing 24 still pictures per second.

Types of Animation

Modern animation techniques can be divided into four basic types:
• Hand-drawn animation
• Stop-motion animation
• Computer animation
• Motion-capture animation

Hand-Drawn Animation

In hand-drawn animation, a series of drawings is photographed. Each drawing makes up one frame of the film. The position of the character or object changes very slightly from frame to frame. This technique is known as cel animation because the first animated characters were drawn on thin plastic sheets of celluloid. See **Fig. 13-2**. Cel animation has been used to create many well-known cartoon and feature films.

Fig. 13-2. In hand-drawn animation, animators draw the characters. Artists then trace the animator's drawings onto clear plastic sheets called cels.

In 1937, Walt Disney released *Snow White and the Seven Dwarfs*. This was the first full-length animated film produced using cel animation. Over the years, the cel animation process has been improved. Many of the techniques used in cel animation are now used in other kinds of animation as well.

Stop-Motion Animation

In **stop-motion animation**, models are photographed on a set one frame at a time. The frames are then played in sequence to create movement. Early stop-motion models were made of modeling clay. A model would be posed, photographed, and then adjusted slightly for the next shot. More recent models have a rubber skin over a moveable metal frame or are created entirely on the computer. Have you ever seen a commercial with the *Pillsbury Doughboy*®? He was created using stop-motion animation. See **Fig. 13-3**. Stop-motion techniques have also been used in major films, such as *Star Wars* and *The Lord of the Rings*.

DOUGHBOY® is a registered trademark of General Mills, Inc. and is used with permission.

Fig. 13-3. In many commercials starring Pillsbury Doughboy®, the model is positioned, photographed, and then reposed to create the illusion of movement. What kinds of small movements might be needed to animate this character?

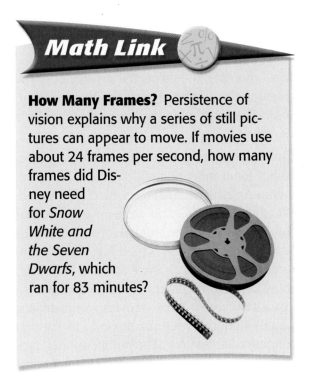

Math Link

How Many Frames? Persistence of vision explains why a series of still pictures can appear to move. If movies use about 24 frames per second, how many frames did Disney need for *Snow White and the Seven Dwarfs*, which ran for 83 minutes?

Computer Animation

The skills of layout, design, and timing are important in traditional animation. These same skills are also important in computer animation. Animators with traditional experience can learn to use the computer as an animation tool.

Computer animation software can be used to create an animated scene, a cartoon, a video game, or even a full-length movie. Software may be used for two-dimensional (2D) or three-dimensional (3D) animation.

Impact of Technology

One Technology's Impact on Another

Each new technology affects the technologies that came before it. Sometimes the new technology helps the older one. For example, cartoons were once shown only in theaters. When television was invented, it provided an additional market for cartoons.

Sometimes a new technology replaces an older one. This is happening now with animation. Computer animation is replacing hand-drawn animation. While hand-drawn animation may not completely disappear, it cannot compete with the speed and special effects of computer animation. Large studios find it more efficient and economical than hand-drawn animation.

Investigating the Impact

Throughout the history of moving pictures, many inventions have come and gone.

1. What preceded the modern version of the movie theater? Did people go to an auditorium and look at still pictures projected on the wall?
2. Research devices such as the praxinoscope, magic lantern, kinetoscope, vitascope, and kinora. Create an annotated time line of motion picture development.

Two-Dimensional Animation. Computers can greatly reduce the time needed to produce a 2D animated feature. A few skilled computer artists can replace the dozens of animators needed to draw and color the individual frames. Some artists prefer to begin by drawing the characters on paper. The drawings are then scanned into the computer. Other animators use the computer both for the initial drawings and the entire animation process.

After the original drawings are stored on the computer, they can be copied and changed as needed to make more frames. There is no need to hand-trace and color each frame. The computer can also be used to add sound effects.

Paint programs can also be used to add color to the drawings. Even inexpensive paint programs can provide hundreds of different colors.

After choosing the color, the artist moves the cursor to the desired area on the object or character. The artist adds shading and texture in a similar way.

Three-Dimensional Animation. Three-dimensional animation adds realism and excitement. As new technologies have developed, 3D computer animation has evolved. The first completely computer-animated movie was *Toy Story*, released in 1995. See **Fig. 13-4**. More recent movies such as *Shrek* and *The Incredibles* show how 3D computer animation improved in just a few years. Today, movies, commercials, games, and architectural designs all use 3D computer animation. Three-dimensional programs can even be used to take a customer on a tour of a building that does not yet exist.

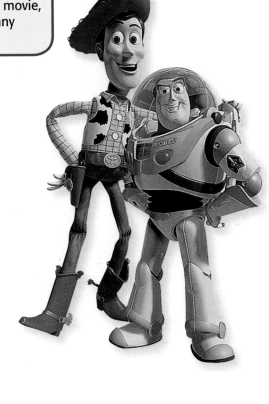

Fig. 13-4. The popularity of *Toy Story*, the first completely computer-animated movie, has spurred the development of many new animation techniques.

Computer animation programs vary in price and quality. Several good programs are available for use with personal computers. These allow you to learn the basics. Professional quality animation requires expensive software and fast computers with a lot of memory.

Motion-Capture Animation

Motion-capture animation is a 3D representation of a live performance. This animation can be done using three different technologies: magnetic, optical, and electro-mechanical.

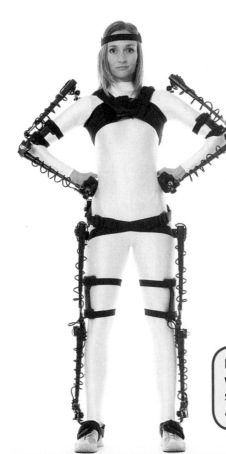

- Magnetic motion capturing uses sensors placed on an actor's body that measure magnetic fields created by a transmitter. The fields are then represented in 3D by a computer.
- Optical motion capturing uses video cameras to track the motion of light-emitting diodes or reflective markers placed on the joints of the actor's body.
- Electro-mechanical devices resemble an exoskeleton that is worn by the actor being recorded. The exoskeleton's movements are recorded without any additional sensors. See **Fig. 13-5**.

Fig. 13-5. The exoskeleton shown here will record the woman's movements. The same movements will then be used to animate a character.

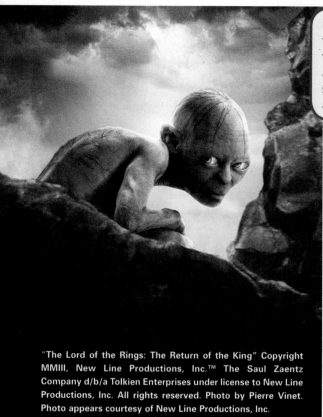

"The Lord of the Rings: The Return of the King" Copyright MMIII, New Line Productions, Inc.™ The Saul Zaentz Company d/b/a Tolkien Enterprises under license to New Line Productions, Inc. All rights reserved. Photo by Pierre Vinet. Photo appears courtesy of New Line Productions, Inc.

Fig. 13-6. More traditional kinds of animation were combined with motion-capture technology to create Gollum. Because an actor's motions were used, Gollum can speak, move, and interact with other characters much more realistically.

TV and Movies

No matter what kind of animation you use, making television shows and movies requires several steps to get from start to finish.

Storyboards. A **storyboard** is a series of sketches that can be used as a guide for making a show. A typical full-length animated feature requires more that 4,000 storyboard drawings to describe the action and dialogue of the film. Storyboards can be fairly simple or very detailed. They may be made by hand or on a computer. See **Fig. 13-7**.

Motion-capture animation is one of the fastest growing animation techniques because it creates such realistic movements. Gollum, a character from *The Lord of the Rings*, had realistic movements because traditional animation was combined with motion-capture technology. See **Fig. 13-6**.

Where Is Animation Used?

Animation is widely used today. If you turn on the television, you may see a cartoon or animated commercial. In the classroom, your teacher may be using PowerPoint® presentations that have animations. (See more about PowerPoint presentations in Chapter 12, "Digital Multimedia.") Movies, games, and Web sites also use animation.

Writing Link

Storyboard. A public service announcement (PSA) serves the public interest and is run by radio or television at no charge. Prepare a storyboard for an animated PSA to be shown on television. Choose a health or safety topic, such as wearing a seat belt, healthy eating, or exercise. Be sure to include sketches, descriptions, and dialogue.

Fig. 13-7. The sketches on the storyboard show the action and dialogue for a television show or movie. An actor or actress then reads the dialogue for the animated character. Animators often study the movements of the actor or actress and transfer those movements to the character.

Modeling. After storyboards are created, the modeling can begin. **Modeling** refers to using computer software to create 3D computer models of characters, props, and sets. The process usually begins with computer-generated shapes such as spheres, cubes, cylinders, and cones. New shapes can then be created by changing the sizes of the shapes and combining them to form the desired objects. Drawing tools similar to those used in CAD programs are used to draw free-form shapes. You can learn more about CAD in Chapter 2, "Computer-Aided Design."

Animation. Once the models are created, animation can begin. A **key frame** shows a beginning or ending point in an action sequence. Key frames are drawn using the 3D models. The computer then generates the "in-between" frames needed to simulate motion.

Imagine a scene showing two characters playing catch with a ball. The key frames might show each character either throwing or catching the ball. The frames in between determine how fast the ball is thrown, how may times the ball goes back and forth, and how long the

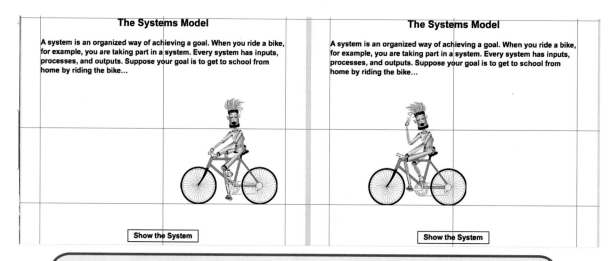

Fig. 13-8. Interactive labs from the Student CD were created using an animation program. Key frames were used to make the robot ride his bicycle and wave. In the screens above, what key frames do you think would be needed to start and stop movement?

game continues. The action in these frames can be controlled by the software. Some animation programs can even show how the ball changes shape when it bounces. See **Fig. 13-8**.

Shading, Lighting & Rendering. Shading, lighting, and rendering are all used to add realism.

- Shading adds colors and textures to objects. A variety of materials can be simulated, including glass, wood, and metal.
- Lighting is used to add illumination and shadows to scenes. Such lighting is similar to stage lighting.
- Rendering combines computer information from the modeling, animation, shading, and lighting steps to create the final images.

Web Sites

Animation can add interest to ordinary Web sites. For example, suppose you want to learn about the internal parts of an automobile engine. Why settle for an ordinary drawing with labels? Instead, look for an interactive drawing showing an engine with all the major

parts in motion. The animated version will be more interesting and show how the parts work together. See **Fig. 13-9**.

There are a number of techniques that Web designers use to create animation. GIF animation was the first successful kind of animation used on the Web. **GIF** stands for Graphics Interchange Format. GIF files are still images that can be animated using software. It is a popular way to store and distribute animations. GIF animations work well with most browsers and are a good way to add a simple animation to a Web site you are designing. You can learn GIF software animation in a few hours. Just remember that it takes up a lot of file space and is crude compared to other animation tools.

Other techniques produce much higher quality Web animations. When Dynamic HTML is used, one still image is moved across the screen instead of using individual frames. Several other kinds of Web animation require plug-ins to be viewed. Plug-ins are small programs designed to play particular media files. QuickTime® and Windows Media Player® are common video plug-ins. They show a sequence of still images

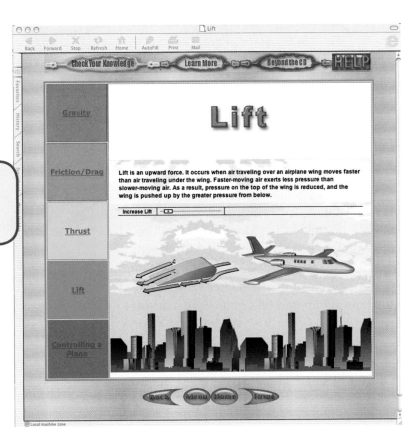

Fig. 13-9. This interactive lab from the Student CD uses animation to show how lift affects an airplane.

Fig. 13-10. Because of streaming animation, you could watch this video of a fireworks show as the video is downloading to your computer.

to create a movie using a process called streaming. **Streaming animation** allows the movie to begin playing before the entire file has downloaded. See **Fig. 13-10**.

Flash™ has become the most popular Web animation program. It can be used to produce high-quality animations and movies. The software needed to design Flash animations is expensive, but the plug-in needed to view them is now included with most browsers or can be downloaded for free.

One reason that we now see so much Flash animation is that a beginner can learn to do basic animations quickly. For example, if you want to show a ball traveling across a surface such as a pool table, you specify where the ball will start and stop and tell the program you want it to roll. The program then creates all the frames needed to show the rolling motion.

Video Games

Video games are now one of the most exciting and financially rewarding parts of the entertainment industry. More than $9 billion of video games are sold each year, and growth is expected to continue. Some of the additional growth will come as the games move beyond individual computers and traditional platforms, like the ones currently produced by Nintendo, Sony, and Microsoft. Much attention is now being focused on games that can be played in real-time by hundreds of players on the Internet and on games designed for mobile phones and other portable devices. See **Fig. 13-11**.

Video game development is the most complex form of animation. See **Fig. 13-12**. It's not unusual for a company to budget several mil-

lion dollars to cover the cost of developing a new game. A key difference between the video games and movie animation is that video games need to be interactive. They need to respond

Fig. 13-12. Shown here is an example of a 2D video game character from more than twenty years ago next to his 3D counterparts that are more common today. Graphics are improving rapidly as competition among video game manufacturers continues to grow.

to commands given by players. Programmers with advanced skills are needed to produce the challenging games that serious video game fans demand. See **Fig. 13-13**. Most of the work is done by teams made up of artists, programmers, and marketing experts.

Like other kinds of animation, video games start with a story. Characters and backgrounds are designed and storyboards are prepared. Early in the planning process, game design teams give artists ideas about the characters that will be needed. As they work, the artists will be concerned about how the characters look and how they behave. Behavior includes factors such as speed, timing, and strength. These details are included on the storyboards.

In addition to planning how the game will look, the design team will also plan levels of difficulty so that players will be able to advance to new challenges as their skills improve.

Once the initial version of the game is complete, post-production begins. Extensive testing of the game takes place before it is released to the public. First, experts help correct major flaws. Then experienced players suggest additional changes. Programmers use the feedback to correct problems and make changes so that the game can be sold for a variety of platforms. Finally, after a year or more of hard work, the game will be ready for marketing and sales. See **Fig. 13-14**.

Fig. 13-14. After testing and feedback, designers program final details into the game. The game is then manufactured and sold to consumers.

Creating a Flip-Book

Identify a Need/Define the Problem

As you learned in this chapter, each picture in a flip-book is slightly different from the pictures on the page before and the page after. See **Fig. A**. When the pages are flipped the images appear to be moving, as if they were animated. See **Fig. B**. In this activity, you will design and build your own flip-book.

Gather Information

The flip-book, or kineograph as it was originally known, has been around for over a hundred years. It has been made in a variety of shapes and sizes. Before designing your flip-book, you may want to research how one works and what makes an effective design for a flip-book.

Develop Possible Solutions

Prepare several different designs for your flip-book. You will need to decide on page size as well as the type of paper. Some paper sizes and types may affect how the book flips through your fingers. You will also need to decide on an image and action that the flip-book will show the user.

Materials and Equipment

Select from this list or use your own ideas.

- paper
- pens
- pencils
- computer system with drawing program
- binding material

Model a Solution

1. Once you have chosen the most effective solution, you will need to prepare the pictures. The flip-book pictures can either be drawn by hand or by computer. (It is relatively easy to draw the pictures on a computer. You can save the image as a separate slide and then modify it slightly, save it again, and so on.)
2. Cut the paper to the appropriate size. Make sure that each image is correctly placed in the sequence you designed.
3. Staple the pages together. If the book is too thick, you may need another possible solution, such as using brush padding compound along one edge. (Padding compound is a substance used to make pads.) You may also want to put the book on a stiff backing, such as cardboard.

Test and Evaluate the Solution

Thumb through your flip-book to see how the images blend together. Share your design with your classmates and ask them for their feedback.

- Do the images appear to move smoothly when the pages are flipped?
- Are the pages of the flip-book fastened tightly enough to allow repeated use?
- Does the flip-book contain enough pages to effectively communicate the story or action?

Refine the Solution

- If the action in the flip-book wasn't clear, maybe you need to add more pages.
- If the animation didn't work, can the materials be improved to help?
- Revise your flip-book as needed.

Communicate Your Ideas

Create a portfolio of the research and work that you did in order to create your flip-book. Include some facts and information about how flip-books work and are made. Document the solutions you came up with and the steps you took to create it. Also include important information on how you tested your solution and any refinements you made. Present your portfolio to the class.

Fig. A

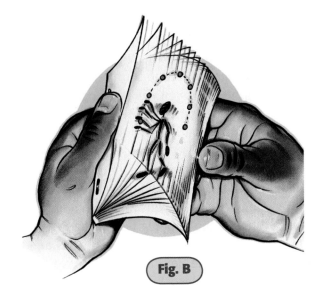

Fig. B

Make Your Own GIF

Identify a Need/Define the Problem

In this chapter, you learned about GIFs. Use a computer and GIF editor software to create an animated GIF.

Gather Information

In order to create an animated GIF, you will need to use a program that can edit several GIF images and then compile them into one animated image. Free trials and shareware are available online. Research these programs. Using "Help" menus can help you review basic functions.

Develop Possible Solutions

Before implementing your design, you will want to decide on a few factors for your animated GIF. You will need to consider the size of the image, the colors you will be using, and the action you will be animating. Begin by developing several storyboards showing possible solutions. This will give you an idea of what you need to do and will serve as a guide while you are editing the individual images.

For example, what if you wanted to animate a stick figure doing jumping jacks? See **Fig. A**. Start out with the figure that has its arms and legs at its side. The next image may have the figure's arms and legs further apart and the third would have them completely apart. Then you would work back towards the original position. By combining these few images together you can create an animated GIF that looks like the figure is doing endless jumping jacks.

Materials and Equipment

Select from this list or use your own ideas.

- computer
- GIF editing software
- online access

Safety Alert

Downloading Software

Before you download free trials or shareware onto a school computer, please ask permission from your school's network administrator.

Model a Solution

1. Create the first image in your animated GIF. Save this as a file named "image 1." If the program you are using requires you to use layers, create the first image on "layer 1."
2. Create your next image, or layer, using the same size you specified for the first one. Modify the image to make it slightly different from the previous one. You will want the change to be subtle. Save the file as "image 2," or "layer 2" if using layers.
3. Repeat the previous step, saving each image as an individual file or layer.
4. Compile the images into one animation using the software.
5. Save the animated GIF and place it in a basic HTML page.

Test and Evaluate the Solution

- Did your animated GIF perform as you had designed it to?
- Are the individual images similar enough so that the animated GIF isn't choppy?
- Is the file too big to upload or save on your storage device?

Refine the Solution

- If your file doesn't perform as you expected, try using different software.
- If the file seems choppy, you will need to create additional images to make it smoother.
- If the file is too large and takes up a lot of room, you may need to modify things such as your color palette or the physical sizes of the images.

Communicate Your Ideas

Document the steps that you took in developing the animated GIF. You may want to include links to any sites that you used in developing the animation as well as the steps you took in creating and refining your design. With permission, post a link to your animated GIF on your school's server for all to see.

Fig. A

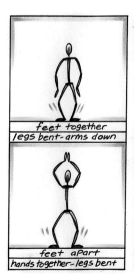
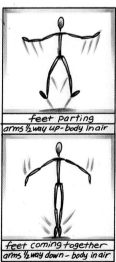

feet together
legs bent-arms down

feet parting
arms ½ way up-body in air

feet apart
hands together-legs straight

feet apart
hands together-legs bent

feet coming together
arms ½ way down-body in air

feet together
arms down-legs straight

Design, Build & Evaluate

Presenting Video Games

Identify a Need/Define the Problem

The video game industry is one of the most popular fields of animation. See **Fig. A**. Use a computer and PowerPoint software to create a presentation about video game animation.

Gather Information

Conduct research on video game animation. Use resources available in your classroom, school library, and online. While conducting the research, gather information on types of video game animation, the people employed in the field, images and videos of the animation, and other interesting informative facts. Since there are some video games that may be unsuitable, make sure that any information you use is appropriate for a school activity.

Materials and Equipment

Select from this list or use your own ideas.

- computer
- PowerPoint
- online access

Safety Alert

Downloading Files

Before you download images or videos onto a school computer, please ask permission from your school's network administrator.

Develop Possible Solutions

Using the information that you have gathered, prepare outlines describing the text that will appear on each slide. You may have more than one outline, depending on the focus of your presentation. Select three or more digital images and video clips to use. Revise the text and images until you are satisfied that your presentation will be interesting to the audience and communicate important information about video game animation.

Fig. A

Model a Solution

1. If you are not familiar with PowerPoint, press the F1 key in PowerPoint for the Help Menu, or you can go online and search for "PowerPoint Tutorial." Search the help menu or complete the tutorial to learn how to use the program.
2. Open PowerPoint and create the presentation using one of these options: "Auto Content Wizard," "Design Template," or "Blank Presentation."
3. If using "Blank Presentation," select "Insert—Text Box" to add the text of your presentation from your outline.
4. Select "Insert—Picture" to add the digital graphics that you have collected.
5. Select "Insert—Movies and Sounds" to add the digital graphics that you have collected.
6. Select "Slide Show—Slide Transition" to automate the presentation and to allow the presentation to play by itself.
7. Create a final slide citing the sources you used for your presentation.
8. Make any changes to eliminate errors and save your presentation.
9. Share your animated presentation with classmates.

Test and Evaluate the Solution

• Did the slide show automatically transition from one slide to the next?
• Did your classmates find your presentation interesting, informative, and visually clear? Can you use their feedback to improve your presentation?
• Did you use PowerPoint and its many features to its full advantage?

Refine the Solution

• Use the feedback you received from your classmates to refine your presentation.
• You may need to clarify a statement or improve the transition of one slide to the next.
• See how the various features in PowerPoint could help you improve the overall effectiveness of your presentation.

Communicate Your Ideas

Create additional slides showing your steps for creating this presentation. This presentation can contain links to Web sites, any outline documents, or additional sources of information. Present your new slides to the class.

Science Link

Video Game Science. Did you ever stop and think how much scientific knowledge is needed to design a video game? To create realistic characters, designers must understand anatomy. To create movements for characters or objects, designers must have an understanding of physics.

Play an all-ages video game and look for demonstrations of science. Can you find examples of biology? How about astronomy or chemistry? Video game designers might use these and other fields of science when designing video games for you to enjoy.

Exploring Careers

Animator

ENTRY LEVEL **TECHNICAL** PROFESSIONAL

Animators draw cartoons and other images, such as special effects. A cartoon is a series of individual drawings displayed at a speed that makes the drawings look like they are moving. Animators work as part of a team to develop the ideas for the drawings and what the final product will look like.

Most animators work in the motion picture and video industries, advertising, and computer systems design services. They create animated images for movies, television programs, commercials, and computer games.

More than 50% of animators are self-employed. Their clients are advertising agencies, publishers, and design firms. Animators who are salaried also work for advertising agencies and for newspapers and other publishers, motion picture and video industries, and computer systems design companies.

Qualifications

The training for animators varies depending on what field they will work in. A bachelor's degree is not always required, but animators with a bachelor's or master's degree usually get the better jobs. Animators who go to college would most likely take classes in art history and studio art as well as general education courses. Animators may also take classes in computer techniques.

Animators must be creative, have artistic ability, a good sense of color, hand-eye coordination, and be detail oriented. Good communication skills and the ability to work well with other people on a team are also important.

Outlook for the Future

The job outlook for animators is good. Because this career is highly desirable, there is a lot of competition for jobs. The more educated and skilled animators will have an advantage.

Respecting Differences

Like most careers, this one may bring you into contact with many different kinds of people. Coworkers will have different ideas than yours, and these ideas may be useful. Working together and listening to others' ideas will make your job easier and your product better.

Researching Careers

Find out about jobs for animators. List five companies that employ animators. What are the education and training requirements? Where in your area can you get this training and education? Make a poster showing what you have learned.

**More activities
on Student CD**

Key Points

- Animation is the creation of simulated movement in a series of still images.
- Modern animation includes hand-drawn, stop-motion, motion-capture, and computer animation.
- Computer animation reduces the time needed to produce an animated film.
- Storyboards are used as "blueprints" to guide the action and dialogue in each scene of animated films.
- There are a number of programs designed for creating Web animation.
- Video game animation requires advanced programming skills.

Read & Respond

1. Identify several uses for animation.
2. Explain persistence of vision.
3. What information do storyboards include?
4. What is the difference between stop-motion animation and motion-capture animation?
5. Identify the main difference between hand-drawn animation and computer animation.
6. What are the three types of motion-capture animation?
7. What is an animation cel?
8. What is a key frame?
9. Describe two different Web animation techniques.
10. Describe the key difference between animations for video games and movie animations.

Think & Apply

1. **Create.** Make a series of drawings that show action in a sport. Use the drawings to create an animated flip-book.
2. **Reason.** Why is a storyboard important in preparing all types of animation?

3. **Relate.** Describe how an artist can use a computer animation program to assist in traditional animation.
4. **Persuade.** Describe a class activity that you think could be clearly explained through computer animation.
5. **Design.** Create a Web page design or layout that includes animation on a topic approved by your teacher.

TechByte

Behavioral Animation. Did you know animated characters can determine some of their own actions? Computing visionary Craig Reynolds developed a simple computer model for flocking, in which he dubbed the flocking creatures "boids." The model has proved so successful that it has helped zoologists understand animal behaviors in new ways. The same technique is now used to create many animated battle scenes, where individuals have separate actions and reactions.

CHAPTER 14

Electricity & Electronics

Objectives

- Explain how atoms are responsible for electricity.
- Describe Ohm's law and how it is used by engineers.
- Identify three sources of electricity.
- Compare and contrast series circuits and parallel circuits.

Vocabulary

- **electricity**
- **voltage**
- **current**
- **resistance**
- **generator**
- **circuit**
- **electronics**
- **transistor**
- **diode**
- **integrated circuit**

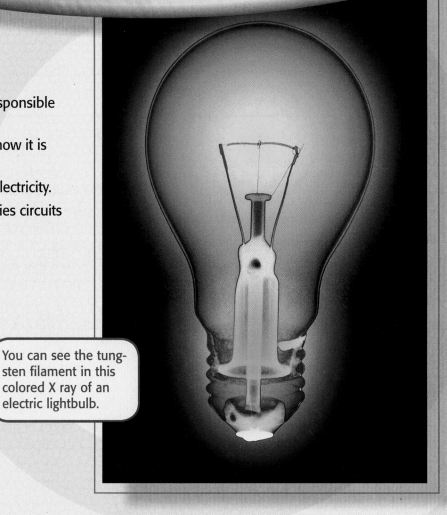

You can see the tungsten filament in this colored X ray of an electric lightbulb.

Activities

- Dim the Lights
- Creating an LED Warning System
- Making an Electromagnetic Crane

Electricity, the Invisible Force

Of all the energy forms we encounter each day, electrical energy is perhaps the most mysterious. Do you know why?

At night we can see the energy of light illuminating a large city. See **Fig. 14-1**. We can feel heat energy radiating from a campfire that warms our hands. Our car shakes from the vibrations of sound energy coming from large speakers pounding in the car next to ours at the traffic light.

When it comes to electrical energy, though, we normally can't see it, hear it, touch it, or even smell it. That's what makes it so mysterious. We know it helps make light energy. We know it powers our computer and it can also create heat to warm our hands. But what is it? That's what this chapter is all about.

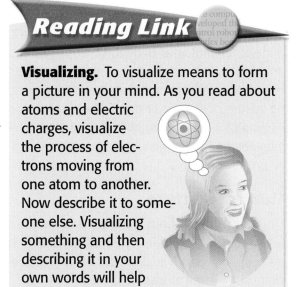

Reading Link

Visualizing. To visualize means to form a picture in your mind. As you read about atoms and electric charges, visualize the process of electrons moving from one atom to another. Now describe it to someone else. Visualizing something and then describing it in your own words will help you learn and remember.

Fig. 14-1. Shown here are the lights of the skyline of Atlanta, Georgia. Just imagine the amount of energy that must be generated for the electricity to run these lights.

A few people would argue that harnessing electrical energy has been the most significant technological event of all time. See **Fig. 14-2**. Modern technologies depend on thousands of devices that rely on electrical energy and electronics. This chapter will help you understand the nature of electricity as an energy resource. It will also help you see how we use electricity and electronics to control complex technological products.

Atomic Structure

Electricity is all about charges. As we learned in Chapter 4, "Materials Science," all matter is made up of atoms. The charges that create electricity come from atoms.

An atom has two parts: a center core, or nucleus, and a cloud of electrons that surround the nucleus. Tightly packed within the nucleus are particles called protons and neutrons.

Science Link

Atoms and the Solar System. This chapter explains that a cloud of electrons revolves around an atom's nucleus. Many sketches of a theoretical atom resemble sketches of our solar system. Compare and contrast the model of an atom with that of the sun and the planets revolving around it. Is it accurate to say that our solar system is similar to an atom, just immensely larger in scale? Research the question and be prepared to discuss your findings in class.

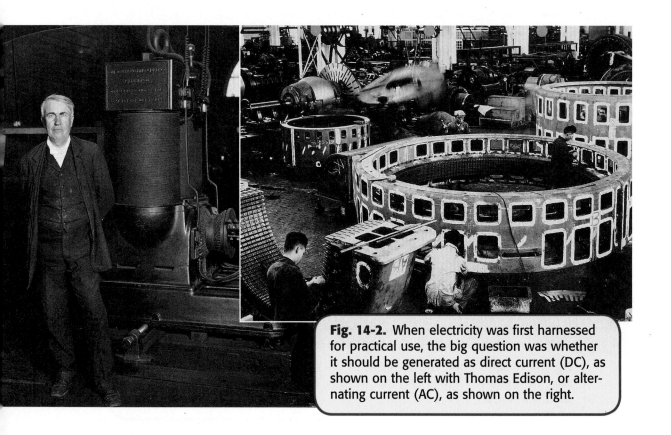

Fig. 14-2. When electricity was first harnessed for practical use, the big question was whether it should be generated as direct current (DC), as shown on the left with Thomas Edison, or alternating current (AC), as shown on the right.

Except for hydrogen, the number of neutrons in an atom is always equal to or greater than the number of protons. (Hydrogen has one proton and no neutrons.)

A neutral atom has one electron for each proton in its nucleus. This makes the atom balanced. Most atoms, however, can lose or gain electrons. This process, which upsets the balance of the atom, is called ionization. Ionization is very important to generating a flow of electricity. See **Fig. 14-3**.

Atomic Charges

Protons and electrons in atoms contain tiny amounts of electrical energy (charges). Protons have a positive charge, while electrons have a negative charge. Neutrons are neutral. They have no charge.

When ionization occurs, the balanced atom loses or gains an electron and now becomes an unbalanced ion. If an electron is lost, the atom becomes a positive ion because it has more protons than electrons. Why do you think atoms that gain an electron are called negative ions?

Why Are Electrons Gained and Lost?

Electrons move around the nucleus at different distances, or levels, from the core. Each level around the nucleus is known as a shell. The electrons in the innermost shell, which is closest to the nucleus, have the least amount of energy. The electrons in the outermost shell have the greatest amount of energy.

Electrons in the outer shell are particularly important in creating electricity. They are called valence electrons. See again **Fig. 14-3**.

Why don't all the electrons fly away from their nucleus? What keeps them bound to the core? As with magnets, there is an attraction between the nucleus of the atom and its electrons. You may have learned that positive and negative poles of magnets attract each other. The positive charges of protons in the nucleus attract

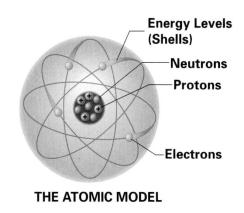

THE ATOMIC MODEL

Energy Levels (Shells)
Neutrons
Protons
Electrons

**LITHIUM ATOM
BALANCED ATOM**

Valence Electron

3 (+)Protons
3 (−)Electrons
———————
0 Charge

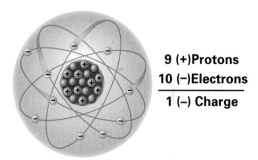

NEGATIVE FLUORIDE ION

9 (+)Protons
10 (−)Electrons
———————
1 (−) Charge

Fig. 14-3. The nucleus of an atom has protons and neutrons. An atom is balanced if there are the same number of electrons as protons. If there is a missing electron or an additional electron, the overall charge of the atom will change.

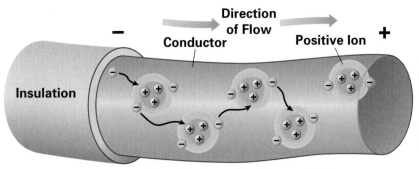

Direction of Flow

− Conductor Positive Ion +

Insulation

Fig. 14-4. Electricity will move through a wire as free electrons move from atom to atom. It is this movement of electrons that creates electricity.

Free electrons travel on until they combine with a positive ion.

the negatively charged electrons in each of the shells. This attraction helps keep the electrons from flying away from the nucleus.

The farther away the electrons are from the nucleus, the less of an attraction the nucleus has on the electron. Because valence electrons are the farthest away from the nucleus and they contain the highest level of energy, they frequently jump out of their orbit. These electrons are called free electrons. These free electrons become attracted to another atom, move toward that atom, and may even join with it. It is this movement of valence electrons that creates electricity. **Electricity** is the movement of electrons from one atom to another. See **Fig. 14-4**.

Static Electricity

Earlier in this chapter you read that the flow of electricity is not usually seen. Static electricity is one exception. Have you ever experienced

touching a doorknob after walking across a thick carpet? Zap! A charge of electricity jumps out and tags your hand. Sometimes you can even see the charge. If this has happened to you, then you have experienced static electricity. Just as an atom can become a negatively or positively charged ion, entire objects can become charged. A neutral object can become electrically charged when it either gains or loses electrons.

As you walk across the carpet, friction causes the carpet to give up electrons. Your body picks up electrons. When you grab the doorknob, electrons jump to the neutral metal. The electrical charge is transferred to the doorknob. See **Fig. 14-5**. Can you identify another example when electricity is visible in the air?

Now you know about that mysterious force that lights our city, warms our hands, and powers our computers. It is simply the flow of free electrons caused by the interaction of charged electrical particles.

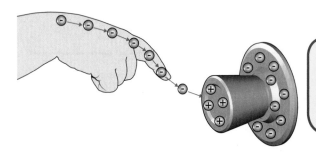

Fig. 14-5. As you walk across a carpeted floor, negative electrons build up over your body. These negative electrons then are attracted to the neutral metal. The movement of the electrons causes the spark.

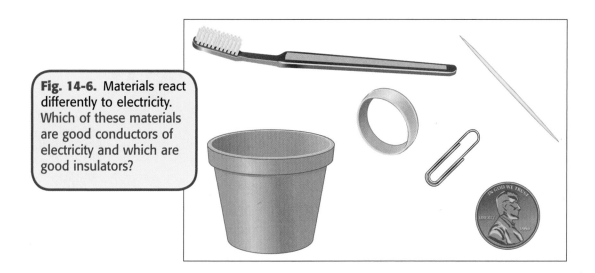

Fig. 14-6. Materials react differently to electricity. Which of these materials are good conductors of electricity and which are good insulators?

Voltage, Current, and Resistance

An electrical charge has stored energy with the potential to do work. For example, the potential energy at the negative terminal of a battery differs from that at the positive terminal. In a complete electric path, this difference in potential causes a charge to move through the circuit. This potential difference, as it is called, is the force that causes electrons to move.

A potential difference is also referred to as a voltage, or electromotive force (emf). **Voltage** is the electric pressure that causes current to flow. It is the force that sets charges in motion.

Voltage is a measurable quantity. The unit of measure for voltage is the volt. The symbol for volts is V, although E for "electromotive force" is sometimes used.

Some materials are made of atoms that do not have a strong hold on valence electrons. These materials are called conductors. Metals are generally good conductors. Copper, aluminum, silver, and gold are all excellent conductors. See **Fig. 14-6**.

When voltage is applied to a wire conductor, the excess electrons at the negative end of the wire travel to the positive end, or the end with too few electrons. The flow of electrons in a wire or other conductor is known as **current**.

Current is measured by the number of electrons that move past a certain point within a wire each second. The symbol for current is I. The unit of measure for current is the ampere, or amp. The symbol for amperes is A.

In contrast to conductors, some materials are made of atoms that have a tight hold on electrons and very few free electrons. Those materials that resist the flow of voltage are called insulators. Plastic and ceramic materials are good insulators. However, if enough voltage is applied, electrons can be forced from atom to atom through the insulator. Insulators have different strengths. The strength of an insulator represents its resistance.

Resistance is the opposition to the flow of electrons. Different substances have different resistances. Conductors have low resistance. Insulators have very high resistance.

The symbol for resistance is R. The unit of measure for resistance is the ohm. The symbol for the ohm is the Greek letter Ω (omega).

Ohm's Law

The unit of measure for resistance, the ohm, was named after the German scientist George Ohm. Through experiments, George Ohm found that the flow of electrons (current) is affected by both voltage and resistance. Ohm's law explains this relationship.

Ohm's law states that current (*I*) is equal to the voltage (*E*) divided by the resistance (*R*). This means that the number of electrons flowing is affected by (1) the amount of voltage pushing them and (2) the amount of resistance to the flow caused by the conductor and other parts being powered.

Ohm's law can be written mathematically as follows:

Current	$I = E/R$
Voltage	$E = IR$
Resistance	$R = E/I$

$$I = \frac{E}{R}$$

$$R = \frac{E}{I}$$

$$E = IR$$

Math Link

Using Ohm's Law. Ohm's law states the relationship between voltage, amperage, and resistance ($E = IR$). If you know any two of the three variables, you can use the formula to figure out the third. Solve the Ohm's law problems below.

1. $E = 120$ volts; $I = 3$ amperes. How many ohms of resistance?
2. $I = 4$ amperes; $R = 1,000$ ohms. How much is the applied voltage?
3. A 9-volt battery is connected to a circuit containing 100 ohms of resistance. How many amperes of current will flow?

When electrical engineers design products, they must be sure that the product gets the correct amount of electricity. Engineers use Ohm's law to ensure the amount of voltage and current is correct so that the product works as designed. See **Fig. 14-7**.

Sources of Electricity

As mentioned, for electrons to flow, there must be a potential difference, or voltage. It can be produced from various primary energy sources. These primary sources take energy in one form, such as chemical or mechanical energy, and convert it to electrical energy. Two voltage sources—batteries and generators—are discussed here.

Cells and Batteries

Have you ever wondered how a battery works? A battery is a device that converts chemical energy into electrical energy. In other words, a battery generates electrical energy (voltage) with a chemical reaction.

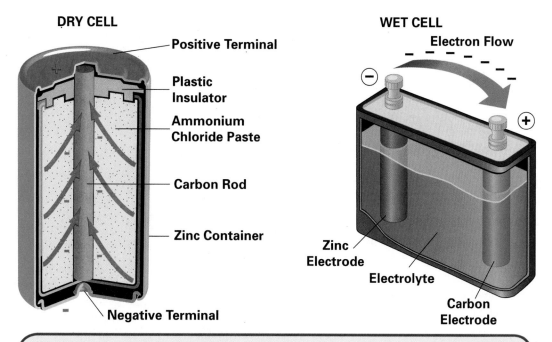

DRY CELL

Positive Terminal

Plastic Insulator

Ammonium Chloride Paste

Carbon Rod

Zinc Container

Negative Terminal

WET CELL

Electron Flow

Zinc Electrode

Electrolyte

Carbon Electrode

Fig. 14-8. Batteries change chemical energy into electrical energy. Voltage is created through chemical reaction. In a carbon zinc cell, electrons travel from the zinc to the carbon rod.

Batteries usually consist of two or more cells. (However, individual cells are often themselves called batteries.) A cell is a device made of two different conducting materials in a conducting solution. The conductors are called electrodes (e-LEK-trodes). The conducting solution is called an electrolyte (e-LEK-tro-lite).

Dry cells have a paste electrolyte made of chemicals. A flashlight battery is a common dry cell. Wet cells contain a liquid electrolyte. A car battery is a wet cell. See **Fig. 14-8**.

Within the cell, a chemical reaction occurs in the cell between the two electrodes. The electrolyte positively charges one electrode and negatively charges the other. In this way, the cell produces voltage.

When connected to a conductor, cells and batteries produce a current that flows in only one direction. This current is called direct current (DC).

Generators

Mechanical energy is the energy of motion. A **generator** is a device that changes mechanical energy into electrical energy. A generator uses electromagnetic induction to force electrons from their atoms.

Many years ago, scientists found that they could produce an electric current by moving a wire through a magnetic field. When a wire cuts across the invisible lines of force of the magnetic field, voltage is induced in the wire. If the wire forms a complete circuit, a current is induced as well.

Generators vary in type and construction. A simple generator consists of a coil of wire wrapped around a metal core and placed between the poles of a magnet. The wire coil and core assembly is called an armature (ARM-uh-chur). As the armature rotates, the coil cuts across the magnetic field. As a result, voltage is induced in the coil.

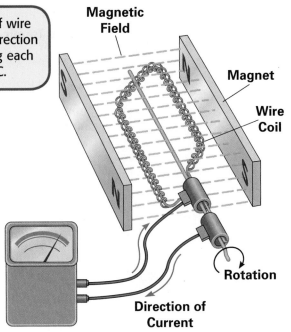

At each half-turn, the two connections at the output of the generator change polarity. First one end is positive and the other is negative, then vice versa. The current induced by such a voltage changes direction each time polarity changes. This electricity is known as alternating current (AC). See **Fig. 14-9.**

Most of the electricity we use is alternating current produced in power plants that use large generators. Mechanical energy turns the rotating parts of these huge machines. This energy comes from turbines. Turbines are bladed wheels that turn when struck by the force of steam or moving water. As the turbine shaft turns, the armature rotates. This generates electrical energy, or voltage.

Electromagnets

If electricity can be induced by cutting across magnetic lines of force, can magnetism be induced from electricity? The answer is yes. An electric current flowing through a wire creates a magnetic field around the wire. The relationship between electricity and magnetism is called electromagnetism.

Electromagnets are powerful magnets created by wrapping wire around an iron core and then passing electric current through the wire. The magnetic field can be controlled by turning the current on and off. The strength of the magnetic field can be increased in two ways: (1) by increasing the number of coils wrapped around the core and (2) by increasing the current flowing through the coil. See **Fig. 14-10.**

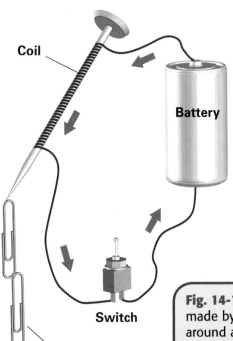

Fig. 14-10. An electromagnet can be made by winding fine, insulated wire around an iron nail and connecting it to a battery.

Electrical Circuits

In electricity, a **circuit** (SIR-cut) is the pathway through which electrons travel. A simple circuit consists of a power source, a conductor, and a load. See **Fig. 14-11**. A circuit that uses direct current as a power source might use a battery or a photovoltaic (foe-tow-vole-TAY-ik) cell. This is also known as a solar cell. (You can learn more about photovoltaic cells in Chapter 8, "Energy & Power Technologies.") The source of voltage for alternating current is usually a generator at the power plant.

Conductors provide a low-resistance path from the source to the load. Typically, copper or aluminum wires serve as conductors. The load is the device that uses the electric energy. The load converts the electrical energy in the circuit into heat, light, mechanical, or other energy forms. The main types of circuits are series circuits and parallel circuits.

Series Circuits

When the components are connected one after another, the circuit is called a series circuit. In a series circuit, there is only one pathway for electrons to follow. See **Fig. 14-12**. A break in any part of the circuit stops all the electrons from flowing. This produces an open circuit. Electrons can flow only in a closed circuit.

Two rules apply to series circuits:
• The current is the same at all points.
• The total resistance of the circuit is equal to the sum of the individual resistance values.

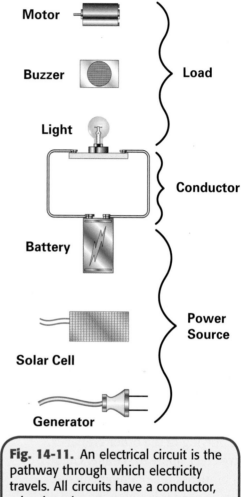

Fig. 14-11. An electrical circuit is the pathway through which electricity travels. All circuits have a conductor, a load, and a power source.

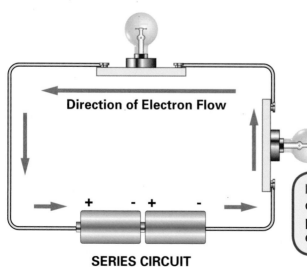

Direction of Electron Flow

SERIES CIRCUIT

Fig. 14-12. A series circuit supplies only one path for the flow of electrons. If that path is broken at any point, the flow of electrons will stop.

Impact of Technology

Lights Out

Can you imagine how things might change if there were no electric power? From heat on a cool morning, to an alarm clock/radio, to lights and a refrigerator, our everyday lives depend to an amazing degree on the use of electricity.

How much electricity do we use? Usage can be measured in megawatt-hours. A megawatt is 1 million watts. A megawatt-hour is the same as using one million watts of electricity for 1 hour. The world uses more than 14 *billion* megawatt-hours of electricity each year.

Investigating the Impact

A blackout is a situation in which electrical power suddenly ceases to flow to a large number of consumers. Investigate a recent blackout.
1. How did people cope without electricity?
2. How might you and your family be able to prepare yourselves for a blackout?

Parallel Circuits

In a parallel circuit, the components are arranged in separate branches. See **Fig. 14-13**. This arrangement provides multiple pathways in which electrons can flow. A break in one branch of the circuit does not prevent electrons from flowing in the other branches.

Two rules apply to parallel circuits:
• All branches are of equal voltage.
• Total current is equal to the sum of the branch currents.

To learn more, try the interactive lab "Series, Parallel, and Combination Circuits" on the Student CD.

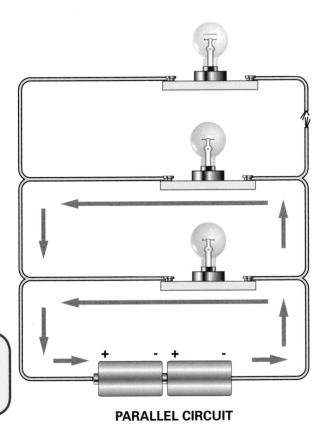

Fig. 14-13. A parallel circuit has many paths along which electrons can flow. A break in one path will not stop the flow of electrons in the other paths.

PARALLEL CIRCUIT

What Is Electronics?

Electronics is the study of the control of electron flow in a circuit. Electronics is an outgrowth of our knowledge and ability to control electricity. However, electricity and electronics are different subjects.

Electricity is about controlling large levels of voltage and current in a circuit. Electronics is about controlling individual electrons in a circuit. Electronic components may increase (amplify) a small signal, limit the amount of current passing through a device, or control the direction of current in a circuit.

Electronic devices often use components that sense changes in light, temperature, and other conditions in the environment. They use this input to adjust the flow of electricity in the circuit. A motion sensor lamp is a good example of this type of electronic circuit.

Transistors

In 1948, engineers at the Bell Telephone Company simplified the precise control of electrons. They invented the transistor. A **transistor** is a tiny device used to control current and amplify voltage or current. It uses very little current. Imagine the transistor as a tiny gate holding back a large flow of current in a circuit. This gate can be opened and closed at lightning speeds to control the flow of current. See **Fig. 14-14.** The transistor acts as an electronic switch in complex circuits.

Transistors are made from layers of semiconductor materials. A semiconductor is a material that can change from conductor to insulator as conditions change. Silicon and germanium are common semiconductor elements. A process called doping adds impurities to the pure semiconductor elements. Doping increases the material's ability to be a conductor or an insulator. The doping of semiconductive materials allows for the precise transfer and control of electrons that in turn can control larger amounts of current.

Current Flowing Through Transistor

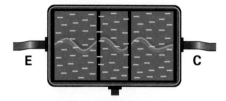

Fig. 14-14. In a transistor, the current flows from the emitter (point E) to the collector (point C). A transistor can be either a conductor or an insulator.

Transistors are often used to amplify an electronic signal. In electronics, a signal is an electric current that carries information. The weak signal enters the transistor. Through the movement of electrons, the signal is amplified (made stronger). Transistors are used in radios, televisions, hearing aids, computers, and calculators to amplify a signal.

Diodes

Diodes are also made from semiconductor materials. **Diodes** are electronic components that allow electrons to flow in only one direction. See **Fig. 14-15.** This characteristic is used

Fig. 14-15. A diode will allow electricity to flow in only one direction. This is used to change alternating current into direct current.

Diode

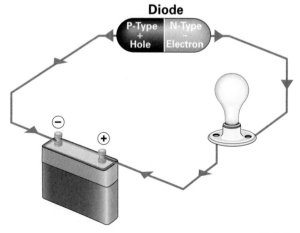

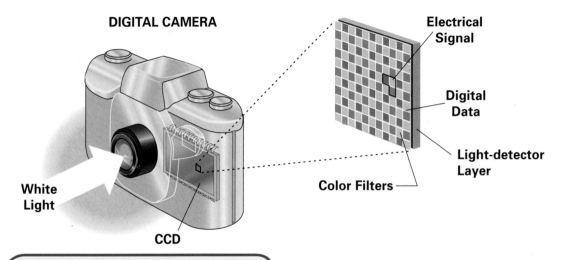

DIGITAL CAMERA

White Light

CCD

Electrical Signal

Digital Data

Light-detector Layer

Color Filters

Fig. 14-16. The light-sensitive diodes in the CCD of the digital camera record the various light levels electronically. The brighter the light is, the greater the electrical charge.

Other Electronic Components

Many times the current in an electronic circuit needs to be reduced. A resistor is a device that acts as a load on the circuit. Fixed resistors have a specific resistance value. Variable resistors can adjust to different resistance values. Volume controls and dimmer switches are examples of variable resistors.

Have you ever wondered how a camera creates the flash of light that brightens up the scene when you take a photograph? Camera flashes use an electronic device called a capacitor. See **Fig. 14-17**.

to change alternating current into direct current. Remember that alternating current changes direction every cycle. Computers, televisions, radios, and other electronic devices are powered by alternating current that is converted to direct current by circuits containing diodes.

Light-sensitive diodes are used in digital cameras to record light energy electronically. The diodes convert light into electrical charges. Groups of diodes form photosites. The brighter the light hitting the photosite, the greater the electrical charge that will accumulate at the site. A computer inside the camera organizes the charges into digital data that can be stored on a variety of media. See **Fig. 14-16**.

CAPACITOR

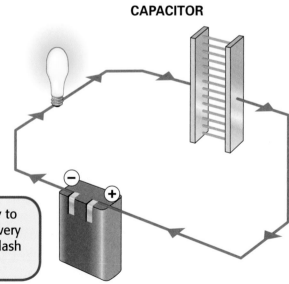

Fig. 14-17. Capacitors have the ability to store electricity and then discharge it very quickly. This is the way an electronic flash works on a camera.

Fig. 14-18. In this microprocessor, the integrated circuit has many miniaturized transistors and diodes.

Capacitors store electrical charges and then release them on command. Lasers and televisions use capacitors because of their ability to deliver a high-speed charge of electricity.

It seems that each year electronic devices, such as computers and music players, get smaller and smaller. This has been made possible by the integrated circuit. An **integrated circuit** is a tiny piece of semiconductor material that contains miniaturized components, such as transistors and diodes, wired into minute circuits. See **Fig. 14-18**.

Writing Link

Troubleshooting. If your toaster, computer, or some other device doesn't work, what do you do? Randomly press buttons? There is a better way. To "troubleshoot" means to follow a systematic method to find the cause of a problem and fix it. Owner's manuals often contain troubleshooting charts to help users find and correct problems. An example is shown in **Fig. 14-19**. Study this example and then make a troubleshooting chart for some other device.

Integrated circuits, sometimes called chips, can have thousands of individual circuits and other components built into a space the size of a pea. The power of modern computers and electronic devices often depends on how many components can be placed into their circuits.

Symptom	Possible Cause	Solution
Computer cannot read CD.	• CD not formatted for your computer system. • CD not properly inserted in drive. • CD damaged. • CD drive damaged.	• Check label of CD for hardware and software requirements. • Reinsert CD into drive, being careful to place it correctly. • Insert a different CD into the drive. • Replace or repair CD drive.

Fig. 14-19. Troubleshooting.

Dim the Lights

Identify a Need/Define the Problem

There are many places where lights need to be dimmed, such as the headlamps on a car or the lighting in a restaurant. See **Figs. A** and **B**.

Design and model a system that either switches between bright and dim lights or can dim the lights gradually. The circuit should include two light bulbs and be powered by one battery.

Gather Information

What type of device can increase resistance in a circuit? Is it possible to make one of these devices?

What type of circuit will be used, series or parallel? Is it possible to use both a series and parallel circuit to solve the problem? Visit the interactive lab "Series, Parallel, and Combination Circuits" on the Student CD. Research the

different kinds of circuits. Experiment with the materials you have. Be sure to take notes on your observations.

Develop Possible Solutions

Based on your research, sketch designs for the ideas that you have. Draw a schematic diagram showing how the electrical circuit of your system will be assembled. Make a detailed drawing of the solution you would like to model. Be sure your solution stays within your instructor's guidelines. Have your instructor approve your solution before you begin to model a solution.

Model a Solution

1. Select building materials to model your idea.
2. Construct the model.
3. Connect the electrical components.

Math Link

Unit Prefixes. As you research circuits and devices that can increase resistance, you may see references to unfamiliar electrical units. For example, you know about volts and amperes, but what are kilovolts and milliamperes?

Research the following unit prefixes and describe what they mean.

giga (G)	micro (µ)
mega (M)	nano (n)
kilo (k)	pico (p)
milli (m)	

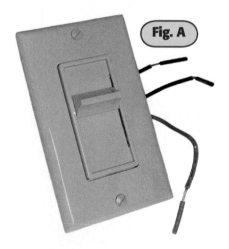

Fig. A

Materials and Equipment

Select from this list or use your own ideas.

- 9-volt battery
- 22-gauge stranded wire
- 1.5v, 3v, and 6v bulbs
- bulb holders
- paper fasteners
- metal paper clips
- aluminum foil
- switches
- photo resistors
- rheostats
- potentiometers
- thermistors
- cardboard
- foam core board
- cardstock
- wood
- metal
- electrical tape
- paint
- plastic
- solder
- soldering iron
- wire cutters/ strippers
- standard material processing tools and machines

For information about the tools listed here, see "Hand Tools" and "Power Tools and Machines" on the Student CD.

Fig. B

Test and Evaluate the Solution

Test your system. Then evaluate your system by asking yourself the following:
- Does your system work as planned?
- What type of circuit did you use?
- Did you have to create a switch?
- What changes can be made to improve the performance of your device?

Refine the Solution

How could the device be improved, based on the knowledge that you have from testing it? After asking the following questions, make any necessary changes to your device.
- If you used a gradual dimmer, would your device work better than switching directly from bright lights to dim lights?
- If you used a switch, would a gradual dimmer work better?
- Was the change in light too sudden? If so, then make adjustments to make the light change more gradual.

Communicate Your Ideas

Demonstrate the operation of your electronic device to the class. Identify the electrical components and describe how they make the circuit work. Be sure to point out any modifications that you made after testing.

Creating an LED Warning System

Identify a Need/Define the Problem

Design a new system or model an existing system that uses a flashing light as an indicator. For this activity, you will be using a special type of diode junction that emits light. These are called light-emitting diodes, or LEDs. The light must be mechanically activated. For example, it might be activated by pulling a lever or turning a crank.

Gather Information

Research mechanisms or simple machines that could be used to activate the light. Visit the interactive lab "Mechanisms" on the Student CD for ideas.

Look around your school, your home, and your community for devices that use flashing LEDs as indicators or warnings. See if you can identify a need for such a device in your school, home, or community.

Safety Alert

Look up "Safety Data Sheets" on the Student CD and prepare a data sheet for this activity. As you work on the activity, be sure to follow all safety rules.

Materials and Equipment

Select from this list or use your own ideas.

- flashing LEDs
- 9-volt battery
- ½-watt resistors
- 22-gauge stranded wire
- switches
- cardboard
- foam core board
- Styrofoam
- wood
- metal
- electrical tape
- paint
- plastic
- solder
- soldering iron
- wire cutters/ strippers
- standard material processing tools and machines

For information about tools, see "Hand Tools" and "Power Tools and Machines" on the Student CD.

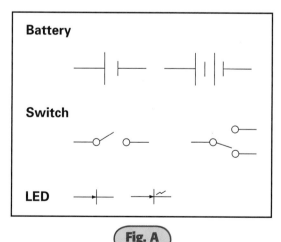

Battery

Switch

LED

Fig. A

Develop Possible Solutions

Sketch designs for the ideas that you have. Draw a schematic diagram showing how the electrical circuit of your system will be assembled. Different symbols you can use in the schematic are shown in **Fig. A**. Make a detailed drawing of the solution you would like to model. Be sure your solution stays within your instructor's guidelines.

Model a Solution

1. Select building materials to model your idea. See **Fig. B**.
2. Construct the model.
3. Install the electrical components.

Test and Evaluate the Solution

Test your system. Then evaluate your system by asking yourself the following:
• Does your LED flash as designed?
• Does your LED serve as an appropriate warning system?
• Does it matter which way the LED is placed in the circuit?

Refine the Solution

How could the device be improved, based on the knowledge that you have from testing it? Make necessary changes.

Communicate Your Ideas

Demonstrate the operation of your electronic device to the class. Be sure to point out any modifications that you made after testing.

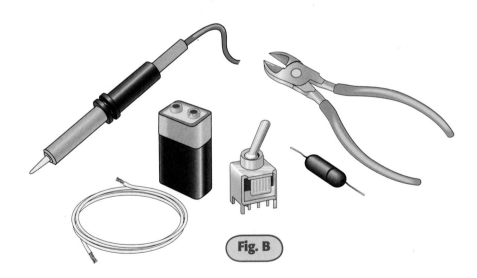

Fig. B

Making an Electromagnetic Crane

Identify a Need/Define the Problem

Electromagnetic cranes are used to move objects made from ferrous metals (metals that contain iron). They can also be used to sort ferrous and non-ferrous metals, since only the ferrous metals will be picked up by the electromagnet. Electromagnetic cranes are often used to pick up scrap metal. See **Fig. A**.

The purpose of this project is to create the strongest electromagnetic crane possible, using a limited amount of material.

Gather Information

How is an electromagnet created? What makes it stronger? See **Figs. B** and **C**. You'll also need to find out what mechanisms are used in the construction of a crane and how the crane is going to be balanced.

Materials and Equipment

Select from this list or use your own ideas.

- 9-volt battery
- 24-gauge solid wire with enamel insulation
- 22-gauge stranded wire
- #1 paper clips
- 6d 2-inch nails
- switches
- cardboard
- foam core board
- string
- wood
- metal
- electrical tape
- paint
- plastic
- solder
- soldering iron
- wire cutters/ strippers
- standard material processing tools and machines

For information about tools, see "Hand Tools" and "Power Tools and Machines" on the Student CD.

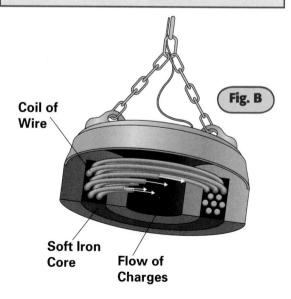

Fig. A

Fig. B

Coil of Wire

Soft Iron Core

Flow of Charges

The red arrows show the direction of the magnetic field.

An electromagnet is made of a coil of wire around a soft iron core.

Develop Possible Solutions

Sketch designs for the ideas that you have. Draw a schematic diagram showing how the electrical circuit of your system will be assembled. Make a detailed drawing of the solution you would like to model. Be sure your solution stays within your instructor's guidelines.

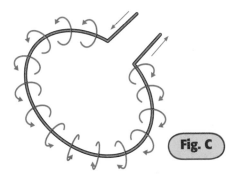

Fig. C

Model a Solution

1. Select building materials to model your idea.
2. Construct the model crane.
3. Install the electrical components.

Test and Evaluate the Solution

- Test the electromagnetic crane to see how many paper clips it can pick up. Does your system work as planned?
- What materials can it pick up, and which ones can it not pick up?
- What changes can be made to improve the performance of your device?

Refine the Solution

- Did your device perform as expected? If not, try using a different kind of circuit.
- Make any other necessary changes to your crane based on your test and evaluation.

Communicate Your Ideas

Demonstrate the operation of your electromagnetic crane to the class. Be sure to point out any modifications that you made after testing.

Science Link

Mooing Magnetics. As you learned in this activity, electromagnets only pick up ferrous metals. Electromagnets are often used to sort large piles of metals. Regular magnets work the same way and can be used for many applications. For example, cow magnets are strong magnets made out of alnico, an iron alloy. Ranchers feed these magnets to their cows. The magnet settles in the cow's first stomach. When the cow eats bits of steel or iron, the magnet attracts the metal bits and holds them in the first stomach. If the sharp pieces of metal were to pass through the cow, the animal would suffer what ranchers call "hardware disease."

Research other ways that magnets can be used. For example, how are electromagnets used in transportation?

Exploring Careers

Electrician

ENTRY LEVEL   **TECHNICAL**   PROFESSIONAL

Electricians install, connect, test, and maintain electrical systems. They connect the main power supply from the outside of a building to the circuit breakers inside the building.

On large projects, electricians refer to blueprints that show them where to install wiring. They use special connectors to join wires and then test the circuits using equipment such as voltmeters. Electricians must follow strict rules to make sure their finished work is safe and functions correctly.

Electricians install wiring in new construction and maintain electrical systems in existing buildings. More than 25% of electricians work at wiring new homes and buildings. About 10% of electricians are self-employed.

Qualifications

Electricians apply the principles of electronics, engineering and technology, and math to do their jobs. An electrician must have a good math and science background as well as good reasoning skills.

If you want to become an electrician, you should take high school courses in math, electricity and electronics, mechanical drawing, science, and technology. After high school, some electricians learn their trade through an apprenticeship program that lasts three to five years. Others take a three-year apprenticeship program through a union. After completing their apprenticeship, workers become journeymen electricians. They take courses throughout their career to learn about new technology.

Electricians must be licensed and pass a test on their knowledge of electrical theory, the National Electrical Code, and local electrical and building codes.

Because electricians match wires by color, they must have good color vision. They should also be in good physical health and be able to work outside and in small spaces.

Outlook for the Future

The job outlook for electricians is good. The population and economy will affect job growth.

Decision Making

As an electrician, you will need to make good decisions by gathering information and asking for advice from experts. Considering all of the consequences will help you make the best decision.

Researching Careers

Research jobs in this field. What does an apprentice electrician earn per hour? What do experienced electricians earn in Seattle and Atlanta? Write a report using what you learn.

More activities on Student CD

Key Points

- The movement of electrons creates electric charges.
- Voltage is the force that sets charges in motion.
- The flow of electrons in a conductor is called current.
- Ohm's law mathematically describes the relationship among voltage, current, and resistance.
- A circuit is the pathway through which electricity flows.
- Electronics involves the control of individual electrons to amplify, direct, or change electric current.
- Integrated circuits can contain millions of miniaturized electronic components.

Read & Respond

1. How are atoms responsible for electricity?
2. Why are valence electrons so important to electricity?
3. When walking across a carpet, what causes someone to gain electrons that can result in static electricity?
4. What do the letters *I*, *R*, and *E* stand for in Ohm's law?
5. How do engineers use Ohm's law?
6. Name three sources of electricity.
7. What is a generator?
8. What is the main difference between parallel circuits and series circuits?
9. Define *diode*.
10. What impact have integrated circuits had on electronic devices?

Think & Apply

1. **Construct.** Using dominos, make models to demonstrate how series circuits and parallel circuits work.
2. **Propose.** Describe the components you would need to design an electronic fan that will turn on when the room temperature reaches 88°F.

3. **Summarize.** Now that you know more about atoms, draw two atoms. Label the different particles and show the movement of electrons from one atom to the other.
4. **Assess.** Write a report on how electronics have affected you over your lifetime.
5. **Extend.** Give an example of how Ohm's law is used in the design of a toaster.

TechByte

Paper Time. Did you know that electronic-paper wristwatches are now possible? Electronic paper is made of semiconductor plastic in which diodes, transistors, and other components of computer chips are etched. Unlike traditional computer displays, it is flexible. An e-paper watch is almost paper-thin. The watch face displays time and other data, while the band changes texture and color to suit the wearer's whim!

CHAPTER 15

Lasers & Lights

Objectives

- Explain how light travels.
- Describe four light sources.
- Name four ways lasers are different from other light sources.
- Describe how a laser works.
- Identify uses for lasers in communication, manufacturing, construction, medicine, and business.

Vocabulary

- **electromagnetic wave**
- **phosphor**
- **light-emitting diode (LED)**
- **laser**
- **monochromatic light**
- **directional light**
- **coherent light**
- **fiber optics**
- **holography**

This laser light show is being displayed on the side of Stone Mountain, near Atlanta, Georgia.

Activities

- Communicating with Light
- Shine a Little Natural Light
- Creating a Fiber-Optic Monogram

The Nature of Light

Have you ever thought about how important light is? Light is all around us. It enables us to see. It creates rainbows and makes it possible for us to enjoy the colors created by nature and technology. Without light from the sun, there would be no life on Earth.

Light is a form of energy. It travels in waves, much like waves of water. Unlike those waves, however, light waves can travel through space where no matter is present.

Light is a type of electromagnetic wave. An **electromagnetic wave** is a wave produced by the motion of electrically charged particles. Scientists often refer to these waves as electromagnetic radiation because they radiate outward from their source. Visible light waves are one form of electromagnetic radiation. Other forms include radio waves, X rays, and gamma rays. Electromagnetic waves are classified by their length. See **Fig. 15-1**.

The color of visible light is determined by its wavelength. Red light has the longest wavelength. Violet light has the shortest. Though it looks white, ordinary light, such as light from the sun, is a mixture of many colors. Another important quality of ordinary light is that its rays travel in many different directions.

 Learn more about electromagnetic waves in the interactive lab "Electromagnetic Spectrum" on the Student CD.

Fig. 15-1. Electromagnetic waves are classified by wavelength on the electromagnetic spectrum. Only a small portion is visible light.

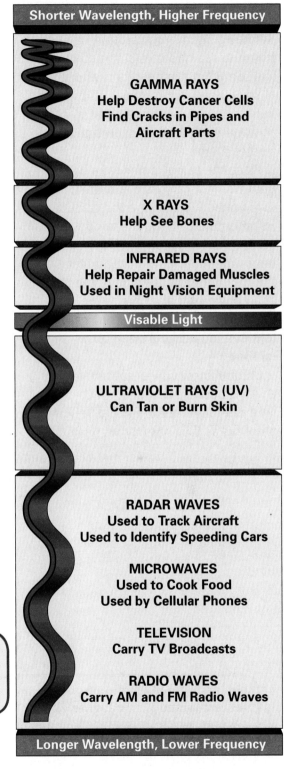

Shorter Wavelength, Higher Frequency

GAMMA RAYS
Help Destroy Cancer Cells
Find Cracks in Pipes and Aircraft Parts

X RAYS
Help See Bones

INFRARED RAYS
Help Repair Damaged Muscles
Used in Night Vision Equipment

Visable Light

ULTRAVIOLET RAYS (UV)
Can Tan or Burn Skin

RADAR WAVES
Used to Track Aircraft
Used to Identify Speeding Cars

MICROWAVES
Used to Cook Food
Used by Cellular Phones

TELEVISION
Carry TV Broadcasts

RADIO WAVES
Carry AM and FM Radio Waves

Longer Wavelength, Lower Frequency

Splitting Light. You can split white light from the sun into colors. The colors you will see will be colors of a rainbow.

You'll need a rectangular baking pan. You'll also need water, a small piece of white cardboard, and a rectangular pocket mirror.

Place the baking pan near a window where it can catch sunlight. Fill it with water. Prop or tape the cardboard between the pan and the window. It will serve as a viewing screen for the rainbow. Place the mirror in the water. Lean the mirror against the side of the pan at an angle so that it reflects the sunlight onto the cardboard. You may have to move the mirror a bit to obtain a rainbow.

Do you know why a rainbow forms? It forms because the different colors of light in sunlight are bent by different amounts as they pass through water.

Light Sources

Engineers continue to work to find new and more efficient light sources. Let's take a closer look at some of the light sources we now have because of technology.

Incandescent Lamps

Most of the lighting in our homes is provided by incandescent bulbs. The basic technology they use has changed very little in the past 100 years.

Take a look at an electric lightbulb. See **Fig. 15-2**. If the glass is clear, you'll be able to see a thin wire inside. This thin wire is called a filament. This filament heats up as electricity flows through it. The filament then gives off photons, which produce the light. This glowing, or "incandescing," is what we see when we look at a lamp that is turned on.

Incandescent bulbs are inexpensive, easy to use, and come in many different sizes and shapes. A major disadvantage is that they radiate a lot of heat along with the light. This heat is wasted energy.

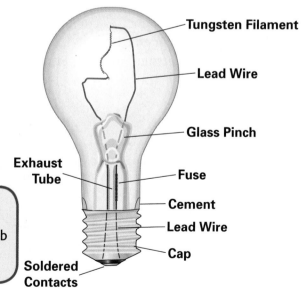

Tungsten Filament

Lead Wire

Glass Pinch

Exhaust Tube

Fuse

Cement

Lead Wire

Cap

Soldered Contacts

Fig. 15-2. The thin filament in an incandescent lightbulb is what gives off the photons, which produce the light. The bulb is hot because the filament heats up as electricity flows through it.

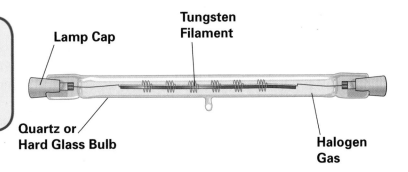

Fig. 15-3. Halogen lightbulbs are filled with halogen gas and are very bright when lit. They also will get very hot and can be a fire hazard if not handled correctly.

Lamp Cap

Tungsten Filament

Quartz or Hard Glass Bulb

Halogen Gas

A halogen lamp is a special kind of incandescent lamp. Halogen lamps use quartz or hard glass bulbs. These bulbs are filled with a halogen gas such as iodine or bromine. (Regular bulbs are filled with nitrogen or argon.) Halogen lamps produce a large quantity of very bright light and last a long time. See **Fig. 15-3**. A disadvantage is that they get very hot during operation. Many halogen lamps have safety switches to help prevent fires. These switches turn the bulb off if the lamp is tipped over.

Fluorescent Lamps

We see fluorescent lamps in schools, stores, and in some homes. Fluorescent lamps use a sealed glass tube filled with a small amount of mercury and an inert gas such as argon. The inside of the tube is coated with a phosphor. **Phosphors** are substances that give off photons when charged with electric current. When you turn on a fluorescent lamp, electrons flow through the tube and some of the mercury inside changes from a liquid to a gas. Photons are released, giving off light. See **Fig. 15-4**.

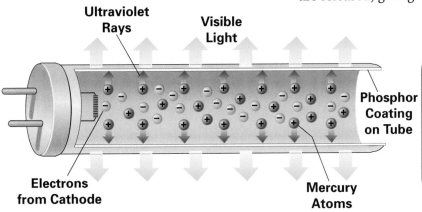

Ultraviolet Rays

Visible Light

Phosphor Coating on Tube

Electrons from Cathode

Mercury Atoms

Fig. 15-4. The electrons from the cathode in a fluorescent lamp cause mercury atoms to emit ultra-violet radiation. The phosphor coating absorbs this to create visible light.

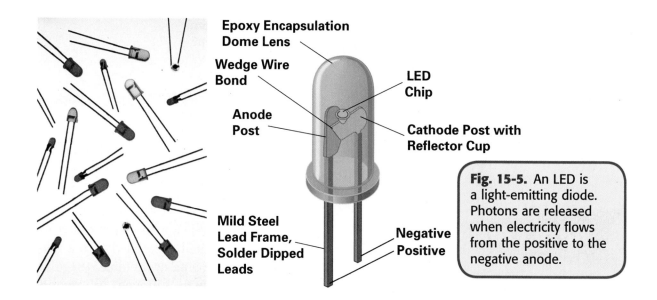

Epoxy Encapsulation Dome Lens

Wedge Wire Bond

Anode Post

Mild Steel Lead Frame, Solder Dipped Leads

LED Chip

Cathode Post with Reflector Cup

Negative
Positive

Fig. 15-5. An LED is a light-emitting diode. Photons are released when electricity flows from the positive to the negative anode.

Fluorescent lights now come in many sizes and shapes. Compact fluorescent lamps can be used to replace standard incandescent bulbs. Although the compact lamps cost more to purchase, they are cheaper to operate because most of the energy they use is converted to light. A 17-watt compact fluorescent lamp will produce as much light as a standard 60-watt incandescent bulb.

LEDs

Light-emitting diodes (**LEDs**) are much different than other lights. They are common electronic components that emit light when connected in a circuit. They allow electricity to flow in only one direction. Inside the LED, electricity flows from the cathode (positive lead) to the anode (negative lead). In the process, photons are released. See **Fig. 15-5**.

LEDs were originally used just to show that electronic devices were turned on. Looking around, you can probably find a number of LEDs in your home. Recent research and development has led to many new applications for LEDs. Solar yard lights, watch displays, exit signs in public buildings, and traffic signals all use LEDs.

When they are used in traffic signals, many LEDs are grouped together in an arrangement called an array. LED arrays have replaced incandescent bulbs in traffic lights. Money is saved because they last for years and use much less energy. See **Fig. 15-6**.

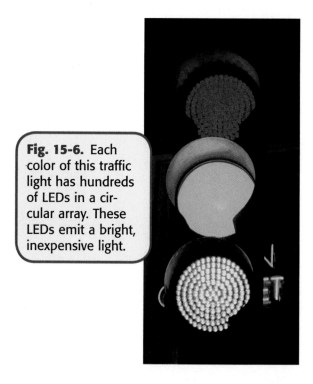

Fig. 15-6. Each color of this traffic light has hundreds of LEDs in a circular array. These LEDs emit a bright, inexpensive light.

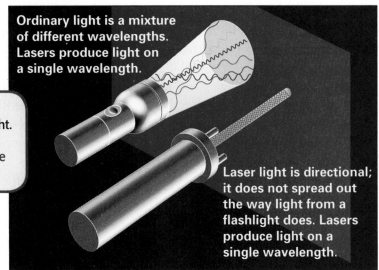

Fig. 15-7. Shown here is laser light compared with ordinary light. Note how one light spreads out and the other does not. What are the advantages of each type?

Ordinary light is a mixture of different wavelengths. Lasers produce light on a single wavelength.

Laser light is directional; it does not spread out the way light from a flashlight does. Lasers produce light on a single wavelength.

Lasers

The **laser** is a light source that sends out light in a narrow and very strong beam. The first laser was built in 1960 by Theodore Maiman. He used a synthetic ruby in his laser. The ruby produced a laser beam when an intense beam of ordinary light was flashed on it. Maiman did not know he had created one of the most important devices of the century.

Why are lasers so important? Laser light is different from other light. Laser light is
• Monochromatic
• Directional
• Coherent
• Bright

Monochromatic light is light that consists of only one color. For instance, most of the laser lights you see are red.

Directional light is light that spreads out very little compared to ordinary light. Because it does not spread out as much, it can be focused on a small spot.

Coherent light is light in which all of the light waves have the same wavelength. They are also "in phase." Waves that are in phase have their peaks and valleys aligned. Waves of laser light are like the members of a marching band, where everyone moves in step. In contrast, waves of ordinary light are incoherent. The waves of incoherent light are like many people leaving a store and walking in different directions. See **Fig. 15-7**.

Reading Link

Acronyms. An acronym is a term made from parts of a series of words. The acronym may be made from the first letters of the series, as in LED (light-emitting diode). Some acronyms are made using the first letters of only the most important words in the series. An example is laser (light amplification by stimulated emission of radiation). If that acronym used letters from all the words, it would be "labseor."

Find out what words were used to make the following acronyms. Hint: They all relate to the electromagnetic spectrum.

AM	ULF
ELF	UV
FM	VLF

Finally, laser light is bright. The light waves from a laser work together to produce a high-energy beam that is brighter than any ordinary light source. A 60-watt lightbulb produces enough light for reading. A 60-watt laser is many times brighter.

Laser stands for **l**ight **a**mplification by **s**timulated **e**mission of **r**adiation. Try to remember this phrase. It helps explain how a laser works.

How Lasers Work

The laser is a system designed to produce a special kind of light. See **Fig. 15-8**. Its subsystems include the following:
• Excitation mechanism
• Active medium
• Feedback mechanism

All lasers need an energy source. This energy comes from the excitation mechanism of the laser. The energy can be produced by electricity, a chemical reaction, or by light from another source.

The active medium changes the energy to light and amplifies (strengthens) it. The active medium is a material that can absorb and release energy. It can be a solid, a liquid, a gas, or a semiconductor.

The feedback mechanism usually consists of a mirror placed at each end of the active medium. The mirrors are used to build the strength of the laser beam. One mirror is made to allow some of the light to escape the active medium. This mirror is an output coupler. The light that is released is the laser beam.

Since light travels at 186,000 miles per second, a laser process occurs in barely a fraction of a second. The process is shown in **Fig. 15-9**.

Laser Safety

Lasers are safe when used properly. Follow the safety guidelines included with any laser you use. Check with your teacher if you have questions.

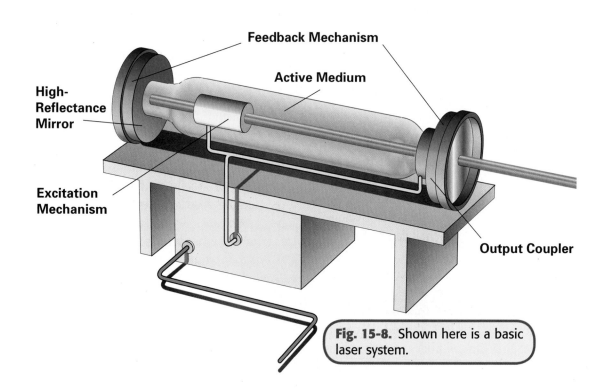

Feedback Mechanism

Active Medium

High-Reflectance Mirror

Excitation Mechanism

Output Coupler

Fig. 15-8. Shown here is a basic laser system.

A. The excitation mechanism pumps energy into the active medium. This releases a small burst of light.

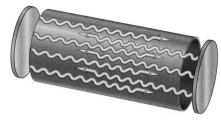

C. Light bounces off mirrors and returns to the active medium, where it is further amplified.

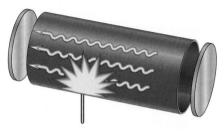

B. The light picks up more energy from the active medium.

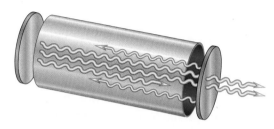

D. Some of the light is released by the output coupler. This light is the laser beam. Other light is bounced back to the active medium, where it is amplified.

Fig. 15-9. The laser process releases energy as a burst of light.

Lasers are grouped into four classes according to the hazard they present. Class I lasers produce no known hazard to people. The lasers used in checkout scanners, laser printers, and CD or DVD players are Class I lasers.

Class II lasers can cause eye damage if not used properly. Lasers used in technology education classes are Class II lasers.

Class III lasers produce a powerful beam that can damage the eyes. Special glasses must be worn when using Class III lasers. Class III lasers might include those used in office scanners or in light shows. Do not stare at these lasers.

Class IV lasers are more powerful than Class III lasers. They can burn the skin. Special glasses are also required with Class IV lasers. These lasers are typically used in cutting, welding, and surgery.

Writing Link

Similes. Writing can be made more interesting and informative through the use of figures of speech.

"Waves of laser light are like the members of a marching band, where everyone moves in step."

"The waves of incoherent light are like many people leaving a store and walking in different directions."

These comparisons are called similes (SIM-uh-lees). A simile is a comparison using "like" or "as." Similes can help us understand ideas more clearly. Write a simile for each of the following:
LED
monochromatic light
fiber optics
UPC

Types of Lasers

The light that comes from a laser can be continuous or released in short pulses. Laser pulses are powerful because their energy is very concentrated.

The strength of a laser is measured in watts. The laser you use in technology education class produces less than $\frac{1}{1000}$ of a watt. Some lasers being used for research can produce millions of watts of energy. These lasers fill an entire building.

Lasers are usually named after their active medium. For example, a solid piece of ruby is the active medium in a ruby laser. Carbon dioxide gas is the active medium in a carbon dioxide laser. Semiconductor lasers use electronic components as their active medium. The helium-neon (HeNe) laser uses a mixture of the two gases. This is the type of laser that is usually used in technology education classes.

Laser Applications

Engineers are finding new uses for existing lasers. They are also working to develop new kinds of laser systems. Lasers are now used in many different applications.

Lasers in Communication

The most important use of lasers in communication is in fiber optics. **Fiber optics** are thin filaments of glass through which light travels. Light pulses are coded as text, audio, and video information.

In a fiber-optic telephone system, a laser changes sound input into a series of light pulses, or bits. See **Fig. 15-10**. They travel along optical fiber at 90 million bits per second. Along their route, the pulses are electronically amplified. When the pulses reach a receiver, they are changed back into sound.

Laser fiber-optic systems are replacing conventional metallic-cable communication systems. They are less expensive and more efficient. Fiber optics are also less subject to static.

Lasers are also used in holography. **Holography** is a photographic process. It uses a laser as well as lenses and mirrors to produce three-dimensional images. These images are holograms.

Holograms are used in art, advertising, business, and industry. Most credit cards now have embossed holograms on them. Such cards are harder to counterfeit.

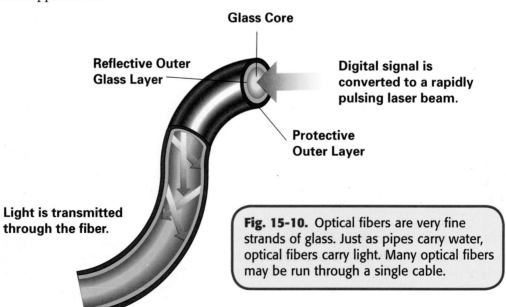

Glass Core

Reflective Outer Glass Layer

Digital signal is converted to a rapidly pulsing laser beam.

Protective Outer Layer

Light is transmitted through the fiber.

Fig. 15-10. Optical fibers are very fine strands of glass. Just as pipes carry water, optical fibers carry light. Many optical fibers may be run through a single cable.

Fig. 15-11. Laser cutters can cut quickly and very precisely. This industrial laser is cutting a block of stainless steel.

Lasers in Manufacturing

Lasers are being used in a wide range of manufacturing applications. They do a good job of cutting many materials, such as steel, plastic, fabric, and wood. In the garment industry, they are used to cut out the pattern pieces for a garment. Lasers can cut through hundreds of thicknesses of cloth at once.

Lasers can cut most metals, even steel that is more than an inch thick. Laser cutting is more accurate than most other methods and the cut edges are clean and smooth. See **Fig. 15-11**.

Lasers can also be used for welding. The automobile industry makes extensive use of carbon dioxide lasers for welding vehicle parts on assembly lines. Another application of laser welding is to join different metals together, such as attaching steel handles to copper cooking pots.

Lasers can drill holes in many materials, such as rubber, wood, and diamonds. They can drill holes more smoothly than standard drill bits can. Laser-drilled items include aerosol spray nozzles and contact lenses.

Tabletop laser systems are now available for cutting and engraving in technology education classrooms. They can produce detailed items and engrave wood, fabrics, and plastic plaques. These systems attach to a computer like a printer does and can duplicate designs created using CAD and graphic design software. See **Fig. 15-12**. For more about CAD systems, see Chapter 2, "Computer-Aided Design."

Lasers in Construction

Surveying is measuring the boundaries of a piece of land. A laser beam can be used as a straight line in surveying. Distances are measured by timing a light pulse from the laser to a mirror and back to a detector near the laser. Laser beams are also being used to align water and sewer pipes and tunnels.

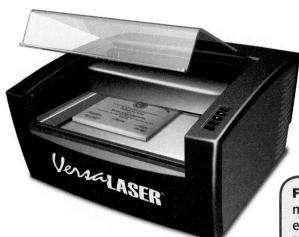

Universal Laser Systems, Inc. manufactures VersaLaser™ as well as a wide selection of laser systems and CO_2 lasers.

Fig. 15-12. Compact lasers can engrave many materials. This model is used for engraving wood plaques, signs, and labels. It operates much like a computer printer.

Fig. 15-13. The laser receiver on this bulldozer will raise and lower the blade to keep it on grade. The receiver accepts the laser beam that has been set at the desired level of the grade.

Lasers can guide the equipment used to level a construction site. See **Fig. 15-13**. For example, a rotating laser sends out a signal that is picked up by a receiver on a bulldozer. This signal contains information that the land level is low, high, or on grade (level). The blade of the bulldozer is automatically adjusted by the laser. Similar systems are used to prepare farmland for planting.

Lasers Used in Health Care

Lasers have changed the way many traditional medical procedures are done. Laser surgery reduces blood loss, risk of infection, and patient discomfort.

Lasers are very useful for eye surgery. They cut more accurately than a scalpel. There is also less damage to nearby tissue. The surgery is often done in the doctor's office. The patient can sometimes even resume activities the same day. **See Fig. 15-14**.

Lasers and fiber optics are used together for some kinds of surgery. Optical fibers are inserted through a natural opening or an incision in the body. Light is transmitted through one fiber and received through another. That way the surgeon can see the problem and work on correcting it. For example, the surgeon can use a laser to cut out diseased tissue and seal nearby blood vessels to reduce bleeding.

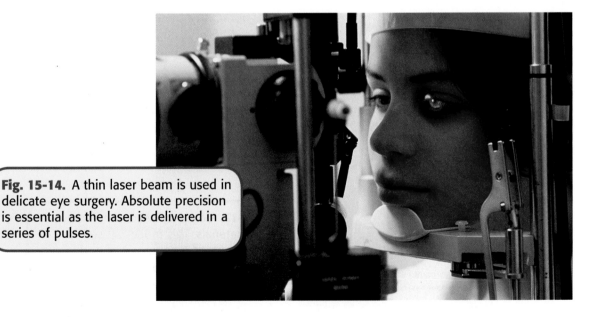

Fig. 15-14. A thin laser beam is used in delicate eye surgery. Absolute precision is essential as the laser is delivered in a series of pulses.

Impact of Technology

Laser Surgery

You have read in this chapter how lasers may be used for surgery. An important advantage of laser surgery is that less time is needed for recovery and/or rehabilitation (such as physical therapy for knee surgery). Because smaller incisions (cuts) are made, there is less damage to surrounding bone and tissue. There is also less bleeding and less risk of a secondary infection. A secondary infection is one that results from a medical procedure and not from the original problem or illness.

Many laser surgeries can be done on an outpatient basis. This means the patient can go home the same day rather than staying in a hospital.

Investigating the Impact

1. Find out whether there are disadvantages or risks to laser surgery. What problems may occur when lasers are used for medical procedures?
2. If the use of lasers for vision correction continues to improve, what may happen to vision health providers in the future?

Lasers Used in Business

Your local grocery or retail store is where you are most likely to see a laser in action. See **Fig. 15-15**. Most of the items sold there have a Universal Product Code (UPC) printed on the package. It consists of a bar code, which looks like parallel lines of varying widths.

How does a UPC work? A laser scanner at the checkout counter reads each symbol. As each item is moved across the scanner, information from the item's bar code is sent to a computer.

Fig. 15-15. This laser is used as a price scanner in the store. The laser will read the UPC so the customer knows the cost of the item before checking out.

The computer identifies the item. It then signals the cash register to print the name and price of the item. At the same time, the information is used to update the store's inventory. This helps managers pinpoint when it is time to reorder. For more on UPCs, see Chapter 5, "Manufacturing Technologies."

Many offices now have laser printers. Laser printers use a low-power laser to form images on a rotating drum. Powdered ink sticks to the image formed on the drum. This image is transferred to paper. Laser printers print quickly and quietly. They produce high-quality copies.

Lasers in the Home

Laser printers are not only used in offices. Many homes now also have laser printers.

Affordable laser levels are also now available for home use. They are easy to use and can help with various tasks. You can make sure that fence posts are all aligned or check to see if picture frames are hung straight. See **Fig. 15-16**.

Compact discs (CDs) are the most common way to record and listen to music. Every CD player has a tiny laser inside it. The laser reads digital code from tiny pits on the CD's surface. (Digital code, also called binary code, is discussed on the Student CD.) A computer inside the CD player turns the code into music. See **Fig. 15-17**.

Digital video discs (DVDs) are also played using lasers. A DVD player works much like a CD player. A DVD has smaller pits than a CD, so DVDs contain more information.

Most DVD players use a red laser. However, DVD players may soon start using blue lasers. Blue lasers can read tinier pits. More pits can be added to the DVDs, increasing storage space. A disc that could once only contain three hours of video might contain thirteen hours instead.

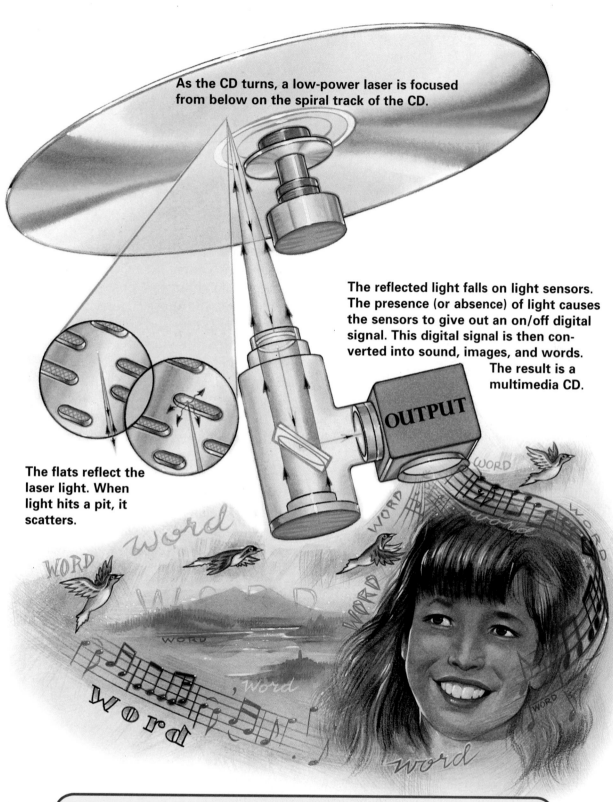

As the CD turns, a low-power laser is focused from below on the spiral track of the CD.

The reflected light falls on light sensors. The presence (or absence) of light causes the sensors to give out an on/off digital signal. This digital signal is then converted into sound, images, and words. The result is a multimedia CD.

The flats reflect the laser light. When light hits a pit, it scatters.

OUTPUT

Fig. 15-17. Compact discs hold information in a spiral track. When you play a CD, a laser beam reads the track to reproduce sounds, photos, or text.

Communicating with Light

Identify a Need/Define the Problem

You are part of an expedition to uncharted territory where the technological communication devices, such as radio, telephones, and the Internet, do not work. Your team needs to develop a communication system that works off a flashlight in order to relay information to a plane that flies overhead once a week to check up on you. (Assume that the flashlight used in the expedition will be strong enough for someone in an airplane to see.)

The communication system must be able to transmit basic information. This information should include the following items:

- The status of the expedition, good or bad.
- Any supplies that are needed, such as food, water, gear, or medical attention.
- The end date of the expedition.

Gather Information

Light systems can be used to communicate many kinds of information. The stoplight shown in **Fig. A** is one example. Research the ways light can be used in sending and receiving messages. For example, can color be used? Can light intensity or duration help convey your message?

Develop Possible Solutions

Brainstorm ideas on different systems of communication with light in your group. Draw sketches of anything that needs to be constructed. Record all of your ideas in a portfolio.

Math Link

Coded Communication. When deciding how to communicate, you may assign different signals for specific words or phrases. The Morse code works the same way. Letters and numbers are represented by a unique pattern of dots and dashes. The code was patented by Samuel F.B. Morse in 1840 and is still being used today.

Did you know the Morse code we are familiar with was not Morse's original idea? Morse's original idea was to use a unique number for each word, and then use a code to send the numbers. Take the information you want to send to the plane in your activity. Assign a number to each word. How many different numbers would you need to send a simple message?

Safety Alert

Look up "Safety Data Sheets" on the Student CD and prepare a data sheet for this activity. As you work on the activity, be sure to follow all safety rules.

Materials and Equipment

Select from this list or use your own ideas.

- aluminum foil
- assorted colored light gels
- cardboard
- cardstock
- clear plastic rod
- two flashlights
- foam core board
- glue
- paper towel tubes
- plastic mirrors
- prism
- ruler
- scissors
- stopwatch
- tape

 If you need help with measuring, see "Measurement" on the Student CD.

Model a Solution

1. Create the device that you feel will communicate most effectively. This will be the ground team's device.
2. Create a duplicate device. This will be the airplane team's device.

Test and Evaluate the Solution

- Divide your team into two groups. Decide which team will be the ground team and which will be the airplane team.
- Position each team on the opposite end of the room or on the other side of a window. The teams cannot communicate by talking.
- The ground team will start the conversation by relaying a message.
- The plane team should respond.
- How long can the conversation continue?
- How easy is it to communicate?

Refine the Solution

- How could the system be improved based on the testing that you have completed?
- If necessary, make any changes and redo the test. Remember to meet the activity's criteria and constraints.

Communicate Your Ideas

Create a portfolio of all of the work that you have completed. Demonstrate your completed system. Be sure to point out any refinements you made to your original device.

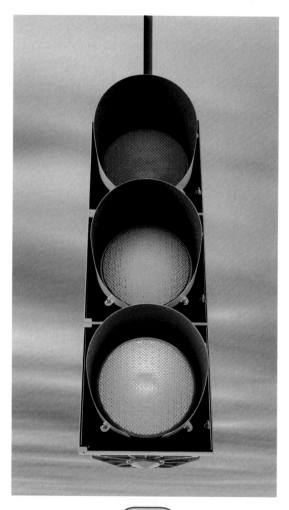

Fig. A

Shine a Little Natural Light

Identify a Need/Define the Problem

Many homes do not have windows in basements or closets. Electrical lights that can be installed cost money when used.

A homeowner has contacted you, a light expert, to design and build a system that will transfer natural light to these hard-to-get-to places through small windows on the roof of the house. See **Figs. A** and **B** for examples. When making your model, the system must be at least two feet long and have two turns in it.

Gather Information

Research the methods for transferring light. What materials can be used to do this? Research what obstacles can hinder your system so that you can plan accordingly.

Develop Possible Solutions

Create sketches of at least three different solutions for the problem. Be sure to label the materials that are used and the dimensions of the parts.

Model a Solution

1. Choose the solution that you feel will work the best.
2. Construct the solution according to the plans you have created.

Materials and Equipment

Select from this list or use your own ideas.

- plastic mirrors
- clear plastic rod
- aluminum foil
- cardstock
- cardboard
- paper towel tubes
- foam core board
- tape
- scissors
- glue
- ruler

Science Link

Light Intensity. Would you agree that the intensity (brightness) of natural light varies with the distance from the source? Try to answer the following questions.

1. If this statement is true, why is this so?
2. How could an experiment be set up to actually measure the light at any particular distance?
3. How could this information influence your design of the natural light system?

Safety Alert

Look up "Safety Data Sheets" on the Student CD and prepare a data sheet for this activity. As you work on the activity, be sure to follow all safety rules.

Test and Evaluate the Solution

- Set up your system to allow natural light to shine through it.
- Does the light flow through the entire system?
- Does the light flow through some areas better than others?
- Is the light as bright at the end as it is at the beginning? If not, then why?

Refine the Solution

How could the system be improved based on the testing that you have completed? Make any necessary changes.

Communicate Your Ideas

Create a portfolio of all of the work that you have completed. Share your completed system with the class. Be sure to include any refinements you made to your original design.

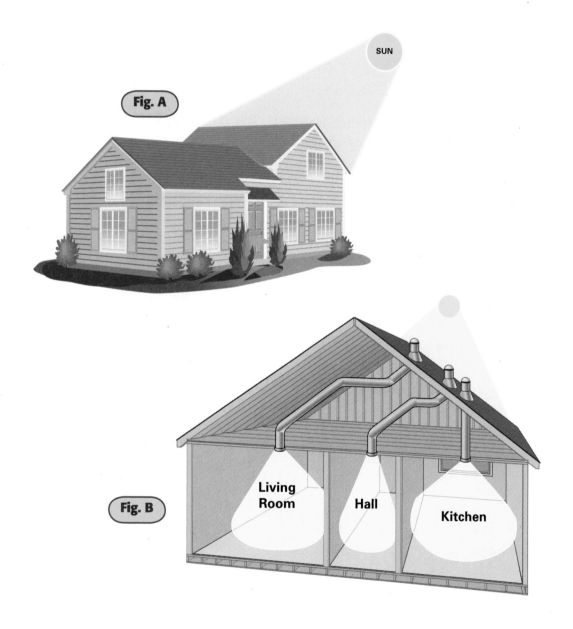

Creating a Fiber-Optic Monogram

Identify a Need/Define the Problem

In this activity, you will design and build a lighted monogram, using the principles of fiber-optic light transmission. Your monogram will be only one letter, such as the first letter of your name.

Gather Information

Research the uses of monograms. How is a single light going to be used to create the monogram?

Develop Possible Solutions

Create a series of sketches for the letter. Avoid tight bends and multiple pieces. Develop a method to secure the letter to a base or wall plaque.

Safety Alert

Look up "Safety Data Sheets" on the Student CD and prepare a data sheet for this activity. As you work on the activity, be sure to follow all safety rules.

Materials and Equipment

Select from this list or use your own ideas.

Monogram
- plastic rod, round ¼" to ½" in diameter
- plastic solvent

Light source
- two "D" cells
- lamp and socket
- cardboard tube

Mounting plaque
- ¾" wood
- ¼" dowels

Bending form
- ¾" wood
- ¼" dowels

Equipment
- plastic material processing equipment
- plastic ovens

Model a Solution

1. Using graph paper, prepare a full-scale drawing of the monogram. This drawing will be the pattern on which you will shape the plastic rod.
2. Tape the graph paper to a ¾" piece of plywood. Place the dowel pins (¼") at strategic locations where the rod has to be bent.
3. Using string or a flexible ruler, calculate the length of rod you will need for the monogram.
4. Heat the rod to the proper temperature.
5. Using gloves, remove the now flexible rod from the oven. Bend it around the form you have created. Hold it in place until it cools. See **Fig. A**.
6. Mount the monogram to the stand or plaque. See **Fig. B**.
7. Install the light source.

BENDING FORM

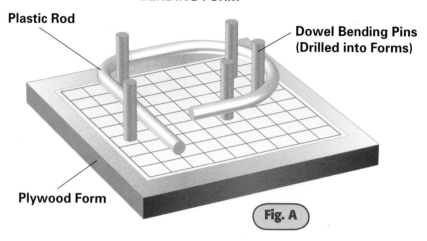

Plastic Rod

Dowel Bending Pins
(Drilled into Forms)

Plywood Form

Fig. A

Test and Evaluate the Solution

- Turn off the lights. Can you see your monogram clearly?
- Does the light flow through the entire plastic tube? Does the light flow through some areas better than others?
- Can your classmates recognize the letter you made?

Refine the Solution

- In what ways could the monogram be improved?
- Would the diameter of the dowel have any impact on how well it works?
- Do you think a brighter bulb makes a difference?
- Make any necessary changes to your monogram. Test and evaluate your improved design.

Communicate Your Ideas

Create a portfolio of all of the work that you have completed. Share your completed monogram with the class. Be sure to include any changes you made after testing and evaluating your monogram.

MONOGRAM DISPLAY

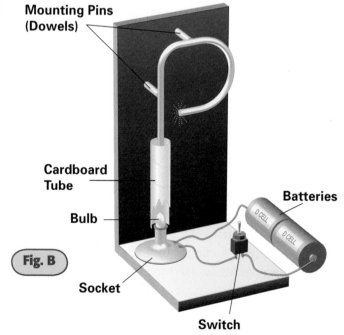

Mounting Pins
(Dowels)

Cardboard
Tube

Batteries

Bulb

Fig. B

Socket

Switch

Exploring Careers

Laser Technician

ENTRY LEVEL | **TECHNICAL** | PROFESSIONAL

Engineers and scientists develop new ways to use lasers. Laser technicians assist the engineers and scientists by building, installing, repairing, and testing laser equipment. Laser technicians also work with fiber optics and laser systems.

Laser technicians use computers in their work and run the controls that fill the laser device with the correct amount of gas. Technicians install optical parts and assemble the parts that make up the laser. They set up electronic and optical equipment to test the laser devices. Laser technicians keep detailed records of their work and must perform the work safely.

Laser technicians also work in the telecommunications field and as field representatives. In the field, they install lasers for the customers at their site. They make adjustments and demonstrate to the customer how the laser equipment works. Technicians also repair and service the equipment.

Qualifications

A laser technician must have a good math and science background as well as good problem-solving skills. An associate's degree in electronics or a laser technology certificate is required. Additional technical courses might include vacuum technology and optics. Useful high school classes include electricity and electronics, advanced math, and computer science.

Laser technicians need to have good vision and the ability to do detailed hand work. Field service representatives need to have good communication skills.

Outlook for the Future

The job outlook for laser technicians is good. The number of available jobs will depend on the economy and the field of laser technology. The more skilled and educated technicians will get the better jobs.

Positive Attitude

In any job, a positive attitude is important. Instead of complaining about a problem, you can suggest solutions to fix the problem. Learning from problems is useful. Keeping a positive outlook is a good way to stay motivated and to do a good job.

Researching Careers

Find out about jobs for laser technicians by looking at want ads in the newspaper or on the Internet. What education is required? What experience? What work would they do? Make a poster about what you find.

**More activities
on Student CD**

Key Points

● Light is a type of electromagnetic wave.

● Incandescent lamps, fluorescent lamps, LEDs, and lasers are all sources of light.

● A laser's subsystems include an excitation mechanism, an active medium, and a feedback mechanism.

● Laser light differs from ordinary light.

● Lasers are grouped into four classes according to the hazard they present.

● Lasers are usually named after their active medium.

● Lasers are used in communication, manufacturing, construction, medicine, business, and the home.

Read & Respond

1. How does light travel?
2. Explain how incandescent lights work.
3. What creates photons in a fluorescent bulb?
4. Why are LEDs being used to replace other kinds of lighting?
5. Name four ways lasers are different from other light sources.
6. What type of laser is usually used in class-rooms, and what is its safety classification?
7. Describe the function of each subsystem of a laser.
8. What is the most important use of lasers in communication?
9. Describe at least three ways that lasers are being used in manufacturing.
10. Identify three advantages of laser surgery.

Think & Apply

1. **Assess.** How do laser scanners help to increase the efficiency of supermarkets?
2. **Connect.** Each chapter in this textbook discusses a different area of technology.

How can lasers be used in areas such as robotics, forensics, materials science, and graphic communications?

3. **Propose.** What other ways could lasers be used in the home besides those discussed in this chapter?
4. **Construct.** Create a simple alarm system using fiber optics.
5. **Classify.** Find lasers in your house or around your school. What do you think is the safety class of each laser?

TechByte

UPLs. Ultrashort-pulse lasers (UPLs) generate pulses of laser light lasting only a few quadrillionths of a second. Invented in the 1960s, UPLs have only recently been miniaturized enough to be useful in practical applications. Because UPLs concentrate power in such tiny bursts, they are extremely precise surgical tools.

CHAPTER 16

Hydraulics & Pneumatics

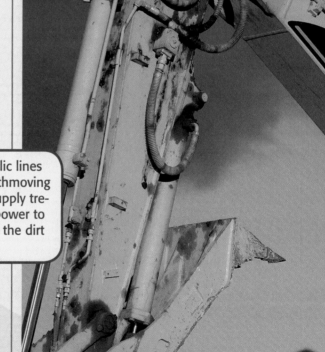

Objectives

- Define *fluid power*.
- Explain the difference between hydraulic and pneumatic systems.
- Identify the basic components of hydraulic and pneumatic systems.
- Give examples of how hydraulic and pneumatic systems are used.

Vocabulary

- **fluid power**
- **hydraulic system**
- **pneumatic system**
- **Boyle's law**
- **pressure**
- **Pascal's principle**

The hydraulic lines on this earthmoving machine supply tremendous power to the jaws of the dirt bucket.

Activities

- Making an Air-Cushion Vehicle
- Creating a Hydraulic Jack-in-the-Box
- Mixing with Pneumatics

Fluid Power

Fluid power is the use of liquids or gases under pressure to move objects or perform other tasks. A fluid is any substance that flows. Liquids and gases are both fluids. When they are not moving, fluids have no power. Fluids can be put under pressure and moved through pipes or hoses to where they are needed. Fluids are extremely helpful when attached to an appropriate machine. See **Fig. 16-1**.

People have used fluid power for thousands of years. Windmills have been used to turn millstones to grind grain. Water from rivers supplied the power for many manufacturing plants. Many of the devices that make our life easier depend on fluid power.

In this chapter you will learn about two types of fluid power systems: hydraulic (high-DRAW-lick) and pneumatic (new-MAT-ick). **Hydraulic systems** use oil or another liquid. **Pneumatic**

Fig. 16-1. Hydraulic-operated pistons transfer the power of the system to the various components of the backhoe.

Dipper Ram

Digger Bucket Ram

Loader Lift Ram

Boom Lift Ram

Stabilizer Ram

systems are fluid power systems based on the use of air or another gas. See the "Fluid Power" interactive lab on the Student CD for an overview.

Fluid power is one of three basic systems used to transmit and control power. The other two systems are mechanical and electrical. Mechanical power moves automobiles and other vehicles. Electrical power gives us light and operates motors.

Fluid Science

All objects are made of matter. There are three states of matter: solid, liquid, and gas. The state in which matter exists depends on how tightly its molecules are held together. Solids have molecules that are strongly linked. The molecules of liquids are loosely held together. Gas molecules are even less tight and can move in all directions.

Because of the space between their molecules, gases are easy to compress. Solids and liquids are not. Solids have a definite shape

and occupy a certain amount of space (volume). Liquids have a definite volume and take the shape of the container they are in. Gases do not have a definite volume. They will fill a container of any shape and size. As the container is made smaller, pressure on the gas increases. The volume of the gas decreases. This is known as **Boyle's law**. See **Fig. 16-2**.

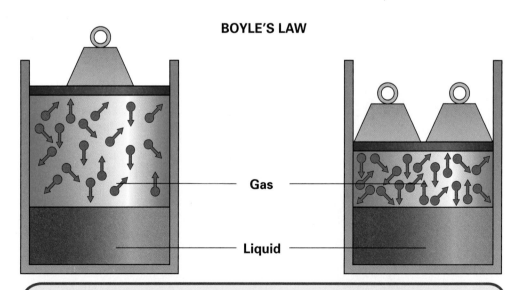

BOYLE'S LAW

Gas

Liquid

Fig. 16-2. Boyle's law states that as pressure increases, the volume of gas decreases. Note that pressure does not change the volume of liquids under constant temperature.

As pressure on the gas increases, its temperature increases. This is why compressed air can ignite fuel in a diesel engine. For more information on how a diesel engine works, see Chapter 9, "Land & Water Transportation."

Pressure is the force on a unit surface area (such as a square inch). Pressure is essential in all fluid power systems. The molecules of both liquids and gases bump into the walls of their containers. This pushing is pressure.

Blaise Pascal, a French scientist who lived in the 1600s, found that when force is applied to a confined liquid, the resulting pressure is transmitted unchanged to all parts of the liquid. His discovery became known as **Pascal's principle**.

Hydraulic Systems

Liquids cannot be compressed. Therefore, they can be used to transfer force. **Figure 16-3** shows this. The force applied to the piston in cylinder A puts pressure on the fluid. The fluid then exerts the same amount of pressure in all directions (Pascal's principle). Thus the 10 pounds of force from cylinder A are transmitted to cylinder B. Note that the two cylinders are the same size.

Math Link

Calculating Pressure. Pressure is the force per unit of surface area. In the customary system, pressure can be measured in pounds per square inch or pounds per square foot. The formula for pressure is $P = F/A$.

On a separate piece of paper, complete the data in the table below. Then prepare a graph to show the relationship between pressure and surface area.

Can you see a pattern? What happens as the surface area gets smaller?

Pressure (psi)	Force (pounds)	Surface Area (sq. in.)
	1	1.00
	1	0.50
	1	0.25
	1	0.10
	1	0.05
	1	0.01

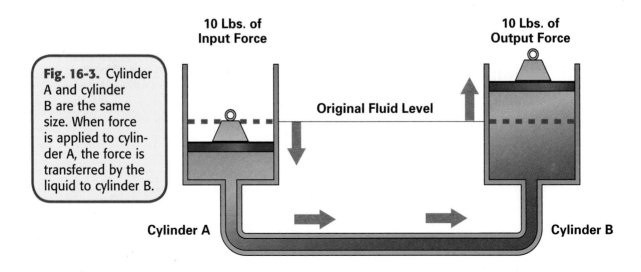

Fig. 16-3. Cylinder A and cylinder B are the same size. When force is applied to cylinder A, the force is transferred by the liquid to cylinder B.

10 Lbs. of Input Force

10 Lbs. of Output Force

Original Fluid Level

Cylinder A

Cylinder B

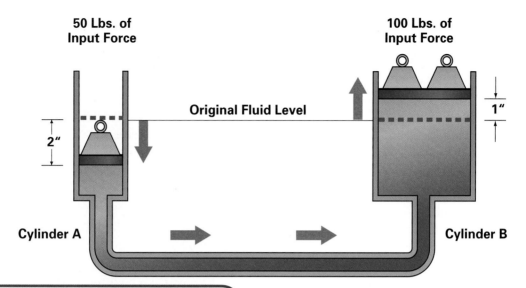

50 Lbs. of Input Force

100 Lbs. of Input Force

Original Fluid Level

2"

1"

Cylinder A

Cylinder B

Fig. 16-4. Force can be multiplied by using pistons and cylinders of different sizes. In the system shown here, 50 lbs. of input force produces 100 lbs. of output force.

Hydraulic systems can also multiply force. **Figure 16-4** shows how cylinders of different sizes can produce a greater force. Note that two inches of input movement were required to produce one inch of output movement. A gain in force results in a loss of distance.

Automobile jacks work this way. The handle must be moved up and down many times on the smaller cylinder to raise the jack just a few inches on the larger cylinder.

Hydraulic systems are ideal where strength and accuracy are required. That is why they are used on heavy construction equipment such as backhoes and bulldozers.

Simple Hydraulic Systems

A simple hydraulic system consists of the following parts.
- Fluid
- Pump
- Reservoir
- Relief valve
- Control valve
- Single-acting cylinder
- Transmission lines

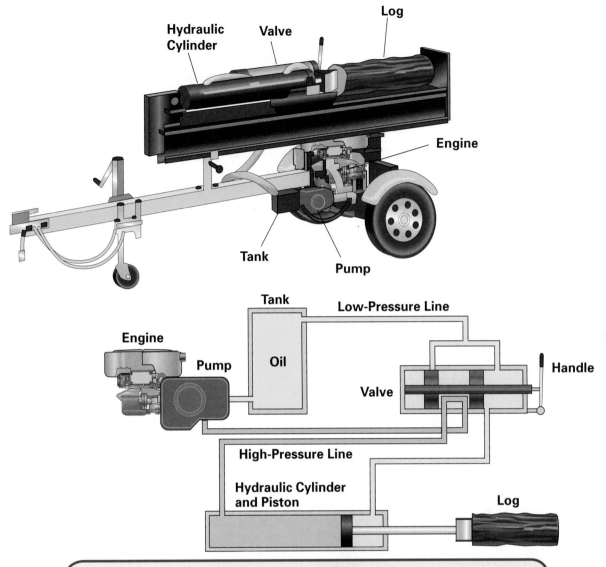

Fig. 16-5. A log splitter is a perfect example of a simple hydraulic system. The pump pushes the fluid through the line at a high pressure, which then moves the piston through the cylinder and applies the force to split the log.

A log splitter has all the components of a simple hydraulic system. Log splitters use great force to split logs into pieces that are suitable for use as firewood. See **Fig. 16-5**.

The fluid in most hydraulic systems is oil. The oil is stored in a reservoir. A motor-driven pump provides the pressure needed to move the oil throughout the system.

A relief valve is used to limit pressure in the system to a desired and safe amount. Control valves open and close to direct the oil to the proper location in the system.

Single-acting cylinders are made up of the cylinder body, piston, and piston rod. When oil is pumped to the cylinder, it causes the piston to move, extending the rod.

Uses of Hydraulic Systems

Today, hydraulic systems perform important tasks in manufacturing, transportation, construction, agriculture, and even the entertainment industry.

Hydraulic power systems are used in almost every manufacturing plant. Many of the material-moving devices on assembly lines are powered by hydraulics. Hydraulic lifts move materials to where they are needed. See **Fig. 16-6**.

Hydraulic systems make automobiles steer easily, ride smoothly, and stop quickly. The controls that pilots use to guide airplanes, helicopters, and even the Space Shuttle rely on very accurate, computer-controlled hydraulic systems.

Most pieces of heavy equipment use hydraulic systems. The lifting, pushing, and digging mechanisms on construction machines are hydraulically operated. Hydraulic devices are also present in the equipment used for planting and harvesting. See **Fig. 16-7**.

Hydraulic systems are also used heavily in the entertainment industry. Movies often feature simulated earthquakes, explosions, tidal waves, or cars that bounce in the air. These effects all use hydraulic systems. See **Fig. 16-8**. Many theme parks use the same hydraulic systems to create similar special effects.

Fig. 16-6. The powerful hydraulic system on this Sizzor® lift is capable of moving workmen and materials up to the roof of this building.

Fig. 16-7. Hydraulics lift up and empty this full bin of harvested cotton into the wagon.

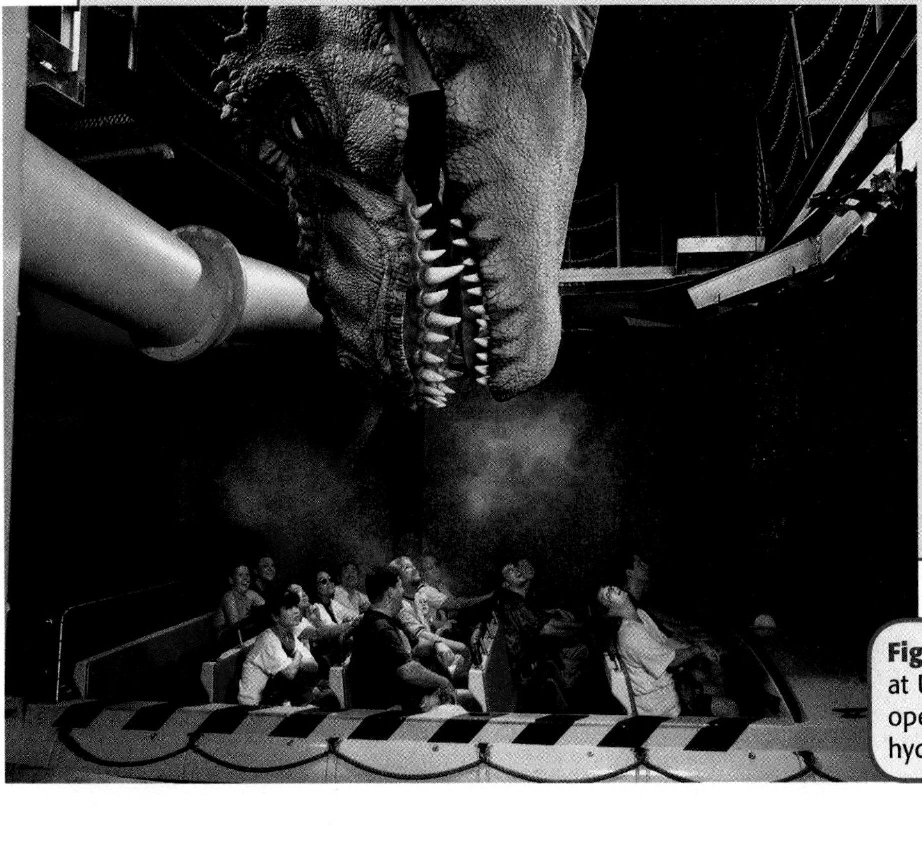

Fig. 16-8. This T-rex at Universal Studios is operated by a hidden hydraulic system.

Pneumatic Systems

Pneumatic systems use air, not liquid. They have some advantages over hydraulic systems. The air they require is readily available. If a pneumatic system leaks, there is nothing to clean up. No hazardous materials are released. Pneumatic systems also have some disadvantages. The energy required to compress air can be expensive. They do not produce as much force as hydraulic systems, and many pneumatic devices are noisy.

Simple Pneumatic Systems

A basic pneumatic system includes the following components:
- Compressor
- Receiver
- Check valve
- Control valve
- Actuator

The compressor is a device that compresses (squeezes) air. The compressor draws air into a chamber. Then the air is squeezed into a smaller space. The compressor can be operated by hand or motor driven. A bicycle pump is not really a "pump." It is actually a manual compressor. Most pneumatic systems use motor driven compressors.

The check valve is a one-way valve that allows pressurized air to enter the system but not to leave it. The valve prevents loss of pressure when the compressor is stopped.

The receiver tank takes the air from the compressor and stores it. Later the air is released as needed.

Control valves direct air to the proper location in the system. They also regulate air pressure and the rate of flow. They can be operated manually, electrically, or by pressure.

Actuators change pressure into mechanical motion. The actuator is usually a cylinder or motor. Single-acting cylinders are designed so that air pressure is applied to only one side of the piston. When the air pressure is released, a spring returns the piston to its original position.

Double-acting cylinders are designed so that air can be applied to either side of the piston. With this type of cylinder, air pressure can be used to extend and retract the piston.

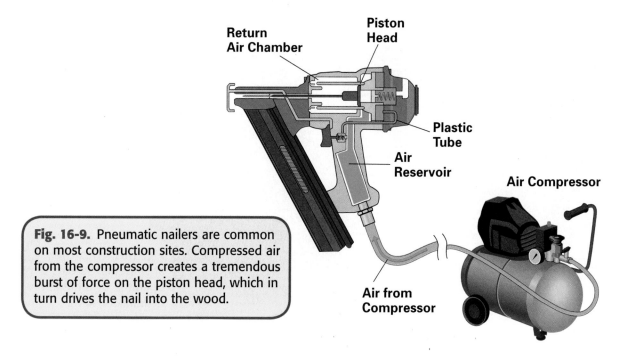

Return Air Chamber · Piston Head · Plastic Tube · Air Reservoir · Air Compressor · Air from Compressor

Fig. 16-9. Pneumatic nailers are common on most construction sites. Compressed air from the compressor creates a tremendous burst of force on the piston head, which in turn drives the nail into the wood.

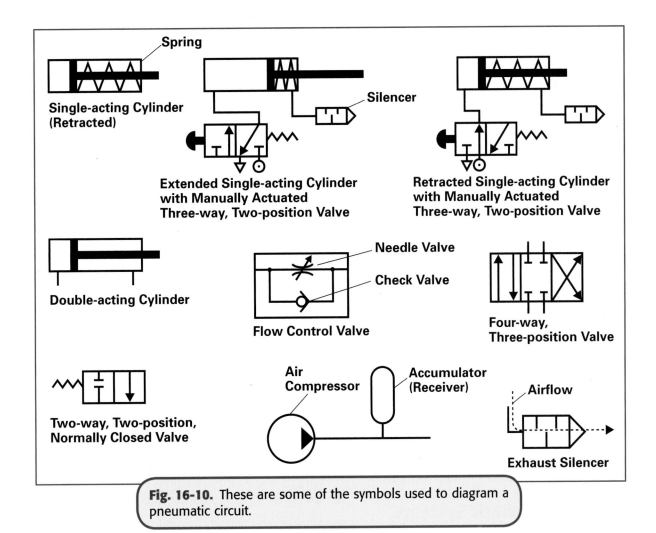

Fig. 16-10. These are some of the symbols used to diagram a pneumatic circuit.

Cylinders produce linear (straight line) or reciprocating (back and forth) motion. Air motors produce rotary (circular or spinning) motion. Rotary motion is produced in a tool such as the air-powered wrench. Complicated industrial equipment often uses combinations of the different types of motion.

A pneumatic nailer is a good example of a simple pneumatic system. **Figure 16-9** shows the interior view of a typical pneumatic nailer. Air comes from a separate electric or gasoline-powered compressor and flows through a hose into the nail gun's air reservoir. When the trigger is pressed, compressed air flows to the piston head to drive the nail out of the chamber. When the trigger is released, compressed air forces the plunger back into place and air above the piston head is forced out of the gun.

Pneumatic System Diagrams. Engineers plan pneumatic systems by drawing schematic circuit diagrams. They use symbols to represent components. **Figure 16-10** shows a few of these symbols. These symbols were devised by the American National Standards Institute

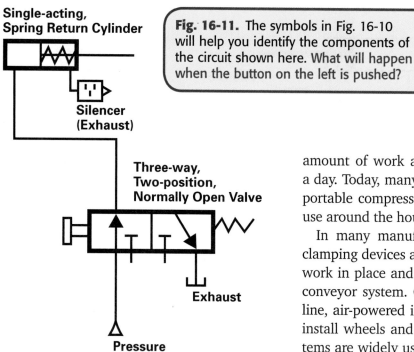

Single-acting,
Spring Return Cylinder

Silencer
(Exhaust)

Fig. 16-11. The symbols in Fig. 16-10 will help you identify the components of the circuit shown here. What will happen when the button on the left is pushed?

Three-way,
Two-position,
Normally Open Valve

Exhaust

Pressure

(ANSI). **Figure 16-11** shows how some of these symbols are arranged to create a circuit. Fluid power diagrams are read from bottom to top.

Uses of Pneumatic Systems

Pneumatic devices have many uses. Examples can be found in construction, manufacturing, packaging, and health care.

You have probably seen jackhammers in use. These large pneumatic drills are important road construction and maintenance tools. For home construction, pneumatic nailers are replacing hammers. Pneumatic framing, roofing, and trim nailers have greatly increased the amount of work a carpenter can complete in a day. Today, many homeowners also purchase portable compressors and pneumatic tools for use around the house. See **Fig. 16-12**.

In many manufacturing plants, pneumatic clamping devices are used on machines to hold work in place and to move items on and off a conveyor system. On an automobile assembly line, air-powered impact wrenches are used to install wheels and other parts. Pneumatic systems are widely used for packaging food products because it is safe to use air around food products. See **Fig. 16-13**.

Pneumatic devices also have important health care applications. Your dentist uses a high-speed pneumatic drill. Medical personnel may now use pneumatic devices to give injections. These devices force vaccines through the skin without piercing it.

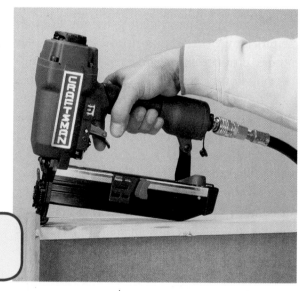

Fig. 16-12. Homeowners use small staplers and nailers, such as this one, to make small do-it-yourself projects easier.

Impact of Technology
Fluid Systems in Automobiles

It is easy to be unaware of the many hydraulic and pneumatic systems around us. As an example, look at the average automobile. It contains a number of fluid power systems that allow it to operate. Could we easily drive cars if there were no such systems? You be the judge.

Investigating the Impact

Research the steering and braking systems in a modern car.
1. Are they hydraulic or pneumatic?
2. Why would car manufacturers opt for such systems? What are the alternatives?

Be prepared to report to your class on how the steering and braking systems function.

Rescue workers use inflatable air bags to lift overturned vehicles in order to rescue people trapped inside. The air bags can be placed in just the right area to lift the vehicle off any victims. Similar airbags are used to lift sections of collapsed buildings.

The fluid power industry will continue to develop new uses for pneumatic and hydraulic control systems. In the future, individual components are expected to be smaller, lighter, and less expensive.

Engineers will continue to develop new uses for fluid power in manufacturing and other industries. Microprocessors will combine fluid and computer control. These devices will play an important role in automated systems.

Fig. 16-13. Pneumatic systems are used in many phases of food processing. Many of the machines used in food canning are powered by pneumatic systems.

Making an Air-Cushion Vehicle

Identify a Need/Define the Problem

Air-cushion vehicles, also known as ACVs, use a pneumatic system to provide lift as they travel. See **Fig. A**. Design and build a model ACV capable of transporting as many pennies as possible.

Gather Information

ACVs are used in a variety of environments. Commonly known as hovercrafts, ACVs are versatile machines which can be used for moving people or cargo across bodies of water, exploring the frozen tundra, or transporting troops across rough terrain. Research how ACVs work and how they are used.

Develop Possible Solution

A simple ACV design is shown in **Fig. B**. You can use a design like this one to build a working ACV. How can you modify this design to lift and move as many pennies as possible? Sketch several design ideas for the model you will build.

Fig. A

Materials and Equipment

Select from this list or use your own ideas.

- corrugated cardboard
- compass
- scissors
- pencil
- hot glue gun
- balloons in assorted sizes and shapes
- empty thread spools
- pennies

Model a Solution

1. Select the design that you think will be the most effective.
2. Make your ACV using a 4" diameter base. Use a pencil to make a hole in the center of the base.
3. Center a thread spool over the hole in the center of the base so the two holes will line up. Attach it to the base using hot glue. Allow it to dry.
4. Inflate the balloon and stretch it over the spool. Pinch the balloon at its base with your fingers so that the air cannot escape.

Test and Evaluate the Solution

- Place the ACV on the floor and release it. Observe the movement of your ACV.
- Add several pennies to the top of the ACV and repeat the test.

- Repeat the test using balloons of various sizes and shapes. You might also use bases of different sizes. Determine how changing these components affects the lifting power.
- Record your results in a table, listing the different sizes or shapes used.
- Which combination of balloon and base worked best?
- Describe one practical use of an ACV in business or industry. How might the design of such an ACV differ from the design of your model?

Refine the Solution

What could you do to improve the control of the ACV? Design a system that will allow your ACV to move in a forward direction. Produce several sketches before implementing your solution.

Communicate Your Ideas

Show your table to the class and explain what you learned about ACV designs. Be sure to include any steps you made from designing possible solutions to refining your model.

Science Link

ACVs in Reverse. Air-cushion vehicles use forced air to resist the force of gravity, thus allowing the vehicle to float above the surface. On the opposite end of this concept would be using air to keep the vehicle in close proximity to the ground. Who would desire such an effect? The answer is race car drivers! Race cars are designed to take advantage of the high wind speed to force the car down against the road surface. This is often called downforce.

1. Why would a speeding car driver desire to be "held down" against the road surface?
2. How could downforce be accomplished using scientific principles such as aerodynamics and the Bernoulli effect?

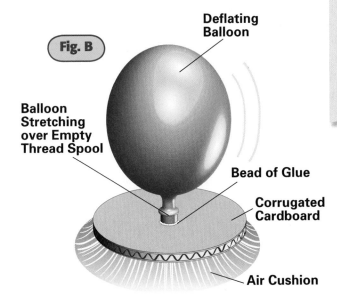

Fig. B

Deflating Balloon

Balloon Stretching over Empty Thread Spool

Bead of Glue

Corrugated Cardboard

Air Cushion

Creating a Hydraulic Jack-in-the-Box

Identify a Need/Define the Problem

Design and build a jack-in-the-box that operates on hydraulic power. The design must meet these specifications:
- The "box" must not exceed the maximum size (e.g., 8" × 8" × 8").
- The object that pops out must be of a historic figure in technology (e.g., an inventor, scientist, or engineer).

Gather Information

Jack-in-the-boxes are simple toys that traditionally use mechanical power. You wind a crank until a clown or other object pops out of the box. Research how a jack-in-the-box works and take notes on its design and operation.

Develop Possible Solution

Create a design that will allow all the internal workings of the hydraulic systems to be hidden from view. The action that the jack-in-the-box produces may be either linear or rotary. Sketch several design solutions for your jack-in-the-box. An example is shown in **Fig. A**.

Materials and Equipment

Select from this list or use your own ideas.

- strips of wood
- poster board
- plastic tubing
- plastic 20 cc syringe
- hot glue
- water
- food coloring

Model a Solution

1. Select the design that you think will be the most effective.
2. Make a simple cube shape using strips of wood and paper gussets.
3. Create a flap on the top of the cube using a piece of poster board and a wooden frame.
4. Create an image of a historical figure and place it on the top of the plunger of the syringe. Fill the syringe with colored water.
5. Attach the syringe that holds your figure to a piece of vinyl tube. Squeeze the plunger down to the bottom so that the syringe is empty but the tube is filled.
6. Take the remaining syringe and fill it with the colored water. Attach it to the other end of the vinyl tube.

Test and Evaluate the Solution

- Push the plunger of the filled syringe in. Take notes on the movement of the figure at the other end.
- Retract the figure. Again take notes on the movement of the figure at the other end.
- Empty the system of all liquids and replace it with air.
- Repeat the test and observations.
- What is the difference between the hydraulic and pneumatic jack-in-the-boxes? Is one system smoother than the other?
- Is there anything you can do to make the figure pop out quicker?

Refine the Solution

Develop a method that will allow your jack-in-the-box to retract once the plunger is released. Think about what causes the plunger to go up and what could be used to make it retract.

Communicate Your Ideas

Develop a promotional poster for your new design. The product will be sold as an educational toy. Include information about the historical figure as well as how the jack-in-the-box operates. Include terms and definitions about the power system used. Be sure to also explain the principle of hydraulics.

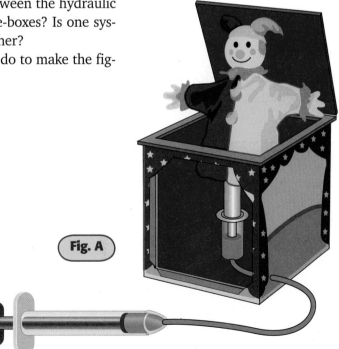

Fig. A

Mixing with Pneumatics

Identify a Need/Define the Problem

Pneumatic devices are often used more than traditional mixers in manufacturing pharmaceuticals. Why is this so? Producing certain products can be delicate and dangerous if other chemicals, such as hydraulic fluid, or ignition sources, such as electricity, are in the area.

In this activity, you will develop a pneumatic mixer that meets the following specifications:
• The device must use different speeds.
• The device must be handheld.
• The device can be operated up to 4' away from the air source.
• The device must mix two liquids of different thicknesses.

Gather Information

Investigate air motors and how they operate. Also, research designs for pneumatic mixers.

Materials and Equipment
Select from this list or use your own ideas.

• beverage and food containers
• plastic tubing
• plastic 20 cc syringe
• dowels
• low pressure air source (5–15 psi)

Safety Alert
Look up "Safety Data Sheets" on the Student CD and prepare a data sheet for this activity. As you work on the activity, be sure to follow all safety rules.

Develop Possible Solution

Develop possible ideas for a handheld pneumatic mixer. One possible design is shown in **Fig. A**. This design is presented only to give you a general idea of what will be needed. Do not copy this design.

Model a Solution

1. Select the design that you think will be the most effective.
2. Using various materials, construct your mixer.

Safety Alert
Pneumatic Safety
Always wear safety glasses around pneumatic systems and do not point air sources at yourself or others.

Test and Evaluate the Solution

- Attach your pneumatic mixer to an air source.
- Place your mixer in the liquid solution (e.g., milk and chocolate syrup).
- Turn your mixer on at the slowest speed. Observe how your device operates.
- Slowly increase the speed of the mixer.
- How did your mixer handle the load (liquid solution)? Was the motor able to operate under the stress of the load?
- Was the air source exhausted?
- Was there any of only one liquid (e.g., syrup) left at the bottom of the container?

Refine the Solution

Refine your design so that there is a specific location for the leftover air to escape to once you are done mixing. Have other classmates try your design and tell you if it was easy to operate. Use their input to refine your device further.

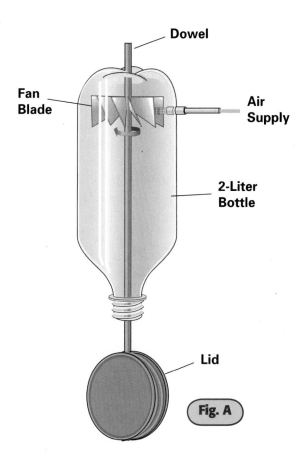

Fig. A

Air Power. Air may seem like a harmless form of power, but it is really a strong force. Wind is a good example of air power.

If a building 100 feet square and 500 feet tall is buffeted by a wind averaging a force of 25 pounds per square foot, how much force is the building resisting?

Communicate Your Ideas

Present the various steps you took in creating your pneumatic mixer. Be sure to include all steps of the design process from the problem to the final refined solution. Key components should include the various ideas that you developed and the observations from the testing.

Exploring Careers

Elevator Installer and Repairer

ENTRY LEVEL   **TECHNICAL**   **PROFESSIONAL**

Elevator installers and repairers install, repair, and maintain elevators, escalators, and moving walkways. Installers set up new machinery and replace older equipment when needed. Repairers work on machinery when it breaks down. Both make sure the machinery works properly and safely.

Elevator installers and repairers work for construction companies, elevator manufacturers and distributors, maintenance and repair companies, or businesses. They use principles of mechanics, engineering and technology, building construction, and public security and safety in their jobs.

Qualifications

Installers and repairers must be at least 18 years old, have a high school diploma, and pass an aptitude test to work for a union company. They learn to install and repair elevators through on-the-job training and classroom instruction. Knowledge of hydraulics, electronics, and electricity is needed. Elevator installers and repairers would most likely take math, electricity, and physics in high school.

Elevator installers and repairers need to maneuver in tight spaces and work with heavy equipment. Therefore physical strength, stamina, and flexibility are important. Night and peripheral vision, depth perception, and good hearing are also important. Installers and repairers must be able to take instructions and then work alone.

Outlook for the Future

The job outlook for elevator installers and repairers is good. However, compared with other careers, available jobs will be limited because fewer people work as elevator installers and repairers.

Listening Skills

Listening skills are important in any job. If you don't listen, you will not know what you are expected to do. To be a good listener, make sure you know the main ideas when someone is speaking. Pay attention to body language and to spoken words. Learn to tell the difference between facts and opinions. To help you remember instructions, take notes.

Researching Careers

Find out about jobs for elevator installers and repairers. Interview someone responsible for elevators. What is the person's education and experience? What are the best and worst parts of the job? Give a talk to the class about what you learn.

More activities
on Student CD

Key Points

- A fluid is any substance that flows. Fluids can be a liquid or a gas.
- Fluids can be put under pressure and moved through pipes or hoses to where they are needed using hydraulic or pneumatic systems.
- Pressure is an essential component to all fluid systems.
- Hydraulic and pneumatic systems are used in many fields of technology.

Read & Respond

1. What is fluid power?
2. What is the main difference between a hydraulic and a pneumatic system?
3. What is Boyle's law?
4. Explain Pascal's principle.
5. A simple hydraulic system consists of what parts?
6. Name three uses for a hydraulic system.
7. A simple pneumatic system consists of what parts?
8. What are three advantages of pneumatic systems over hydraulic systems?
9. What are three disadvantages of pneumatic systems?
10. Name three uses for a pneumatic system.

Think & Apply

1. **Plan.** Create a system diagram for a simple pneumatic device. Be sure to use the proper symbol for each component.
2. **Research.** Find out how fluid power devices are used in your community and nearby areas.

3. **Organize.** Make a time line showing different inventions related to hydraulics.
4. **Relate.** You learned that there are many uses for fluid power. How do hydraulic and pneumatic systems relate to other chapters in this textbook?
5. **Extend.** Does a simple squirt gun use hydraulics, pneumatics, or both? Explain your answer.

TechByte

A Pneumatic Hospital. Did you know that pneumatic systems can be used in hospitals? In 2003, Queen's Medical Center in Honolulu, Hawaii, installed pneumatic tubes to transport lab samples, drugs, and other items throughout the huge hospital. Now the hospital's staff can take the time to care for patients instead of running all over the hospital to transport goods.

CHAPTER 17

Robotics

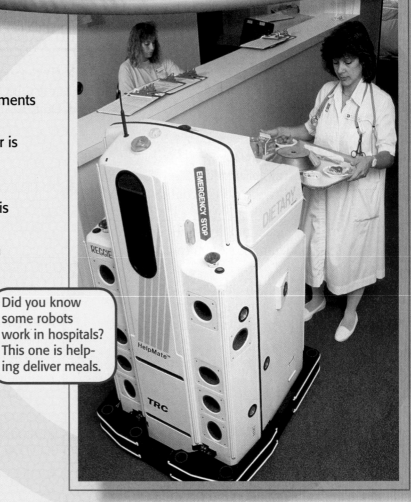

Did you know some robots work in hospitals? This one is helping deliver meals.

Objectives

- Identify technological developments that led to modern robotics.
- Explain how the stepper motor is used in robotics.
- Define *work envelope*.
- Explain how feedback control is used.
- Describe four types of modern robots.

Vocabulary

- robot
- robotics
- controller
- computer program
- manipulator
- end effector
- power supply
- degree of freedom
- work envelope
- feedback control

Activities

- Trial and Air
- Arming Yourself with Hydraulics

How Robotics Developed

A **robot** is a machine that does complicated tasks and is guided by automatic controls. **Robotics** is the design, construction, and operation of robots. We think of robots as marvels of modern technology. However, the idea of machines designed to imitate human actions existed over 3,000 years ago. The ancient Egyptians made puppets on strings.

During the 1700s, automata were popular in Europe. Automata were mechanical devices that imitated the actions of people or animals. One of the most famous automatons was a duck built by the French engineer Jacques de Vaucanson. This duck could walk, flap its wings, and even "eat." While the automata had no practical purpose, building them did encourage the development of technology. For example, de Vaucanson was the first European to make hoses from a "new" South American material—rubber.

Once computers were developed that could control robotic systems, robotics began to grow rapidly. The automata, computers, and other technological developments have led to today's modern robots.

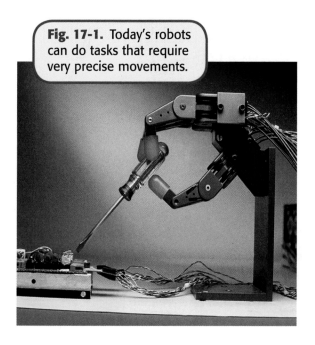

Fig. 17-1. Today's robots can do tasks that require very precise movements.

Robot Generations

The first generation of robots was designed for factory work. Known as steel collar workers, these robots did simple tasks that were dangerous or unpleasant for human workers. Early robots were used to handle hot metal, weld metal parts, spray paint, move parts, and load pallets. These early robots were large and not very flexible.

The second-generation robots used today can perform more complex tasks and simulate many human functions. They move, sense their surroundings, and respond to changes in the environment. Today's robots are flexible, and they can quickly be taught to do several different operations. With movements accurate to a fraction of a millimeter, robotic arms can assemble intricate electronic circuits. They can solder wires as thin as a human hair. See **Fig. 17-1**.

While many robots are mechanical arms attached to a base, some robots are independent. The robot shown on the opposite page is able to move through rooms and hallways and can even use elevators. It is controlled by a central computer and wireless radio. The robot can transport meals, medicine, lab samples, medical records, and supplies.

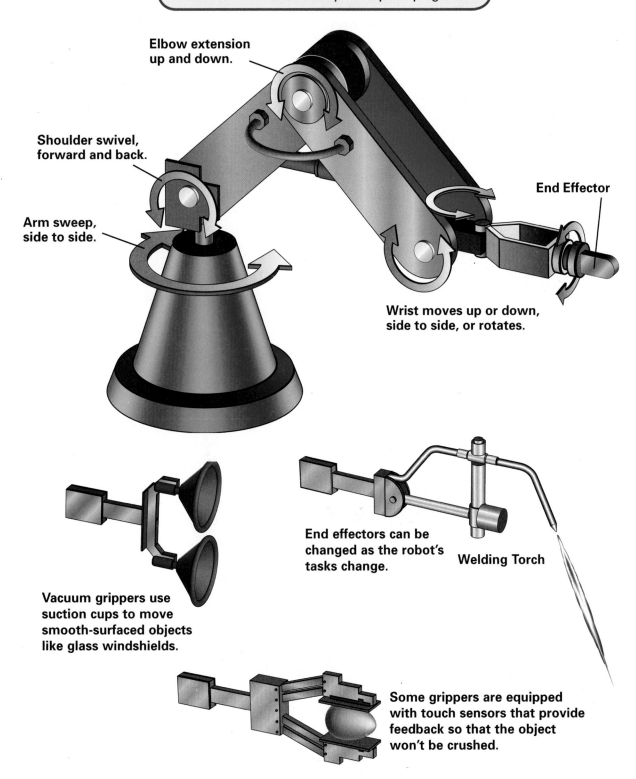

Fig 17-2. The manipulator uses cables, motors, gears, and pneumatic cylinders to move within a space. The movements are controlled by a computer program.

Elbow extension up and down.

Shoulder swivel, forward and back.

Arm sweep, side to side.

End Effector

Wrist moves up or down, side to side, or rotates.

End effectors can be changed as the robot's tasks change.

Welding Torch

Vacuum grippers use suction cups to move smooth-surfaced objects like glass windshields.

Some grippers are equipped with touch sensors that provide feedback so that the object won't be crushed.

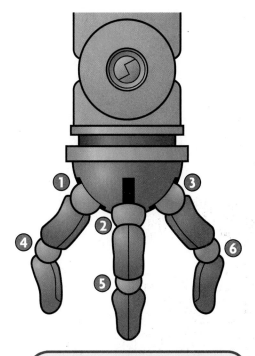

Fig. 17-3. This robotic hand has six degrees of freedom, and the wrist can turn 360 degrees.

Robotic Systems

In many ways, robotic systems model human systems. In order for you to move, your brain has to send signals to your arms and legs. In order for a robot to move, a computer sends signals to it. Computers are the brains of modern robotic systems. A robotic system includes the following:

- The **controller** is a tiny computer that acts as the robot's brain and contains the computer program.
- The **computer program** is a set of coded instructions the robot must follow.
- The **manipulator** is the robot's mechanical system. It is often jointed so the robot can move, and it may resemble a human arm or torso. See **Fig. 17-2**.
- The **end effector** is the robot's hand. It may be in the form of a gripper, or it may hold an attachment, such as a welding torch or a paint sprayer nozzle. See **Fig. 17-2**.

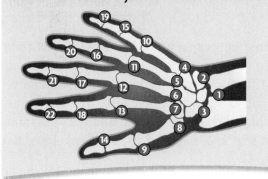
- The **power supply** provides power to the robot. It may supply electricity, hydraulic power, or pneumatic power. These different sources all affect the type of work a robot can accomplish.

Degrees of Freedom

A robot's **degree of freedom** is its ability to move in a particular direction. Each degree of freedom requires a separate joint. Most robots have at least six degrees of freedom, and some have many more.

While a robot's freedom of movement is more limited than a person's, its range of movement is greater. Your wrist can bend only about 165 degrees, but a robot's wrist can spin 360 degrees. See **Fig. 17-3**. Just think how easy it

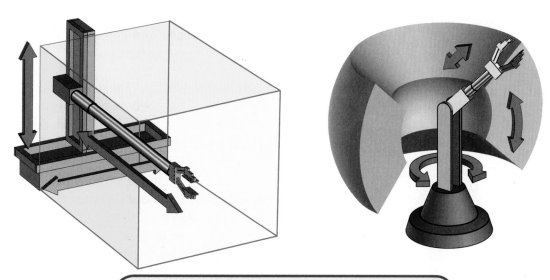

Fig. 17-4. The area a robot moves within is called its work envelope. The size of the robot and its degrees of freedom determine the size of the work envelope. What is the size of your work envelope?

would be to spin an object if our hands could turn completely around!

The space within which a robotic arm moves is called its **work envelope**. The design, or architecture, of the robotic arm will determine the size and shape of its work envelope. See **Fig. 17-4**.

Math Link

Working Out Your Work Envelope.
The formula for the volume of a sphere is $V = \frac{4}{3}\pi r^3$. Radius (r) can represent the reach of your arm, while a sphere represents your work envelope. Have a classmate measure your reach.
You should then figure your arm's work envelope.

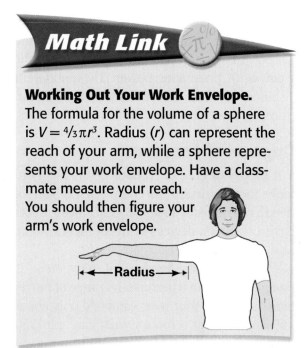

|←—— Radius ——→|

Power for Robotic Movements

Moving robotic parts can be powered in various ways. The selection of a power source, or actuator, depends on what the part has to do.

An electric motor known as a stepper motor is commonly used as an actuator when robotic movement has to be fast and accurate. See **Fig. 17-5**.

One complete rotation of a stepper motor can be divided into hundreds of individual steps. Each step represents a fraction of a degree of movement. A stepper motor can rotate a small amount, or a step, each time an electrical signal is sent to it. Waist, shoulder, elbow, and wrist joints may each be powered by separate motors. The motor shaft transmits the mechanical energy through gears, shafts, and pulleys to the robotic joint. Computer programs control the precise movements of each joint by controlling the steps of the motor.

At times, robotic arms must lift heavy objects. Pneumatic and hydraulic actuators use compressed air or hydraulic fluids to transfer

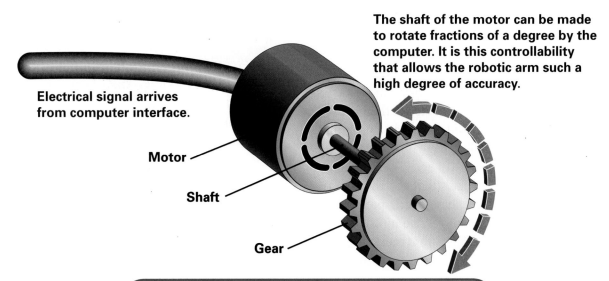

Electrical signal arrives from computer interface.

The shaft of the motor can be made to rotate fractions of a degree by the computer. It is this controllability that allows the robotic arm such a high degree of accuracy.

Motor

Shaft

Gear

Fig. 17-5. Stepper motors provide the force to move the robotic arm. Gear trains and chain drives transfer mechanical energy from the motor to the moving part of the robotic arm. Gears are used to adjust the speed of the motors.

power to the joints and grippers. Pneumatic and hydraulic systems are made up of cylinders and pistons. The piston pushes on the fluid in the cylinder. A second piston on the other end of the cylinder moves as the fluid presses on it. The moving piston can make the robotic arms move forward and back. The pistons are controlled by electrical switches connected to computers. The switches open and close valves controlling air or hydraulic fluids. See **Fig. 17-6**. To learn more, see the "Fluid Power" interactive lab on the Student CD.

Controlling Robotic Systems

How does a robotic arm know what movement to make? How much pressure should the gripper apply? How can a robotic arm "remember" the patterns needed to paint an automobile? Just as your brain controls your every movement, computers control the movement of robotic systems. The computer uses a series of instructions known as a program. Robotic

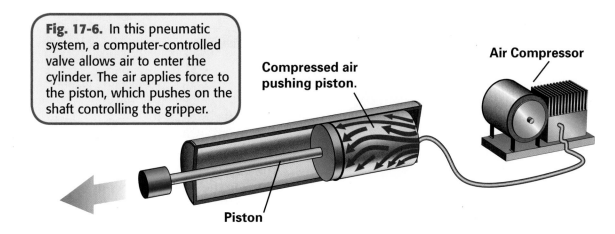

Fig. 17-6. In this pneumatic system, a computer-controlled valve allows air to enter the cylinder. The air applies force to the piston, which pushes on the shaft controlling the gripper.

Compressed air pushing piston.

Air Compressor

Piston

programs are very complex. They must list in logical order all the steps needed for the robot to perform a task.

Robotic software (program) designers prepare flowcharts that list the basic movements of the robot. These movements are then broken down into finer detail and written in a programming language. That language is converted into binary code, a series of 0's and 1's. The binary code contains instructions that tell the robotic arm how far to travel, how much pressure to apply, or how to move a tool to perform a task. You'll find more information about the binary system on the Student CD.

The instructions travel through cables of wire from the computer to the robotic interface. The interface links the computer to the robotic motors. Inside the interface are electronic switches that turn the motors on and off. The interface reads the binary code and produces signals that rotate stepper motors or open pneumatic valves, causing the pistons to travel. See **Fig. 17-7**.

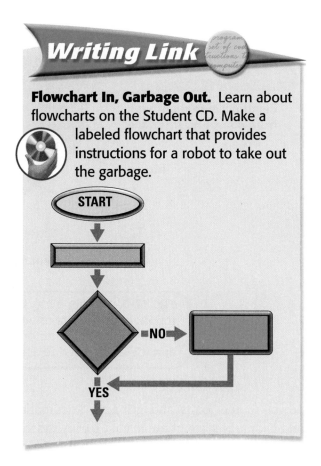

Writing Link

Flowchart In, Garbage Out. Learn about flowcharts on the Student CD. Make a labeled flowchart that provides instructions for a robot to take out the garbage.

START

NO

YES

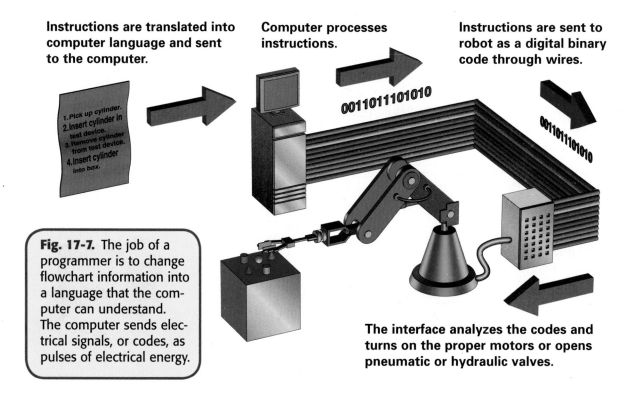

Instructions are translated into computer language and sent to the computer.

Computer processes instructions.

Instructions are sent to robot as a digital binary code through wires.

1. Pick up cylinder.
2. Insert cylinder in test device.
3. Remove cylinder from test device.
4. Insert cylinder into box.

0011011101010

0011011101010

Fig. 17-7. The job of a programmer is to change flowchart information into a language that the computer can understand. The computer sends electrical signals, or codes, as pulses of electrical energy.

The interface analyzes the codes and turns on the proper motors or opens pneumatic or hydraulic valves.

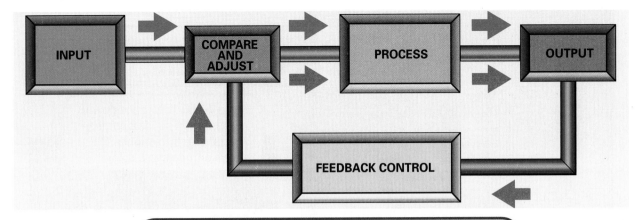

Fig. 17-8. The feedback control process. On robotic systems, sensors monitor what the system is actually doing. If the input to the system does not match the output to the system, the system is adjusted.

Lead-through programming is another way of controlling a robot. Instructions are created by guiding the robotic arm through a sequence of movements and programming the computer to remember the pattern of motion.

Robots can also be programmed using keyboards or teach pendants. The pendant and keyboard give the robot direct instructions to move up, down, left, and right. Each movement is remembered by the computer and repeated as often as required.

Feedback Control

How does a robot detect where an item is? Humans have sensory organs such as eyes, skin, and ears. These allow us to track changes in our environment. Robots also have sensors so they can keep track of what's going on around them.

Imagine that you are going to touch the handle of a saucepan on a stove. Your brain sends signals to the muscles and tendons in your hand, and you grasp the pot handle. Information is quickly sent to your brain through nerve bundles. The message is that your hand has grasped the handle and is ready for the next command.

What if the handle is too hot? Signals are returned to your brain and translated as pain. Your brain sends new signals to your hand and your grip is released. The process of sending signals, interpreting received signals, and adjusting through signals is called **feedback control**. Robots use feedback control constantly. It allows them to detect where they are and to adjust their actions. See **Fig. 17-8**.

Your fingers have over 17,000 sensors (nerve endings) that send data to your brain as you touch things. Robots use touch, or contact, sensors. These touch an object and send electrical signals to the computer. The data might include information on the shape of the object and how much pressure the grippers are placing on it. The computer can then adjust the actions of the robot if changes are needed.

A robot may be equipped with cameras so that it can view objects. The image picked up by the camera is sent to the computer for analysis. Using that data, the computer outputs directions to the robot.

Fig. 17-9A. Cameras and sensors in this robot help scientists study how the human brain does visual tasks and controls movement.

Fig. 17-9B. This robot flower uses infrared sensors to detect movement. In response, it sways and changes color.

Fig. 17-9C. ASIMO's visual and kinesthetic (movement) sensors enable it to detect and respond to movement.

Fig. 17-9D. This robot's eyes and brain mimic human vision.

Impact of Technology

Robots in Industry

Automated manufacturing plants can operate with very few people. Robots can usually work faster, at lower cost, and more accurately than human workers. They don't even need a building's lights on to work!

Remember that robots are not paid a salary, are never late, never call in sick, never need health insurance, and never take vacations. Since using robots saves time and money, manufacturers can produce products more cheaply, allowing products to be sold for less and enabling the manufacturer to become more competitive in the world market.

Investigating the Impact

You've just read about advantages of having robots in manufacturing plants. However, impacts of technology can be both positive and negative.

1. What are some disadvantages of using industrial robots?
2. If robots are used instead of people, what problems might occur, both for the employer and for the employees?

Imagine freshly baked cookies moving down a conveyor line. Using its attached camera, the robotic arm looks for burnt and broken cookies. When one is sighted, the computer instructs the grippers to remove the bad cookie from the line.

Robotic arms that perform detailed work such as welding and painting use sensors to track the arm's movement. Disks displaying markings are placed at each joint in the robotic arm. As the joint moves, optical scanners (like cameras) read the markings. They send this information to the computer. The computer interprets the information, calculates the angle of the joint, and outputs needed commands to the arm.

There are many types of sensors. Robots may use microphones to sense sounds. They may use sonar to measure distances. The more complex the task, the more sensors a robot may need. See **Fig. 17-9.**

Modern Robots

Robots today come in a variety of shapes and sizes and are used for many purposes. Robots work in many places, from within the home to outer space. Some of them are mobile, while others remain stationary. Four useful types of modern robots include industrial, medical, assistive, and household robots.

Fig. 17-10. Tug is a robot that can be trained to pull hospital carts. An onboard computer contains a map of the building. Users choose a destination and send Tug on its way.

Surgeries are often delicate procedures. In some joint replacement operations, a hole must be drilled into the bone to accept the artificial joint. The accuracy of this hole is critical. Some surgeons now use robotic systems to position the drill and make the actual hole. The accuracy and steadiness of the robotic arm are hard to beat. Robotic hands also can work in areas that are too small for human hands.

Robots are also being used to fill hospital prescriptions. Errors in filling prescriptions are a leading cause of preventable deaths in hospitals. Robots are more accurate and have helped reduce prescription mix-ups.

The Tug® robot is used to transport goods throughout the hospital. In areas where health care workers are in short supply, Tug can carry medical records, food, and medications to where they are needed. See **Fig. 17-10**.

Robots can even be used to fill a void left by absent doctors. While doctors are often needed in many places at once, a robot can be remotely controlled by the doctor to visit patients. The top of the robot has a display screen that shows the doctor's face, so that patients can still interact with their own physician rather than a stranger.

Assistive Robots

People with disabilities may have a difficult time with everyday tasks. Assistive robots can help with tasks such as eating, cleaning, and grasping or reaching for objects.

For instance, a robotic arm attached to a wheelchair could pour a glass of milk for someone who is paralyzed. See **Fig. 17-11**. A robot with an extendable arm could reach objects on high shelves. Robotic arms can also easily hold a tray of food and even feed people who have difficulty feeding themselves.

Household Robots

Have you ever read a story or seen a show with a robotic butler or maid? In reality, robots already exist in some households, doing every-

Industrial Robots

Most industrial robots are robotic arms. They can do many factory jobs, such as welding, spray painting, and assembly. Industrial robots are often used to do jobs that are boring or unsafe for humans. The majority of robots today are industrial robots.

Medical Robots

Robots have become very valuable in the medical field. They can assist in surgeries, transport hospital materials, dispense medicine, and much more.

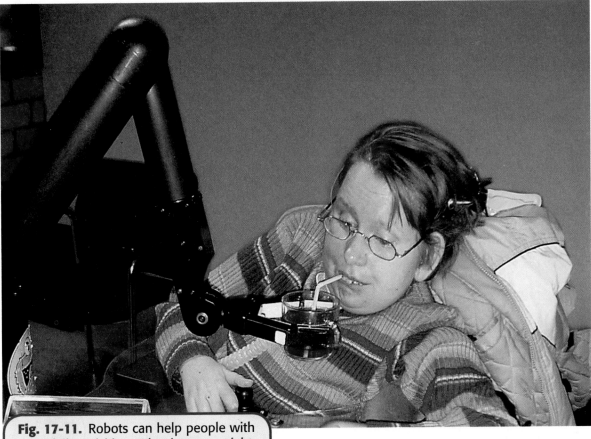

Fig. 17-11. Robots can help people with their daily activities. What impacts might assistive robots have on society and the economy?

day tasks. The Roomba vacuum cleaner was the first affordable house-cleaning robot. Roomba vacuums your floors when you have other commitments, such as watching football on television! The vacuuming robot can sweep under furniture, and it uses sensors to detect trouble spots that might need more thorough cleaning.

Do you find mowing the lawn to be tedious? Robomowers can mow your yard while you relax nearby. All you need to do is push a button on a remote control to operate it. See **Fig. 17-12**.

Fig. 17-12. Robomowers can cut grass in perfectly straight lines.

 # Trial and Air

Identify a Need/Define the Problem

Design and build a model of a lifting device that converts the pneumatic force of an expanding balloon into linear motion.

Gather Information

Research information on pneumatic devices and how they are used in robotics. Be sure to also read the discussion of pneumatic power in this chapter.

Develop Possible Solutions

Develop sketches for devices that can capture the expanding pneumatic force of a balloon and convert it into an up-and-down motion that can be used to move an object. See **Fig. A**.

Select one of your possible solutions. You will build a model of this solution. Determine what materials and equipment you will need.

Materials and Equipment

Select from this list or use your own ideas.

- ¼" plastic tubing
- squeeze bottle, turkey baster, or similar item
- cardboard
- dowels of assorted diameters
- wood scraps
- tape
- glue
- string
- rubber bands
- balloon
- tools as needed

Safety Alert

Look up "Safety Data Sheets" on the Student CD and prepare a data sheet for this activity. As you work on this activity, be sure to follow all safety rules.

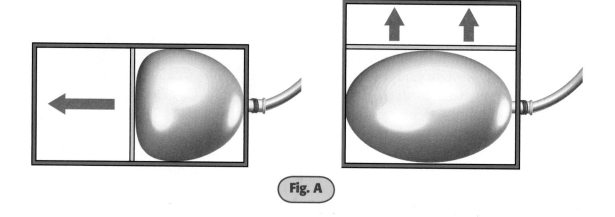

Fig. A

Model a Solution

1. To make the pneumatic system for your lifting device, attach the balloon to the plastic tubing with a rubber band as shown in **Fig. B**. Attach the other end of the tubing to the squeeze bottle by forcing it through the hole in the cap. A tight fit is needed.
2. Experiment with the pneumatic system that you just created. Get a feel for its power. Try lifting a book.
3. Now incorporate this pneumatic system into your lifting device design by creating a model, such as a cylinder or box.

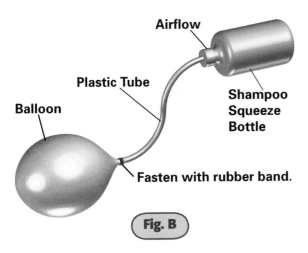

Airflow

Plastic Tube

Balloon

Shampoo Squeeze Bottle

Fasten with rubber band.

Fig. B

Test and Evaluate the Solution

Test your model and make changes if needed. Then evaluate your model using the following questions:
- Does your device create up-and-down motion?
- Can the device lift an object into the air?
- In a traditional lifting device, what mechanism would replace the balloon as a power source?
- If the air in the balloon were replaced by water, what kind of system would it be?

Science Link

Expanding on Pascal. Pascal's principle says that when air is blown into a balloon the pressure in it will be transmitted equally all around the balloon's interior. This is why a balloon expands evenly in all directions when you blow it up. How will Pascal's principle affect the design of your device?

Equal Pressure

Refine the Solution

- Did your design prove successful through testing and evaluation?
- How might you improve your design?

Communicate Your Ideas

Prepare a presentation for your class. Be sure to include all your information, such as design notes and sketches. Relate problems you may have encountered and how you dealt with them.

Design, Build & Evaluate

Arming Yourself with Hydraulics

Identify a Need/Define the Problem

In this activity, your team will design and produce a robotic manipulator. Some examples are shown in **Fig. A**.

The number of axes a robot arm has depends on the type of work it is designed to do. In manufacturing, robots commonly have six axes. Your robot will need at least two axes. It must be hydraulically controlled. The robot must also be able to pick up, move, and put down an object. This means that your design must allow for both horizontal and vertical movement.

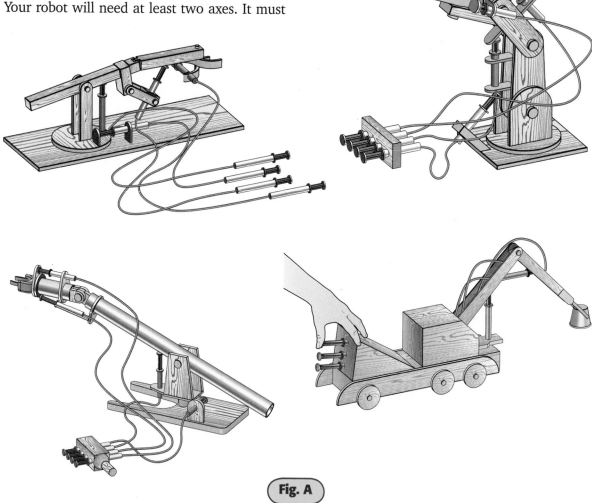

Fig. A

Materials and Equipment

Select from this list or use your own ideas.

- plywood, assorted sizes
- scrap lumber
- dowel rods, various sizes
- PVC plastic pipe
- sheet metal
- metal cans
- clear plastic tubing, ⅛" diameter
- syringes, 6–20 cc
- ball bearings
- springs
- adhesives
- fasteners
- abrasive paper
- drafting equipment or CAD system (for making working drawings)
- tools for cutting, drilling, and sanding

Safety Alert

Look up "Safety Data Sheets" on the Student CD and prepare a data sheet for this activity. As you work on this activity, be sure to follow all safety rules.

Gather Information

You will need to research information on robotic arms powered by hydraulics. First, look at the material included in this chapter. For additional information, you could research books, magazines, or the Internet. On the Internet, you will find more information if you use key terms such as hydraulics, leverage, prototype, and robotics.

Science Link

Mechanical Advantage. Using two different-sized pistons within connected cylinders filled with oil, like in a hydraulic jack, is a great mechanical advantage. A large movement of the small piston creates a small movement of the large piston, but the larger piston acts with much greater force. The opposite is also true; a small movement of the large piston can create a large movement of the small piston. What other devices besides a hydraulic jack might work in this way?

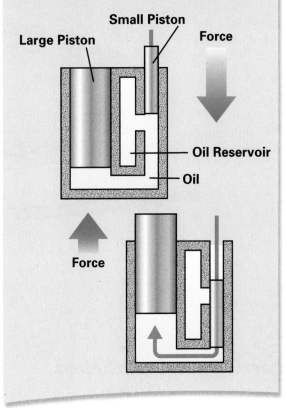

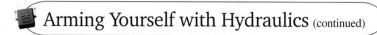

Arming Yourself with Hydraulics (continued)

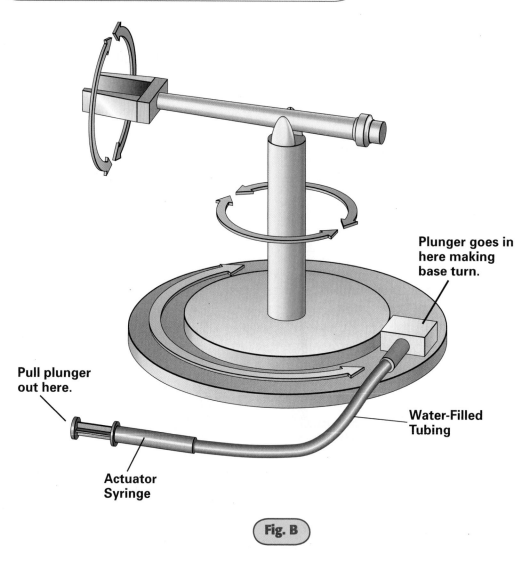

Plunger goes in here making base turn.

Pull plunger out here.

Actuator Syringe

Water-Filled Tubing

Fig. B

Develop Possible Solutions

Remember your robot is to be a prototype; that is, a working model. Brainstorm possible solutions with your fellow team members. Discuss pros and cons of each solution. Make several sketches of possible robot designs and then combine your group's best ideas into a final sketch.

Model a Solution

1. Prepare working drawings for the solution you've chosen.
2. Create a bill of materials for making your robot.
3. As in the manufacturing industry, your production must be well-planned and organized. Prepare a plan of procedure for making your robot.

Math Link

Cylindrical Math. The formula to calculate the area of a cylinder head is $A = \pi r^2$. The formula to calculate the volume of a cylinder is $\pi r^2 h$. Using these formulas, calculate the area and volume of the cylinder illustrated below.

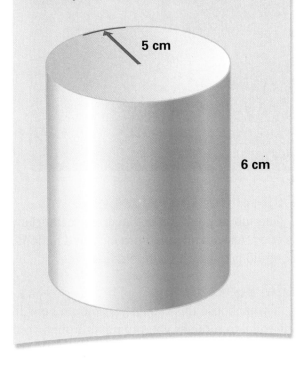

5 cm

6 cm

4. Following your teacher's instructions for working in the lab, build your robot. Pay special attention to the syringes and plastic tubing, which will act as the power/control system for the robot. One syringe should act as part of the actuator, while the other syringe will be mounted on the part of the arm you wish to move. See **Fig. B**.

Test and Evaluate the Solution

Test your model and make changes if needed. Then evaluate your model using the following questions:

• Does the robotic arm have two or more axes?
• Does your robotic arm move both horizontally and vertically?
• Can the device lift an object into the air and put it back down again?
• How might your robot be used on a production line in your lab?

Refine the Solution

• Did your design prove successful through testing and evaluation?
• How might you improve your design?
• If you were to make another robot, what new or additional features would you include?

Communicate Your Ideas

Prepare a presentation for your class. Be sure to include all your information, such as sketches, working drawings, the bill of materials, and the plan of procedure. Demonstrate what your robot is capable of doing.

Relate problems you may have encountered and how you dealt with them. Discuss some of the design ideas you considered but did not include. Explain how your robot might be used on a production line or how a computer could be incorporated into the design. Pose the question of the positive and negative impacts of robots in manufacturing as well as in our everyday life.

Exploring Careers

Engineering Technician

ENTRY LEVEL | **TECHNICAL** | PROFESSIONAL

Engineers need help to design and build robots. Engineering technicians are key members of the team. They help the engineers build prototypes, collect and record data, and design the robot—often using computer-aided design.

Engineering technicians work in many kinds of jobs, not just robotics. Most specialize in one area, such as civil or chemical engineering. More than 40% of engineering technicians work in the electrical and electronics area.

Qualifications

A technician must have a good math and science background as well as good problem-solving skills. An associate's or bachelor's degree is required. The technical courses required will depend on the engineering area. For example, students who want to become electrical engineering technicians may need to take classes in microprocessors, electric circuits, and digital electronics. A mechanical engineering technician would most likely take courses in thermodynamics, mechanical design, and fluid mechanics.

Because many engineering technicians are helping with design, creativity is a good trait to have. Good communication skills and the ability to work well with other people on a team are also very important.

Outlook for the Future

The job outlook for engineering technicians is good. As with all jobs, both local and national conditions affect the outlook.

Lifelong Learning

If you choose this career, you should be willing to learn more as you progress through your work life. Employers want people who are willing to learn more about their chosen field and want to improve their skills.

Researching Careers

Find out about jobs for engineering technicians in the field of robotics. What are the educational requirements? Where in your area can you meet those requirements? Write a brief summary of what you find.

**More activities
on Student CD**

Key Points

- A robot is a machine that does complicated tasks and is guided by automatic controls. Robotics is the design, construction, and operation of robots.
- Robots use joints to move. These joints give robots degrees of freedom.
- The stepper motor is a common power source for robotic movement.
- Computers are used to operate a robotic system.
- Robots depend on feedback control.
- Modern robots are used for a variety of purposes.

Read & Respond

1. How are robots similar to humans?
2. What important technological developments had to take place before robotics control system technology could become a reality?
3. How does a robotic arm achieve degrees of freedom?
4. Explain how the stepper motor is used in robotics.
5. Define the work envelope of a robotic arm.
6. List three end effectors commonly used on robots.
7. Explain how feedback control is used to adjust a robotic arm's movements.
8. Where are robots used the most?
9. What is a robot's manipulator?
10. Name four types of modern robots.

Think & Apply

1. **Plan.** You are the owner of a company that manufactures small appliances. You have decided to add robots to the assembly process. Describe the plans you have for the workers who will be displaced by your actions.

2. **Organize.** Prepare a list of tasks usually performed by people in their homes that could be done by robots.
3. **Design.** Prepare sketches of a robotic arm that could turn pages in a book for a person who is paralyzed. Label the parts.
4. **Summarize.** Describe the factors that affect the adoption of robotics technology by manufacturers.
5. **Evaluate.** While in a hospital, you are visited by a robot that has a display screen showing your doctor's head. Describe the pros and cons of this robot.

TechByte

Learning to Walk, Robot Style.
Scientists have created a robot that can learn to walk, just like people do. This robot uses gravity, along with springs and motors that imitate muscles, to move without using as much energy. While it starts walking by falling and catching itself just like a toddler might, the robot learns to walk in 20 minutes!

CHAPTER 18

Bioengineering

Objectives

- Define *bioengineering*.
- Give three examples of functional prostheses.
- Identify two ways of diagnosing sicknesses or conditions.
- Describe how ergonomics can help someone work more efficiently.
- Explain how bioprocessing is used to help improve the environment.

Vocabulary

- **bioengineering**
- **prosthesis**
- **myoelectric signal**
- **implant**
- **diagnosis**
- **medical imaging**
- **ergonomics**
- **bioprocessing**

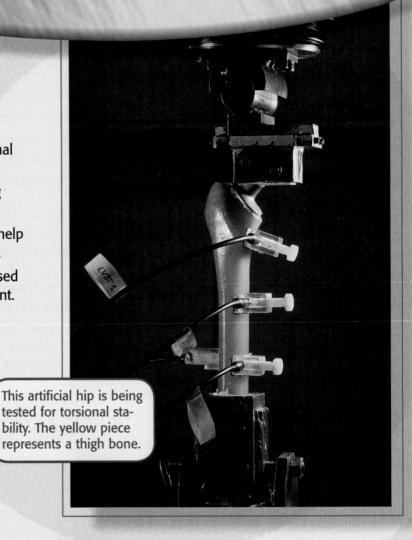

This artificial hip is being tested for torsional stability. The yellow piece represents a thigh bone.

Activities

- Making a Heart Beat
- Controlling Your Comfort
- Creating a Joint Replacement

What Is Bioengineering?

It was a sunny July day in 2002. Eleven hundred athletes representing 75 countries gathered in Lille, France, for world-class track and field competition. Expectations were high that the United States team would break many standing world records.

In an early event, the women's 800-meter run, Kelly Bruno set a new record of 2 minutes and 46 seconds. Marlon Shirley bested his own world record time in the 100-meter dash with a time of 11.08 seconds.

On the last day of competition Jeff Skiba, another American athlete, set a new world record in the high jump, leaping over the bar at 6'10".

Sounds like an ordinary track and field event, doesn't it? This event was far from ordinary. Each of the athletes in this competition was disabled and competing in the paralympic games.

Kelly Bruno, Marlon Shirley, and Jeff Skiba are each leg amputees. Fitted with lightweight, high-performance carbon fiber legs and feet, these athletes brought home the gold for the United States. See **Fig. 18-1**.

How do you design artificial limbs that work like the ones nature gave you? Who creates these technological marvels? The answers are in the fascinating field of bioengineering.

The prefix "bio" means life or living organism. In Chapter 1, "engineering" was defined as the use of math, science, communication, and technology to develop design solutions for technical problems. **Bioengineering** is the use of engineering to solve problems in biology, medicine, human behavior, health, and the environment.

Bioengineers design replacements for body parts. They develop new materials that can be used inside a living organism. Bioengineers design devices and machines to diagnose and treat disease and to help rehabilitate patients with impairments.

Fig. 18-1. Technology has made it possible for athletes with all kinds of physical disabilities to compete. This athlete is running with a carbon fiber leg and foot.

What else do bioengineers do? They develop biological systems for use in cleaning up the environment. Bioengineers also study the workplace, trying to create a better match between the human worker and the work environment to increase productivity and safety.

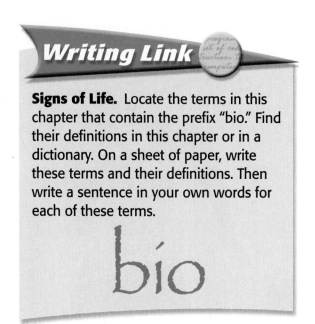

Designers of replacement limbs start by studying the natural movements of the body. For example, they measure the amount of flex and the energy returned by a knee or ankle joint. They examine all the ways a limb moves. This field of study is known as biomechanics. Prostheses are designed to move like natural limbs.

Early prosthetic limbs were mechanical devices that required pulleys, levers, and cables to operate. The user needed a lot of energy to move the limb, and the amount of motion was limited. Modern devices use motors and computers to give the limb a more natural movement. See **Fig. 18-2.**

Prostheses

We usually give very little thought to everyday actions like walking, running, throwing a ball, or lifting a glass. For many people with disabilities, these tasks can be very difficult or even impossible to perform. A person with a disability may be missing a body part or have a faulty biological control system that does not allow the body part to work properly.

Specialists in biomedical engineering create replacement body parts for people with disabilities. Such parts are called **prostheses**. Arms, hands, legs, and feet are common prosthetic devices. Some prosthetic limbs are purely for looks, but most are functional. Modern prosthetic limbs work more like natural limbs than ever before.

Fig. 18-2. Motors and computer chips in this prosthetic leg enable this athlete to play tennis.

Myoelectric Devices

The brain controls muscle movement in the human body. Commands from the brain are sent through nerve endings as electrical energy to the muscles. The electrical impulse causes the muscle fibers to contract. This electrical impulse is called a **myoelectric** (MY-o-ee-LEK-trick) **signal**.

Biomedical engineers have designed prosthetic devices that use these signals to control the limb. For example, myoelectric signals can be picked up from an arm and relayed to a prosthetic hand. Electrodes placed on the skin of the arm conduct the myoelectric impulse to amplifiers that increase the voltage. A computer built

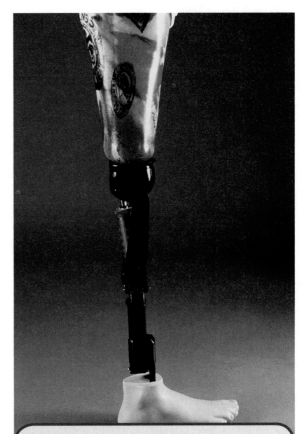

Fig. 18-4. An artificial leg is highly successful in duplicating the movements of a natural leg. The foot moves by reflex action in relation to the leg using a spring-loaded mechanism.

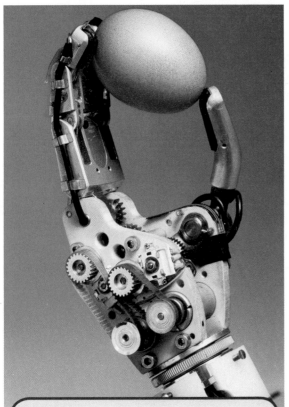

Fig. 18-3. A built-in computer in this artificial hand interprets electrical nerve signals from sensors under the skin of the arm. This opens and closes the hand. The grip is controlled by sensors on the thumb and fingers.

into the prosthetic hand reads the signal and controls motors in the hand, commanding it to open or close the grip. Myoelectric hands can apply twenty pounds of pinch force to grasp an object. See **Fig. 18-3**.

A similar system is used in leg designs. Some prosthetic devices have hydraulic cylinders that provide the force for bending, walking, and running. Computers control the hydraulics, and the leg can be programmed to the user's natural gait (walk), helping to maintain stability. See **Fig. 18-4**.

 For more on fluid power, see the "Fluid Power" lab on the Student CD and Chapter 16, "Hydraulics and Pneumatics."

Sensory Systems

If a prosthetic hand can move like a natural hand, can it also feel heat and cold?

The answer is yes. Temperature sensors placed in the prosthetic hand respond to heat and cold. The sensors send a signal to a computer within the hand. The computer then sends a signal through wires placed on the user's skin. The wires relay the sensation of heat and cold directly to the user's skin. Myoelectric hand users can now know the temperature of an object before they pass it to another person. How might this ability prevent injury?

Implants

Implants are prosthetic devices placed inside the body. Teeth, joints, tissue, and even heart valves can be replaced with implants. Materials and biomechanics are key to implant design.

The human body is a harsh environment for foreign (outside) materials. Chemicals that make up body tissue and organs react badly with many foreign materials. Some materials may even be toxic (poisonous) to our natural systems. Others may be rejected by the body's natural defense system, called our immune system.

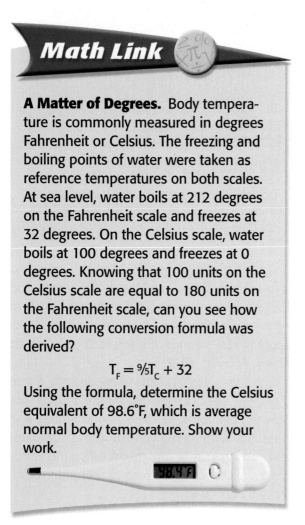

Math Link

A Matter of Degrees. Body temperature is commonly measured in degrees Fahrenheit or Celsius. The freezing and boiling points of water were taken as reference temperatures on both scales. At sea level, water boils at 212 degrees on the Fahrenheit scale and freezes at 32 degrees. On the Celsius scale, water boils at 100 degrees and freezes at 0 degrees. Knowing that 100 units on the Celsius scale are equal to 180 units on the Fahrenheit scale, can you see how the following conversion formula was derived?

$$T_F = \tfrac{9}{5}T_C + 32$$

Using the formula, determine the Celsius equivalent of 98.6°F, which is average normal body temperature. Show your work.

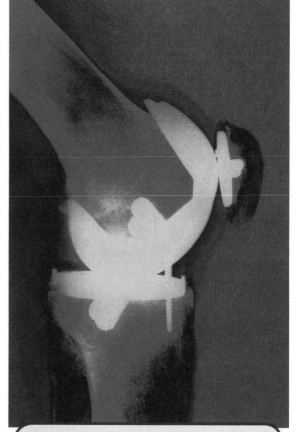

Fig. 18-5. Titanium is a metal that is used in artificial joints because it is strong, lightweight, and does not react with body chemistry.

Impact of Technology
New Materials for the Human Body

As medical treatment continues to progress, unanticipated concerns often arise. In the development of artificial body parts, such as heart valves and knee or hip joints, there arose the problem of immunorejection. This is a situation in which the body rejects foreign matter.

Medical researchers and materials engineers are designing alloys and compounds that do not trigger a rejection response. Parts made from these materials remain strong and durable enough to provide long-lasting service.

Investigating the Impact

Research the Internet or other sources to learn how the problems of immunorejection are being overcome.

1. What alloys and materials are used to reduce rejection?
2. What role is nanotechnology playing in the development of new materials for medical use?

Titanium is a metal that is commonly accepted by the human immune system. Titanium is strong and lightweight. It is used in prosthetic hip and knee joints. See **Fig. 18-5**. Plastics such as woven acrylics are also accepted by the body's chemistry. These materials are used to make flexible artificial veins.

Bioengineers working with doctors, electrical engineers, and speech science specialists have created an implant that has changed the lives of many hearing-impaired people. A cochlear (ko-KLE-er) implant is a device that is placed into the inner ear. It can restore some hearing to people with hearing impairments. See **Fig. 18-6**.

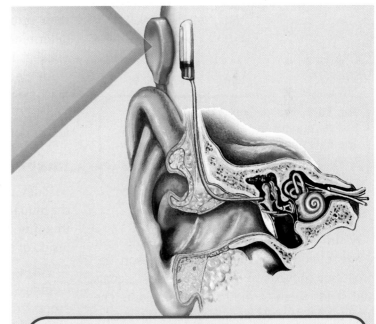

Fig. 18-6. This cochlear implant transmits electrical impulses directly to sensory cells in the cochlea to restore some hearing in the hearing-impaired.

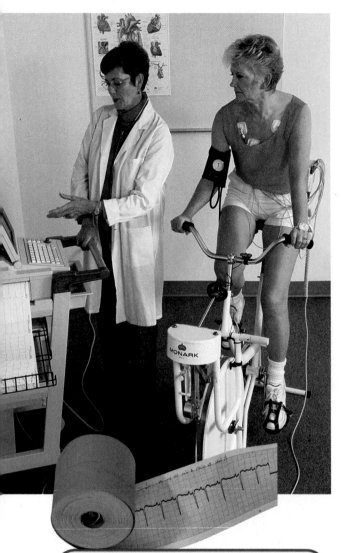

Fig. 18-7. This person is taking a stress test to monitor the efficiency of her heart. The EKG printout shows how well her heart is reacting to the stress. The arm cuff measures her blood pressure.

A tiny microphone worn outside the ear sends sound energy to a speech processor. The processor converts sounds to electronic signals and sends them to the cochlear implant. The implant delivers the signals to nerve fibers connected to the brain. The brain can then interpret the sounds. For more on cochlear implants, see the Student CD.

Diagnosis

Diagnosis is the process of examining a patient and studying symptoms to find out what illness or condition the patient has. Like engineers, doctors use many resources to solve medical problems. Machines developed by bio-engineers are a primary tool used by doctors for the diagnosis and treatment of illness.

Medical equipment can measure, monitor, and image the body's natural functions as well as illnesses and diseases. Using this equipment, doctors can measure your body temperature, blood pressure, pulse rate, breathing rate, blood chemicals, and electrical body signals as well as image your entire body.

Measuring Human Electrical Signals

An electrocardiograph (ECG or EKG) is a machine designed to record the electrical currents of the heart. Electrodes placed on the skin of the chest pick up and amplify the tiny electrical waves produced by the heart. The machine then prints out a graph that doctors use to determine if the heart is working as it should. See **Fig. 18-7**.

An electroencephalograph (EEG) is another machine that reads the body's electrical signals. The EEG measures and records the electrical activity of the brain.

Imaging the Human Body

Today doctors rely on the ability to see into the human body to help with diagnosis. **Medical imaging** is the process of taking pictures of the inside of the human body.

X Rays. Many different machines are used to image the body. X-ray machines have been around for a long time. X-ray machines take photographs of the inside of the body using film that is sensitive to X-ray light. X rays are a type of electromagnetic wave. They have a shorter wavelength than visible light, and they

Fig. 18-8. X rays can take pictures of various parts inside the human body. This X ray of the hand can be taken because the bones absorb more of the X-ray light than the soft tissue.

Science Link

The Body Electric. In the late 1700s an Italian doctor and scientist, Luigi Galvani, theorized that nerves conduct electricity and that this electricity is generated within the body. He devised an experiment in which the nerve of one frog touched the muscle of another, causing the muscle to move. Galvani's discovery of bioelectricity paved the way for inventions such as the EEG and EKG. Research Galvani's work. What other discoveries and inventions did he make that relate to electricity?

can penetrate deep into materials like bone, tissue, fat, and muscle. Each of these materials absorbs the X rays differently and produces a different image on the film. Bones absorb more of the X rays than the tissue, which is why bones are more visible on X-ray pictures. See **Fig. 18-8**.

MRI. Engineering advancements in imaging technology have created machines that can take even better pictures of the human body. Magnetic resonance imaging (MRI) makes three-dimensional images of areas as small as one cubic millimeter.

The MRI machine uses magnetic energy, radio waves, and computers to image the human body. The patient is placed inside the bore, or tube, of the machine. A large circular magnet creates a magnetic field around the patient. The magnetic energy makes hydrogen atoms in

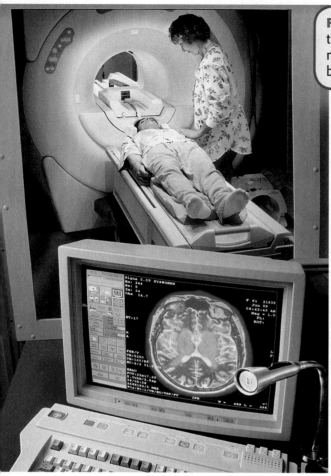

Fig. 18-9. An MRI machine can make three-dimensional images as small as one millimeter. You can see the image of the brain on the monitor.

that can be detected. A computer analyzes the data and generates an image. PET scans are often used to diagnose brain disorders.

Ultrasound. Ultrasound imaging uses sound waves. The sound waves are bounced off structures inside the body, creating echoes. A computer interprets the echoes and generates an image on the computer monitor. Ultrasound is often used to check how a baby is developing inside the mother's body.

Treatment

After an illness is diagnosed, treatment usually follows. Treatment may take the form of drug therapy or even surgery. Here again the machines designed by bioengineers are used to monitor and record body functions. They may even help perform surgery. For example, they may help repair vision.

LASIK Surgery

Do you know anyone who has had his or her vision corrected by LASIK surgery? A patient wearing eyeglasses walks into the eye doctor's office. After a short procedure, the patient leaves the office. Sometimes the patient never has to wear glasses again.

LASIK surgery corrects people's vision by reshaping the cornea of the eye. A computer-controlled laser cuts a tiny flap into the clear covering of the cornea. The flap is lifted up and the laser is used to reshape the cornea to achieve better focus. Lasers, computers, and computer software make the procedure possible. See **Fig. 18-10.**

each cell of the exposed area align with each other. Radio waves from the machine cause the protons in the hydrogen atoms to spin. As the protons spin, they give off a faint signal that is picked up by the machine and amplified. A computer arranges the signals into a 3D picture of the cells. The image is so detailed that even slight changes in the body's structure can be detected. See **Fig. 18-9.**

CT Scan. Computerized tomography (CT) is another method of imaging the body. Unlike an MRI, a CT scan uses a rotating X-ray machine, rather than magnetic fields, to create an image on the computer.

PET Scan. In positron emission tomography (PET), a radioactive substance is injected into the patient. The substance emits gamma rays

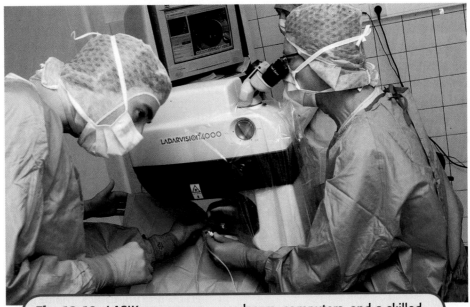

Fig. 18-10. LASIK eye surgery uses lasers, computers, and a skilled surgeon to perform eye corrections.

Ergonomics

Why is a dentist's chair different than other chairs? It has been designed to support your body in a comfortable way. A dentist's chair is designed ergonomically.

Ergonomics (erg-oh-NOM-icks) is the study and design of equipment and devices that fit the human body, its movement, and its thinking patterns. Ergonomics is sometimes called human factors engineering.

Ergonomic designs help people work more efficiently. Human factors engineers must consider the anatomy (structural makeup) and psychology (mind and behavior) of people when they design. See **Fig. 18-11**.

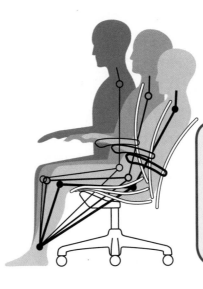

Fig. 18-11. The ergonomic design of this chair is based on human needs. It provides comfortable support. Note the contoured seat and back.

Suppose you are a human factors engineer. You are asked to design a workstation for your classroom. What factors would you consider?

You might start with the design of the chair. You would want the chair to have an adjustable height. This would allow a person's feet to comfortably touch the floor. A lumbar support would brace the lower back. The seat would be contoured to the thighs and lower body.

The desk height would have to be adjustable to match the distance from your other shoulder to your elbow. This would allow people to type comfortably. The position of the computer monitor would be important to avoid eye fatigue and neck strain. The new workstation would be more comfortable. The more comfortable you are when you work, the more efficient you will be.

Environment Design

Human factors engineers design environments as well as individual items. For example, they design living areas and working areas. Again, human anatomy and psychology must be considered. Lighting, noise, temperature, and air quality must also be considered.

What changes would you make to your classroom to improve ergonomic design? What factors would you have to consider in the design of an operating room? How would these factors differ from the design of the welding area of an assembly line?

Reading Link

Graphic Organizer. Making a graphic organizer can help you see the relationships among topics. Make a graphic organizer that shows the topics discussed in this chapter. **Figure 18-12** shows how to start.

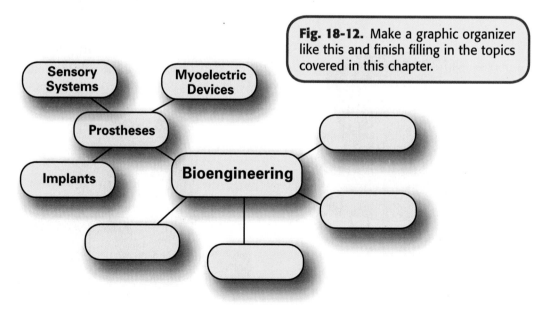

Fig. 18-12. Make a graphic organizer like this and finish filling in the topics covered in this chapter.

Fig. 18-13. This large sewage treatment plant uses bioprocessing to purify the sewage through several different stages.

Bioprocessing

A technological process consists of a series of steps. These steps are used to change materials from one form to another. Try to imagine all the processes needed to manufacture an automobile. Materials need to be separated, joined, formed, and conditioned.

Bioprocessing also processes materials. **Bioprocessing** is bioengineering technology that uses living microorganisms or parts of organisms to change materials from one form to another. Microorganisms are living creatures too small to be seen by the unaided eye.

What products are created through bioprocessing? You probably eat one each day—bread. In breadmaking, live yeast cells are added to the dough. The cells digest the sugar and starch in the dough and also release carbon dioxide. The carbon dioxide forms pockets of gas in the dough. This causes the dough to rise.

Cheese, yogurt, sour cream, vinegar, and sauerkraut are other foods made using bioprocessing microorganisms.

Bioprocessing and the Environment

Do you want to clean up oil spills? Do you want to filter pollutants from rivers? Do you want to clean pesticides from farm fields? Bioprocessing techniques can help.

Algae are microorganisms. Algae are being used to clean up all kinds of messes. For example, mats made from fermenting grass clippings are seeded with blue-green algae. These mats are then placed in ponds. There they form a slimy cover on the surface. The algae eat the pollutants in the pond—as well as the grass mats. The pollutants are turned into less harmful compounds, such as carbon dioxide. Such mats have been used to digest acid runoff from coal, uranium, and manganese mines.

Microorganisms are also used in sewage treatment plants to break down human waste. The waste is converted into methane gas, carbon dioxide, and a fertilizer-type material. The process uses acid-producing bacteria. See **Fig. 18-13**.

Making a Heart Beat

Identify a Need/Define the Problem

Artificial hearts are prosthetic devices. They are often used for people who need a heart transplant but are waiting for one to become available. An artificial heart can be used until a transplant can take place.

In this activity, you will design and build a model of an artificial heart for the human body. An example of an artificial heart is shown in **Fig. A**.

Gather Information

Biomechanical engineers are responsible for designing replacement body parts. Their job requires an understanding of human anatomy and how our natural systems work. Examine reference material on the human circulatory system. Research details on the heart. For example, how many inputs and outputs does the heart have? Does blood move through every part of the heart every time it beats? How much blood is moved with each beat? How does it attach to other parts of the body?

Develop Possible Solutions

Develop drawings for three models for an artificial heart based on the information that you have gathered. Be sure to label the dimensions and different parts according to names used to define parts of a real heart. Select modeling materials.

Materials and Equipment
Select from this list or use your own ideas.

- assorted size plastic syringes
- bucket
- cardboard
- charts on the circulatory system
- hoses
- plastic
- metal
- modeling clay
- plastic models of a human heart
- standard material processing equipment
- Styrofoam
- valves
- water
- wire
- wood

Model a Solution

After choosing the best design, build a working model of an artificial heart.

Test and Evaluate the Solution

- Compare the workings of your design to the way the natural heart works.
- Measure the freedom of movement and the ease of movement.
- Measure the volume of water pumped with the artificial heart.
- How will your design fit into the body?

Refine the Solution

- How could the device be improved?
- Modify the design based on the knowledge that you have gained from testing it.
- What materials would be used to construct an actual artificial heart so that it would work inside the body?

Communicate Your Ideas

Demonstrate the operation of your artificial heart. Be sure to point out any modifications that you made after testing. Show comparisons between your model and charts or simulations of how an actual heart works.

Math Link

Heartbeats. Your heart is a muscle that is about the size of your fist and weighs about a pound. It works like a pump and it is always pumping blood throughout the body. How many gallons do you think a pump that small can output? Try to calculate the following:

1. If your heart beats an average of 75 beats per minute, how many times would it beat in a full day?
2. If each beat pumps about .015 gallons of blood, how much blood would be pumped in one minute? In one hour? In one day? In one year?

Fig. A

Controlling Your Comfort

Identify a Need/Define the Problem

Human factors engineers design products so that they fit the human body comfortably. In this activity, you will design and build an ergonomic video game controller. Your controller must have the following:
- Buttons for character actions
- A directional pad or joystick
- A "start" button
- A cord

Gather Information

What should be taken into consideration when designing a video game controller? What size should it be? How many buttons and controls will it have? See **Fig. A**. Play several appropriate video games and take note of the actions of characters. You should have enough buttons to operate a character's movements. Take note of any problems you have with the controller you are using while playing the video games. You can incorporate improvements into your model.

Develop Possible Solutions

Develop drawings for three models of a video game controller based on the information that you have gathered. Be sure to label the dimen-

> ### Materials and Equipment
> **Select from this list or use your own ideas.**
>
> - cardboard
> - material processing equipment
> - metal
> - modeling clay
> - plastic
> - Styrofoam
> - wire
> - wood

sions of the controller. Try to come up with a design unique from existing video game controllers. Select modeling materials.

Model a Solution

After selecting a design, build a full-scale model of the controller.

Test and Evaluate the Solution

- Compare your design to existing video game controllers.
- How well does your design fit in a user's hands?
- Can all the buttons or controllers be reached easily?
- Allow classmates to handle the controller and provide feedback on its ergonomics.

Refine the Solution

How could the device be improved? Modify the design based on the knowledge that you have from testing it.

Communicate Your Ideas

Demonstrate the ergonomics of your video game controller to the class. Be sure to point out any modifications that you made after testing.

Science Link

Ergonomically Aware. In this activity, you should have learned how human factors engineers design video game controllers. However, engineers must design more than just controllers for a video game user. Research such items as chairs, TVs, and computer monitors to see how else engineers help users play video games comfortably.

Engineers have also developed alternatives to controllers. Find out how these alternatives have fared and discuss any failures or successes.

Fig. A

Creating a Joint Replacement

Identify a Need/Define the Problem

Design and build an implant to replace a joint found in the human body. Common joint replacements focus on the hip, elbow, knee, or finger joints.

Gather Information

Examine reference material on the human skeletal system. See **Fig. A**. Select a body joint for which you wish to design an artificial replacement. Research details on the joint. What type of joint is it? How does it move? What muscles control it? How does it attach to other parts of the body? What size is the joint? Answering these questions will help you construct your model.

Develop Possible Solutions

Develop drawings for three models of the same joint based on the information that you have gathered. Be sure to label the dimensions of the joint. Select modeling materials.

Materials and Equipment
Select from this list or use your own ideas.

- wood
- metal
- wire
- modeling clay
- plastic
- cardboard
- Styrofoam
- material processing equipment

Safety Alert
Look up "Safety Data Sheets" on the Student CD and prepare a data sheet for this activity. As you work on the activity, be sure to follow all safety rules.

Model a Solution

After selecting your design, build your model of an implant.

Test and Evaluate the Solution

- Compare your design to the way the natural joint operates.
- Measure the freedom of movement and the ease of movement.
- Measure the degree of freedom.
- How will your design fit into a body?

Refine the Solution

How could the device be improved? Modify the design based on the knowledge that you have gained from testing it.

Communicate Your Ideas

Demonstrate the operation of your artificial joint to the class. Be sure to point out any modifications that you made after testing.

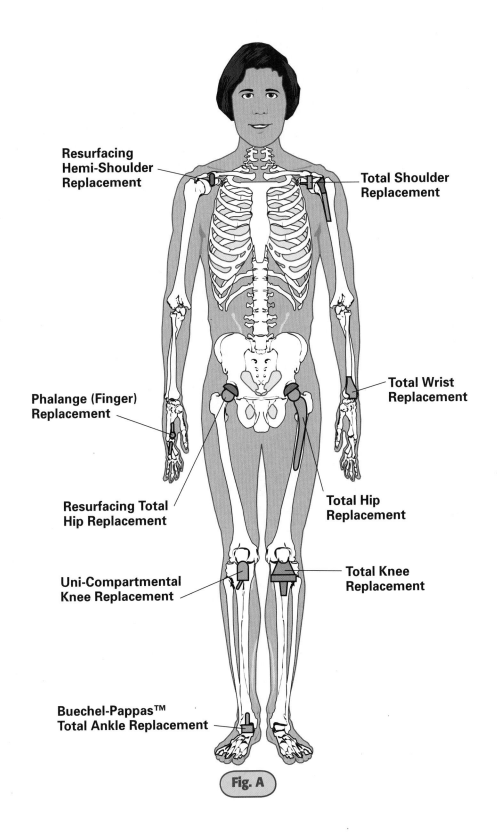

Resurfacing
Hemi-Shoulder
Replacement

Total Shoulder
Replacement

Phalange (Finger)
Replacement

Total Wrist
Replacement

Resurfacing Total
Hip Replacement

Total Hip
Replacement

Uni-Compartmental
Knee Replacement

Total Knee
Replacement

Buechel-Pappas™
Total Ankle Replacement

Fig. A

Exploring Careers

Biochemist

ENTRY LEVEL | TECHNICAL | **PROFESSIONAL**

Biochemists study the chemistry of living things. The chemical combinations and reactions that biochemists study in people, plants, or animals are metabolism, reproduction, growth, and heredity. Biochemists study the complex chemical and biological systems that make up life.

Biochemists are just one type of biological scientist. Nearly 50% of all biological scientists work for federal, state, or local governments. The others work for hospitals, companies that make drugs and medicines, and scientific research and testing labs. They may also teach in colleges or universities.

Qualifications

Biochemists must have a good science and math background as well as good written and verbal communication skills. A biochemistry student usually takes courses in chemistry, biology, math, physics, and computer science. A bachelor's degree is required to do jobs that are not research related, such as testing and inspection or sales. Biochemists with a bachelor's degree may also work as research assistants or technicians. A master's or doctoral degree is required for research or college teaching jobs.

Patience is a good trait for a biochemist to have to be able to work for long periods of time and solve complex problems. Good communication skills and the ability to work well with other people on a team are also important.

Outlook for the Future

The job outlook for biochemists is good. Long-term research projects are sometimes safe from economic downturns because the money is already slotted for the project. On the other hand, new research funding can affect how many new jobs will be available.

Writing Skills

Biochemists should be able to write, rewrite, and proofread their writing until it is clear and accurate. Correct spelling and grammar are important.

Researching Careers

Find out about jobs for biochemists. Research the education and experience of a famous biochemist. What are her or his achievements? Write a one-page paper about what you find.

CHAPTER **18** Review

More activities
on Student CD

Key Points

- Bioengineers design replacement body parts, diagnose and treat patients, and help improve the workplace and the environment.
- Today's prosthetics work more like natural limbs.
- Medical equipment can measure, monitor, and image the human body.
- Human factors engineers design environments as well as individual items.
- Bioprocessing uses microorganisms or parts of organisms to transform materials.

Read & Respond

1. What is bioengineering?
2. What do modern prostheses use to allow for more natural movement?
3. What are three different kinds of functional prostheses?
4. Name two ways to diagnose an illness or condition.
5. Identify five ways to image the human body.
6. What does an EEG do?
7. What is a cochlear implant?
8. How can ergonomics help improve work efficiency?
9. What makes a LASIK procedure possible?
10. How can bioprocessing be used to help clean the environment?

Think & Apply

1. **Design.** Create a simple ergonomic device that can be used in front of a computer.
2. **Connect.** Use the information on lasers in Chapter 15 to provide further details on how LASIK surgery works. For example, what type of laser is probably used during LASIK surgery?

3. **Extend.** Research the Americans with Disabilities Act. What types of changes have public places and private businesses had to make to comply with this law?
4. **Compare and Contrast.** Describe the similarities and differences among forms of medical imaging.
5. **Hypothesize.** How do you think bioengineers obtain microscopic organisms used to clean up waste?

TechByte.

Better Chips. Engineers are currently researching the use of neuromorphic microchips. These chips mimic the organizational structure of the human brain. Like the brain itself, the new chips are faster and more efficient than conventional computer chips. Researchers are using the chips to design artificial retinas for eyes, improved cochlear implants for ears, and other potential prostheses.

CHAPTER 19

Forensics

Objectives

- Define *forensics*.
- Describe the work of the forensics team members.
- Name two major types of evidence and give examples.
- Discuss how science and technology can be used to solve crimes.

Vocabulary

- **forensics**
- **crime scene investigator (CSI)**
- **evidence**
- **latent print**
- **forgery**
- **toxicology**

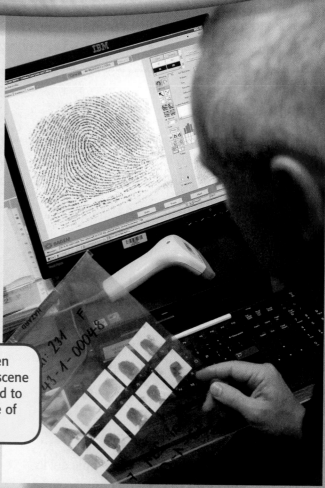

Fingerprints taken from the crime scene can be compared to a huge database of fingerprints.

Activities

- Fighting Forgery
- Name That Ink
- Something Is Afoot

What Is Forensics?

Forensics is the application of science and technology to the law and the solution of crimes. Possibly the first forensic scientist was the fictional character Sherlock Holmes. Holmes used fingerprinting, document examination, and blood analysis to solve crimes. Today forensics is more complex. In addition to the physical sciences of chemistry and physics, it involves behavioral science, which seeks to understand why people behave as they do. For example, chemistry is used to analyze substances found at crime scenes. Physics helps to explain exactly what happens when a gun is fired. Biology is also important. It is used to analyze evidence left by people to determine how a person died.

Modern forensics makes use of many technologies to gather evidence at the scene of a crime and to analyze that evidence in the laboratory. See **Fig. 19-1**. Still and video photography are used to record crime scenes for further analysis and courtroom use. Special lights are used to discover biological evidence, such as blood, even when it is not visible to investigators. In the laboratory, different types of microscopes are used to examine small pieces of evidence. Specialized equipment is used to analyze unknown substances. Computer databases are used to compare fingerprints taken from a crime scene to millions that are on file.

In this chapter, you will read about people who work in forensics. You will learn how they use science and technology to solve crimes.

The Forensics Team

The first person to arrive at a crime scene is usually a uniformed police officer. The police officer will secure the scene to preserve evidence and determine who else needs to be called. An ambulance may be needed to treat the injured. Detectives with special skills may be called. Some detectives specialize in areas such as burglary or homicide.

Fig. 19-1. Science and technology are used in gathering information at the scene of a crime.

Writing Link

Leadership and Teamwork. In the workplace, most people do their jobs as part of a team. Being an effective team member or team leader requires skills, and these skills can be learned. In doing the activities in this book, you have many chances to work in teams and practice leadership. What skills are needed to become a good leader or team member? Communication skills are very important. For example, a leader needs to be able  to explain things clearly. Make a list of other skills that good leaders and good team members need. One source of information is the Technology Student Association.

CSI

Crime scene investigators (CSIs) play an important role at major crime scenes. They use a variety of tools and techniques to find and gather evidence. CSIs record the scene by taking photographs and making sketches. Everything they gather is taken to the crime lab for analysis.

The CSI Toolkit. Crime scene investigators carry toolkits containing the items needed to gather evidence. Items typically found in the toolkit include:

- Special gloves for the CSIs to wear so they don't contaminate the evidence.
- Eye protection to wear when using special chemicals.
- Still and video cameras to record the crime scene.
- Chalk to mark the position of objects by outlining them before they are removed.
- Bags, envelopes, and tubes so that evidence can be taken to the lab for analysis.
- Special tapes and powders for recovering fingerprints. See **Fig. 19-2**.
- Lights to enhance viewing the scene and for discovering bodily fluids.
- Magnifying glasses for locating evidence such as fibers.

- Rulers and tape measures for recording dimensions and distances.
- Crime scene tape to mark off and protect the crime scene.

Medical Examiner

When a crime involves the death of one or more individuals, a medical examiner is usually called to the scene. Medical examiners are physicians who specialize in forensics. Their work begins at the crime scene and then continues in the laboratory. They determine the cause of death, when death occurred, and gather evidence from the body. Medical examiners also perform autopsies to gather additional evidence. Once a case reaches the courtroom, medical examiners testify as expert witnesses.

Lab Technician

Crime lab technicians analyze the evidence found at crime scenes. They perform a variety of tests on blood and other bodily fluids. They use specialized microscopes to examine hair, paint powders, and all other physical evidence found at the scene. See **Fig. 19-3**. Some technicians specialize in areas such as analyzing DNA, discovering fingerprints on objects, or determining who left the prints.

Evidence

Evidence is anything that helps establish the facts of the crime. Items that are very small are called trace evidence.

Two major categories of evidence are physical and biological. Physical evidence is nonliving. It includes such things as fingerprints, shoe

Fig. 19-2. Dusting a suspected area for fingerprints is one way to find evidence at the scene of a crime.

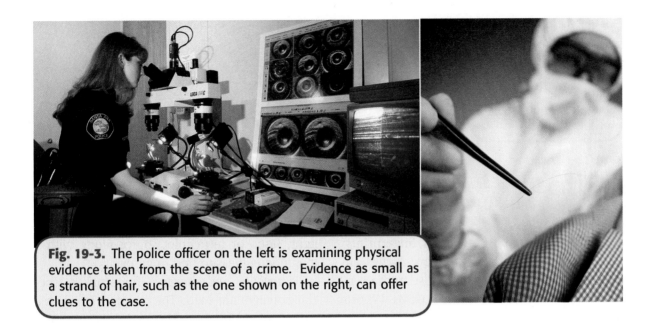

Fig. 19-3. The police officer on the left is examining physical evidence taken from the scene of a crime. Evidence as small as a strand of hair, such as the one shown on the right, can offer clues to the case.

impressions, tire impressions, fibers, firearms, and documents. Biological evidence includes such things as blood, hair, and skin. The findings of toxicologists and medical examiners also are biological evidence.

Fingerprints

The use of fingerprints for identification goes back to prehistoric times. Artists put fingerprint impressions in their clay pottery. Around 1,000 B.C.E. the Chinese used fingerprints as signatures on legal documents. Since around 1880, fingerprints have been used in criminal investigations. To learn about fingerprinting and other forms of physical identification, see the "Biometrics" interactive lab on the Student CD.

Almost every time we touch something, we leave a fingerprint. Our hands are covered with sweat pores. The sweat mixes with body oils and dirt, and this mixture leaves an imprint when we touch a surface.

Fingerprints provide a reliable means of personal identification. The outer layer of skin on our fingers is made up of ridges that are unique to each person. No one else has fingerprints exactly like yours.

Reading Link

Reading Clues. Evidence can provide clues that help law enforcement officials solve mysteries. Did you know that sentences and paragraphs provide clues that help you solve the meaning of unfamiliar words? If you don't know what a word means, look at its context (setting). Other words in the sentence or paragraph may provide clues about what the unfamiliar word means.

Consider the sentence "A shoeprint found at the scene of the crime matched that of Professor Plum's Wellingtons." What do you think *Wellingtons* are?

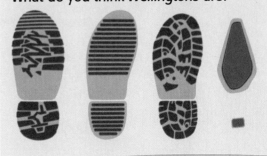

Fig. 19-4. While there are no two fingerprints alike, there are three common types of fingerprints.

Plain Arch   Plain Whorl   Loop

There are three basic types of fingerprints: the arch, the whorl, and the loop. Arches are patterns that start on one side, rise to the center, and then come down like a wave. Whorls resemble circles and sometimes form a bull's-eye pattern. Loop pattern prints are the most common. They start on one side, rise to the center, and return to the same side they started on. See **Fig. 19-4**.

Fingerprints that can be seen with the naked eye are called visible or patent prints. Most prints are not readily visible and are called **latent prints**. The best places to find latent prints are on hard surfaces such as furniture, tools, and weapons. CSIs use light from a flash-

light, laser, or ultraviolet source along with a magnifying glass to find fingerprints.

To recover and preserve fingerprints, special powders are used. The powders come in a variety of colors. Investigators choose a color that contrasts with the surface. The powder is applied with a fine brush. The fingerprint can then be photographed or lifted. See **Fig. 19-5**.

Lifting is done by placing clear tape over the print. As it is carefully peeled off, the pattern sticks to the tape. The tape is placed on a card for further examination and matching.

Chemicals can also be used to help recover fingerprints. For example, a substance similar to super glue is used in a device called a fuming chamber. The object under investigation is placed in the chamber. Fumes from the super glue stick to the prints, which are then photographed.

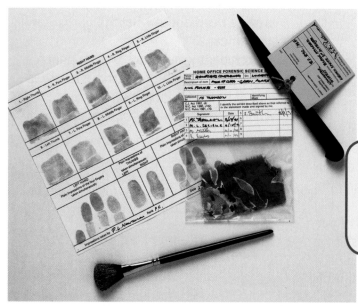

Fig. 19-5. One way to preserve and eventually lift a fingerprint is to use a colored powder that will contrast with the surface. Shown here are a fingerprint record, brush, and forensic evidence.

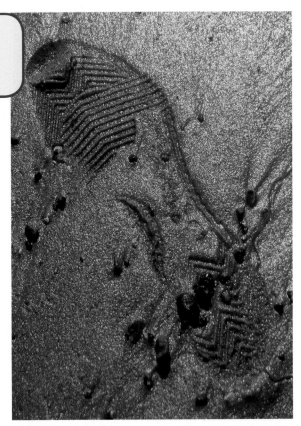

Fig. 19-6. A shoeprint can be a valuable piece of evidence in a crime case. Any defect or unique wearing pattern in the shoe should show up in the print.

AFIS. Fingerprints collected at a crime scene need to be compared to fingerprints on file. For many years technicians needed lots of time and patience to review thousands of fingerprint cards to find a match. The automated fingerprint identification system (AFIS) technology has greatly speeded up and improved this process. To use AFIS, the technician scans the evidence print. A computer searches the millions of fingerprints on file and identifies possible matches. It takes about two seconds to search through a million prints on file. When AFIS finds a match, the technician must verify that the match is correct. An example is shown on page 422.

Shoeprints

Shoeprints are found at many crime scenes. Prints made from soil tracked into a crime scene can help identify the point of entry. Wet, muddy, or bloody shoeprints can help track the movement of a criminal around the crime scene. When the same shoeprint is found at several different crime scenes, it indicates that the same person may be involved in a number of different crimes.

Investigators first try to identify the size and manufacturer of the shoe. The FBI has a database of shoe patterns to help with this task. Next the investigator looks for unique features, such as damaged areas or unusual wear patterns.

Some shoeprints, such as a muddy shoeprint left on a hard surface floor, can be obtained using photography. To record a shoeprint left in soil, a casting can be made using a material similar to plaster. Later the photograph or

casting can be compared to a print made from a suspect's shoe. The investigator will use a magnifying glass or microscope to compare the tread design and wear patterns of the two samples. See **Fig. 19-6.**

Tires

Since automobiles are used in many criminal activities they are an important part of many investigations. Investigators collect evidence from the interior of vehicles and also try to determine if a vehicle's tires have left behind useful evidence. For example, when a vehicle goes through mud, it can leave an impression in the mud and may transfer mud to nearby pavement.

Through databases maintained by the FBI and other law enforcement agencies, it is relatively easy to identify the size and the manufacturer of a tire that left a particular impression.

Fig. 19-7. Tire tracks can show unique characteristics. A suspected tire can then be matched to the tracks.

leave marks. Analysis of the marks will help determine the size of the screwdriver that was used. Unfortunately, it's not easy to match a tool mark to only one screwdriver, but it does provide useful information.

At the crime scene, tool mark evidence can be obtained through photography or by making castings. In the laboratory, microscopes will be used to further examine the tool impression and compare it to marks made by particular tools.

Fibers

We are surrounded by fibers. A fiber is a small strand of fabric. Fibers from clothing, carpeting, seat covers, and other sources are easily transferred from other people or objects that we come in contact with.

There are two major kinds of fibers: natural and synthetic. Cotton, wool, and silk are natural fibers. Cotton is the most common natural fiber, and it has a distinctive look when examined under a microscope. Nylon and polyester are examples of synthetic fibers.

At a crime scene, fibers can be gathered from the victim's or suspect's body using tape or tweezers. In the lab, fibers may be examined under a microscope or tested to determine chemical composition. The chemicals used to dye the fiber can help identify the manufac-

Analysis of the tracks left by each of the vehicle's tires also helps to narrow down the size and type of vehicle.

Tire track evidence can be gathered by photography. A ruler may be placed next to the impression to help record its exact size. Castings similar to those made of shoeprints can also be made. When a suspect vehicle is located, its tires are inked and then rolled across a long sheet of paper to record the impressions. See **Fig. 19-7.**

Tools

Many crimes involve the use of tools. Tools often leave unique marks that can be studied and identified. For example, when a screwdriver is used to pry open a window, it will

Fig. 19-8. This fabric, found at a crime scene, can be tested to see if it matches a tiny piece of fiber found on the suspect.

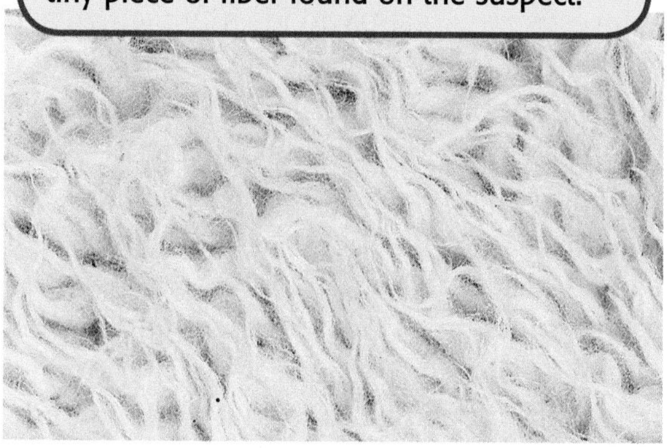

Fig. 19-9. This gun is protected within a casing as it is tested for the ballistics of the bullet. The infra-red radar measures the speed with which the bullets are released from the gun.

turer. This information is helpful for determining whether the victim and suspect were in contact with the same material, such as carpeting. See **Fig. 19-8**.

Firearms

Most people who own guns use them properly and safely. However, guns are used in many criminal activities including robbery, assault, and murder. Firearms examiners are specially trained technicians who study the weapons, bullets, and shell casings from crime scenes.

When the trigger of a gun is pulled, the firing pin strikes the shell casing and the gunpowder inside explodes. The explosion causes the bullet to leave the gun at great speed. The shell casing remains behind. Both bullets and shell casings are collected at crime scenes. Sometimes bullets are removed from floors and walls. Other times they are removed from victims during surgery or during autopsy.

Firearms investigators examine bullets to determine the caliber of weapon used. Caliber refers to the size of the weapon. For example, a 9 millimeter handgun has a barrel with an inside diameter of 9 millimeters. The size of the bullet and the type of bullet help to identify the type of weapon used. Shell casings also provide useful evidence because of the distinctive marks left by firing pins.

As bullets travel through the barrel of a gun, they pick up marks (called striations) that may enable them to be matched to a particular brand of weapon. Guns have special grooves cut into their barrels to increase their accuracy. The high parts are called lands. The grooves cause the bullet to spin to improve its accuracy. Manufacturers design their guns with a certain number and style of grooves and lands.

When a weapon that may have been used in a crime is recovered, it can be fired in a testing chamber. See **Fig. 19-9**. Then a microscope can be used to examine the bullet and compare it to those recovered at the crime scene. Computerized databases also help firearms investigators compare the bullets found at a particular crime scene to those recovered from previous crimes.

There is another way to connect suspects to the use of firearms. When a gun is fired, gases and particles from the explosion end up on the skin and clothing of the shooter. Investigators can swab the hands, arms, and clothing of a suspect and then use a series of chemical tests to determine if the person has fired a gun. Generally this needs to be done within a few hours of using the gun.

Science Link

Spin City. The spiral grooves inside the barrel of firearms are called rifling. The rifling causes the projectile (bullet) to spin as it exits. Similarly, the feathers on an arrow's shaft often have a spiral twist to them, causing the arrow to spin in flight. And of course nearly everyone is familiar with the spiral pass thrown by almost every football quarterback.

Why should these things spin? What advantage might spinning yield? Research the science behind it and let the class in on the secret.

Documents

Document examiners specialize in determining the source and authenticity of documents. They study documents to determine who produced them, when they were produced, and whether they have been altered.

We all have individual handwriting. No two people write alike. In fact our own handwriting changes quite a bit. To verify this, gather a piece of paper, two different pens, and a pencil. Sign your name five times with each instrument and compare your signatures. Are the 15 signatures identical? You will probably find that all your signatures are similar but not identical.

When document examiners compare two written documents, they look for similarities in several areas. They look at the size and shape of individual letters and the speed and pressure used by the writer(s). They also examine the spacing between words and overall spacing on the page. Content is also important. Writing

Fig. 19-10. A person's handwriting has characteristics that are repeated every time he or she puts a pen to paper. These characteristics can be used when examining suspicious documents in a crime case.

style, spelling, and grammar are compared. After comparing these features the examiner can usually determine if the two samples were prepared by the same person. To assist in the process the FBI maintains a database of documents, such as notes used in bank robberies and forged checks. See **Fig. 19-10**.

Forgery refers to preparing or changing a document with criminal intent. Some forged documents involve the copying of signatures. Handwriting analysis can determine whether this has occurred. Other forgeries involve changing things such as the dollar amount on a check. Writing that has been erased or changed can usually be detected by using chemistry, a microscope, or photography.

Document examiners also study papers and inks to determine if any of the pages within a document have been changed or if changes were made using a pen different from the original one. Copy machines and personal computers also produce images that can be compared to other images to determine if they match.

Blood

Bloodstains help solve many crimes because they provide a great deal of useful information. Blood evidence can help determine what happened and when it happened, and it can be used to identify the victim and possible suspects.

Sometimes bloodstains are easy to find. At other times only a small amount of blood is present, or the criminal may have attempted to remove blood evidence by scrubbing the walls or floor. Blood in tiny amounts can be detected through the use of luminol, a fluorescent chemical. Luminol is sprayed where blood is thought to be present. The area is darkened and an ultraviolet light source is used. If bloodstains are present, they glow and become clearly visible. See **Fig. 19-11**.

Identifying the source of bloodstains is challenging. First the material collected must be tested to determine whether it really is blood. If

Fig. 19-11. Hidden bloodstains can be detected by using a special chemical and light to expose the suspected area.

it is blood it must undergo additional testing to determine if it is human blood. Then additional tests are conducted. The investigator will want to determine if the crime scene includes blood from more than one person and the blood type of each sample.

Blood typing provides useful evidence. However, DNA testing, which has been in use for about twenty years, provides a much more accurate way of matching evidence to criminals and victims.

Impact of Technology
Technology vs. Privacy

Forensics can be used both to identify a suspect and to clear one. Knowing this, do you think it is reasonable to expect everyone to submit to testing in a criminal investigation?

Many people feel that their privacy is one of their most cherished rights. They do not think that government agencies (local, state, or national) should be allowed to keep records that might intrude upon that privacy.

Investigating the Impact

Research this issue. What kinds of information do privacy advocates want to keep out of government records? What are the pros and cons?

1. Do you think this conflict of interests can be reconciled?
2. Seek opinions from others and summarize their thoughts on the matter.

DNA

Almost every cell in the human body contains DNA. At a crime scene, blood, saliva, and other bodily fluids can be collected. In the laboratory, DNA molecules can be extracted from these samples.

The results of DNA testing can be compared with DNA evidence maintained in computerized databases or obtained from possible suspects. In recent years, DNA testing has been used to reverse the convictions of a number of prisoners. **Figure 19-12** compares DNA evidence obtained at a crime scene to that taken from two suspects.

Because most people's DNA is unique, DNA testing is often called DNA fingerprinting. However, it does have limitations. The process is complex and requires special equipment and highly trained personnel. Identical twins have the same DNA and family members have similar DNA patterns.

Math Link

DNA Differences. DNA testing is one of the newest tools in forensics. A DNA molecule is like a very long ladder. Each rung of the ladder contains a pair of chemical building blocks called bases. Human beings have many, many base pairs in their DNA, but only about 6 million base pairs vary from person to person (about $\frac{1}{10}$ of 1 percent of all the base pairs available). How many base pairs are there in the entire human genome?

Fig. 19-12. This forensic scientist is inspecting two specimens of DNA evidence to see if there is a match.

Toxicology

Harmful substances are called toxins. **Toxicology** is the study of how drugs, poisons, and other substances affect people. Legal drugs can be toxic if consumed in excess, and so can ordinary substances that we consume every day if used in excess. For example, drinking far too much water can lead to serious health problems and even death.

In the crime lab, toxicologists review evidence to find toxins and determine how they affected a person who consumed or came in contact with them. In the case of a possible homicide, the medical examiner collects bodily fluids and other samples for analysis. Blood is the most important substance studied because it can provide information about what was going on in the body at the time of death. Stomach contents can be analyzed to test for the presence of drugs and poisons. Tests of the kidneys, brain, and liver also provide valuable information. The medical examiner uses information provided by the toxicologist to help determine the cause of death.

Computer Forensics

A computer hacker accesses credit card records and uses them to charge purchases to other people. A con artist sends e-mails to thousands of people, promising quick and easy money. An employee downloads product design files and sells them to a competitor. You hear about these and other computer crimes almost every day. Evidence that helps solve these crimes is collected by CSIs who specialize in computer forensics.

While other CSIs work with physical and biological evidence, computer forensics CSIs search for electronic evidence. They must be able to recover data from computer systems, even if that data has been deleted, damaged, password-protected, or encrypted (written in a secret code). Once they have collected the evidence, they analyze how the data was accessed and used. Like other CSIs, they are responsible for protecting the evidence, making sure that nobody tampers with it.

Fighting Forgery

Identify a Need/Define the Problem

Counterfeiting is one of the largest forms of document fraud. One of the most forged documents is the $20 bill. To fight off potential criminals, the federal government has invested significant amounts of money, time, and effort to create a forgery-proof bill. See **Fig. A**.

In this activity, you will design and produce a forgery-proof banknote (currency bill). The banknote must meet the following criteria:
• Incorporate three or more security features
• Must use three or more colors
• Be no smaller than 4" × 1.5"

Gather Information

Many of the security features found in the new $20 and $50 bills are also found in other international banknotes. You can research and review many of these features at The Bureau of Engraving and Printing's Web site.

Materials and Equipment

Select from this list or use your own ideas.

• paper
• colored pencils
• markers

Math Link

Math and Money. The Bureau of Engraving and Printing produces 37 million notes a day. Of the notes printed, 95% are used to replace notes already in circulation. Nearly half of the notes printed are $1 notes because the average life of a one dollar note is only 22 months. Here are the average times in circulation for the other notes:

$5	24 months
$10	18 months
$20	25 months
$50	55 months
$100	60 months

Did you know that if you had 10 billion $1 notes and spent one every second of every day, it would require 317 years for you to go broke? If you spent 10 billion dollars at the same rate, how long would it take to go broke if you used $5, $10, $20, $50, or $100 notes?

Develop Possible Solutions

Brainstorm multiple ideas for security features and designs for your banknote. Think about the various methods that can be used to reproduce documents and how you can prevent a forgery. Research what other countries do to protect their currency. Also think about the various things that are represented on currency, such as historical figures, national treasures, and pastimes.

Model a Solution

1. Select the solution that you think will best solve the problem.
2. Using white drawing paper and colored pencils, begin to develop your design.
3. Cut out your design and add any remaining security features.

Test and Evaluate the Solution

- Exchange your bill with a classmate.
- Examine the bill and try to identify all the security features.
- Using available resources, attempt to duplicate the banknote.
- Were you able to identify the security features? What was the role of each feature?
- Where you able to replicate the banknote?
- What technologies can criminals use to try to counterfeit money?
- How does technology help us prevent counterfeiting from occurring?

Refine the Solution

After testing, return the bill you examined. Was your banknote forged? Gather the information from the classmate that analyzed your banknote and refine any security features that need improvement.

Communicate Your Ideas

Develop a brochure for your new banknote highlighting the security features and the various technologies used in the banknote. Describe the testing that your design went through as well as any refinements.

Fig. A

Name That Ink

Identify a Need/Define the Problem

When a criminal leaves written evidence, such as a ransom note, document examiners can use the note to identify things such as handwriting and the ink that was used. This information can then be used to identify who was involved in the crime.

One method used to identify the type of ink is called chromatography. Chromatography is the process of separating mixed solutions into individual components. In this activity, you will use the principles of chromatography to identify the type of ink found in a fake ransom note.

Gather Information

Research chromatographs and how they are used in forensics.

Develop Possible Solutions

Prepare a table that will allow you to record the results of your chromatograph tests. Decide on the items that should be included in the table, such as diagrams depicting the individual chromatograph patterns.

Materials and Equipment

Select from this list or use your own ideas.

- assorted blue or black felt pens
- coffee filters
- plastic cups/bowls
- water
- isopropyl alcohol (rubbing alcohol)
- ransom note

Model a Solution

1. Gather several varieties of blue or black felt-tip pens.
2. Obtain a sample of the evidence. (Have a classmate use one of the pens to write a brief ransom note.)
3. Cut several coffee filters into ½" strips.
4. Set up a testing apparatus similar to the testing diagram shown in **Fig. A**.

Test and Evaluate the Solution

- Cut two ½" wide strips of the evidence or ransom note, making sure to select samples having ink on them.
- Hang one strip of the evidence in each solution so that the tip just touches the solution. Make sure the ink does not get immersed.
- Use each of the assorted pens to make a mark ¼" to ½" above the edge of the end of a coffee filter strip. Make two samples for each pen.
- Hang the samples from the dowel so that the very edge of the coffee filter just touches the solvent of each container.
- Examine the banding patterns and determine which of your pens matches the pen used to write the ransom note.
- Did all the samples create a banding pattern in the solvents?
- How can such distinct banding help identify pens used in written evidence?

Refine the Solution

Create a method of testing the pen sample in an alternate fashion. Are there additional solvents that will help you develop more classifications of ink? How would gel ink pens behave when tested? Does inexpensive copy paper produce different banding? Record your answers.

Communicate Your Ideas

Create a reference sheet that can be used to identify the various types of pens used. Be sure to include information on the type, color, and solvency of the pen that you tested. Also include diagrams of the banding pattern so that others can reference their results.

Science Link

Chromatography. As you learned in this activity, chromatography is the process of separating mixed solutions into individual components. In addition to being used in forensics, it can be used in the field of biotechnology to separate various proteins. Different proteins have distinct rates of absorption so that they separate from each other. Research to find out how separating proteins is useful in biotechnology. What other uses does chromatography have?

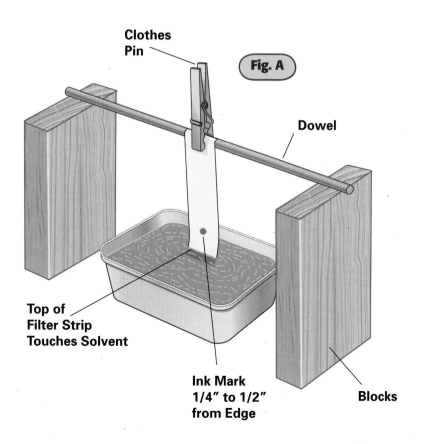

Fig. A

Clothes Pin

Dowel

Top of Filter Strip Touches Solvent

Ink Mark 1/4" to 1/2" from Edge

Blocks

Something Is Afoot

Identify a Need/Define the Problem

Criminals often leave signs of their presence at a crime scene. In this chapter you have read about fingerprints, tire tracks, and shoeprints. See **Fig. A**. Each one of these has unique identifiers. In this activity, you will develop a method of identifying shoeprints.

Gather Information

Research the various types of shoeprints that are available. (You probably have a variety of prints in your class already.) Take notes on the various tread patterns and wear marks that make each print unique.

Develop Possible Solutions

Develop ways to distinguish shoeprints and to categorize them. Be sure to also develop terminology for a variety of wear marks, such as the inside of the shoe heel or the ball of the foot. Develop several sketches and alternate classification formats.

Materials and Equipment

Select from this list or use your own ideas.

- heavy paper
- colored chalk dust
- rulers
- magnifying glass
- shallow trays

Model a Solution

1. Each student in the class should place his or her right shoe in a container of colored chalk dust.
2. Carefully place the shoe on a piece of heavy construction paper and press firmly.
3. Shake off excess powder and spray with a sticky adhesive. Write the name of each shoe's owner on the back of the corresponding print. You have enough information to make a shoeprint database.
4. Repeat steps 1 through 3 for a second set of prints, but have your classmates participate in a different order. Do not write any names on the second set of prints.

Test and Evaluate the Solution

- Pick a shoeprint from your second set of prints. Can you match the shoeprint to the print in the database based on the criteria and identification system you have chosen?
- What elements did you use to help you make your decision?
- Did the unique wear features identified make matching the shoe and its print easier?

Refine the Solution

- If your method was unsuccessful, modify your system to include more identifiers.
- Once you have modified your system, create another set of shoeprints using a different medium. Make sure the medium is something that won't harm anyone's shoes.
- Does your new system work better? Why or why not?

Communicate Your Ideas

Create a forensics report listing all your findings. See if you can make your report as realistic as possible and have categories listed such as location, date, and time of an imaginary minor crime. In the report, describe the process you took in matching prints.

Math Link

A Footwear Problem. Story problems can help you learn about any given topic. Take shoes, for example. Shoes are made in thousands of different shapes and sizes. As the shoe is worn, the shoe's traits change. Shoes acquire cuts, scratches, and other random characteristics.

How unique is a footprint? Pretend that there are 200 different shoe manufacturers. Each manufacturer has 25 different shoe styles available and each style comes in 20 different sizes. Keep in mind that any shoe could be a left or right. Using this data, how unique is any one shoe?

Fig. A

Exploring Careers

Crime Scene Investigator

ENTRY LEVEL | TECHNICAL | **PROFESSIONAL**

A crime scene investigator (CSI) studies a crime scene and its evidence to help solve the crime. CSIs, also called forensic scientists, collect, study, preserve, and document evidence. The evidence might include tissue samples, such as blood or hair; chemical substances, such as drugs; physical materials, such as a vehicle; and ballistics evidence, such as guns or bullets.

CSIs go to the crime scene to photograph and collect evidence. They usually examine it in a laboratory. CSIs are also responsible for storing the evidence. They write reports of what they find and may be called to testify at trials.

Most CSIs work for state and local police departments. CSIs may also work in morgues, hospitals, or universities.

Qualifications

A CSI must have a good science and math background. A high school degree is required, but those with an associate's or bachelor's degree in science or criminal justice will get better, higher-paying jobs. Some employers require specialized training, such as classes in crime scene and fingerprint processing, photography, and blood spatter and arson investigation. Certification may be required in some jobs.

Students who want to become CSIs should take math, chemistry, biology, English, and computer classes in high school.

Because many CSIs are helping solve crimes, reasoning and problem solving are good traits to have. Good communication skills and the ability to work well with others on a team are also very important.

Outlook for the Future

The job outlook for CSIs is good. CSIs with relevant experience and education will have the best opportunities.

Being Ethical

It is very important to be honest and follow professional standards. Your testimony at a trial could help convict or free an accused person. Be conscientious and treat others how you want to be treated. When you say you'll do something, follow through with that action.

Researching Careers

Find out about jobs for CSIs. Where do CSIs work in your area? What education do they have? What experience do they have? Make a poster of what you find.

More activities on Student CD

Key Points

- Forensics involves physics, chemistry, and behavioral sciences.
- CSIs, lab technicians, and medical examiners are all part of the forensic team.
- Evidence is anything that helps establish the facts of the crime.
- Forgery refers to preparing or changing a document with criminal intent.
- Studying how drugs, poisons, and other substances affect people is called toxicology.

Read & Respond

1. What is forensics?
2. What do CSIs do?
3. What do medical examiners do?
4. What role do lab technicians play in forensics?
5. Name two categories of evidence and provide an example of each.
6. What is AFIS?
7. Can fingerprints be invisible?
8. How can you record a shoeprint left in the soil?
9. What is the main reason toxicologists study blood?
10. Why is it important to test fiber from a crime scene?

Think & Apply

1. **Formulate.** With group members, play a game of Clue®. As you play, discuss ways in which a forensics team would solve the crime.
2. **Compose.** Write a description of a crime scene. Then have classmates describe how a forensics team might solve the crime.
3. **Extend.** Make a list of what clothes your classmates are wearing. How similar are the colors of the clothes? How hard would it be to match a fiber to a person in your classroom?
4. **Design.** Make your own CSI toolkit, using only materials found in your classroom.
5. **Distinguish.** Take fingerprints of your classmates and record data about how many of each kind of fingerprint (arch, whorl, or loop) are represented in your class.

TechByte

Labs on the Go. The chemistry lab of the future may be a "lab on a chip" that a police detective could pocket to take to the scene of a crime. Researchers today are developing chemical "labs" on tiny silicon chips that will analyze chemical composition, separate microscopic particles out of blood or other fluids, and decode DNA structure. Only a few years from now, the micro-labs will be more efficient, faster, and more accurate than today's traditional labs.

CHAPTER 20

Agricultural Technologies

Objectives

- Identify five technologies that have changed the nature of farming.
- Discuss how plants are changed through genetic engineering.
- Describe controlled environment agriculture.
- Identify advantages of hydroponics.

Vocabulary

- **agriculture**
- **mechanization**
- **global positioning system (GPS)**
- **genetic engineering**
- **irrigation**
- **fertilizer**
- **controlled environment agriculture (CEA)**
- **hydroponics**
- **aquaculture**

Shown here is a Kansas wheat field being harvested by four combines.

Activities

- From Milk to Glue
- Growing with Hydroponics
- Creating a Water Filter

What Are Agricultural Technologies?

Have you ever visited a farm? Farming, or **agriculture**, is a technological system that produces plants and animals for food, fiber, fuel, and other products. Farms are similar to outdoor factories.

Technology has had a tremendous impact on the farming process and farm efficiency. Think about it. Farmers are less than two percent of the population in the United States, yet they provide enough food for all of us and even help to feed the people of other nations.

Some of the technologies that have changed the nature of farming are mechanization, genetic engineering, irrigation, fertilizing, and controlled environment agriculture. Each of these technologies has helped the farmer produce greater quantities of products with higher quality.

Mechanization

The use of machines to make work easier is called **mechanization** (mek-uhn-eye-ZAY-shun). Mechanization is responsible for some of the greatest increases in agricultural production. Farmers now grow plants and animals with more efficiency than ever before.

Crop Farming

Growing crops involves five activities: clearing the soil, tilling the soil, planting the seeds, cultivating the crops, and harvesting the crops. See **Fig. 20-1**. Long ago these tasks were done by hand. When people learned to train animals for work, the farmer's job was made easier.

The plow was one of the earliest tools used to mechanize agriculture. Plows break up the soil so seeds can be planted. Today, cultivators may be used to break up the soil. Planters drop

Fig. 20-1. Shown below are the five main steps in traditional farming. Today, many farmers practice conservation tillage, in which crops are grown with minimum disturbance of soil.

CLEARING	TILLING	PLANTING	CULTIVATING	HARVESTING

Clearing the land gives the farmer a flat surface to plant.	Tilling loosens the soil so that the plants' roots can take hold.	Seeds are dropped into holes and covered with soil.	Cultivating is the process of caring for the plant.	Harvesting removes the grown plants or products for processing.

Fig. 20-2A. A farmer is planting his crop with an eight-row planter.

the seeds and cover the seeds with earth. This is done in one operation. See **Fig. 20-2A**.

Harvesting is the process of removing the grown plants or products for processing. Many crops are harvested with combines. Combines are huge machines. They can cut a path of wheat thirty feet wide, separate the grain, and package the stalks in one operation. Combines have greatly increased the amount of products a farmer can bring to the market. They have also made larger farms practical. During the 1800s, harvesting at an average farm could take days.

Those farms could have been harvested in one hour by today's combines. See **Fig. 20-2B**.

GPS. Have you heard of GPS technology? A **global positioning system (GPS)** is a navigation system that uses receivers, computers, and satellites to determine the location of an object on Earth. Many hikers use GPS receivers to avoid getting lost.

Farmers are also using GPS. Combined with automated steering devices on tractors and combines, GPS guides the equipment with precise accuracy. The result: perfectly straight furrows

Fig. 20-2B A few months after planting, the combine harvests the field in just a few hours.

Impact of Technology

Technology Assessment

For thousands of years, farming was the chief occupation. Agricultural technology has made farms far more productive and has reduced the number of workers needed. The impact has been immense. Worldwide, fewer than half of all people now work on a farm. In the United States, fewer than 2 percent are farm workers.

"Technology assessment" means examining how technology affects culture, society, politics, the economy, and the environment. One way to begin is to consider what life would be like *without* technology.

(rows), less compacting of the soft soil by the machine, and no overlapping when spraying fertilizers and pesticides. All this adds up to more efficiency in the fields.

GPS is also used to help monitor the farm. Using yield monitors, the farmer can create a map of the farm and determine how each acre is doing. For example, some acres may need more water than others. Some may need fertilizer while others don't. By monitoring each acre, the farmer can improve production and avoid wasting money. See **Fig. 20-3**.

Fig. 20-3. Based on earlier soil samples, this GPS system shows the farmer how much fertilizer should be released in various parts of the field.

Productivity. How has technology improved agricultural productivity? Here's some data to plot that will indicate the trend. It shows the hours of labor (from planting to harvesting) needed to produce 100 bushels of wheat at various times in history.

1830: 250–300 labor-hours with the use of a walking plow, brush harrow, hand broadcast of seed, sickle, and flail. Required about 5 acres of land.
1890: 40 to 50 labor-hours
1930: 15 to 20 labor-hours
1955: 6 to 12 labor-hours
1965: 5 labor-hours
1975: 3.75 labor-hours
1987: 3 labor-hours with the use of a tractor, 35-foot sweep disk, 30-foot drill, 25-foot self-propelled combine, and trucks. Required about 3 acres of land.

Make a bar chart to show this data. Be prepared to explain or interpret your chart.

Dairy Farming

Agricultural mechanization is not limited to just crop production. Labor-saving machines have also changed how animal products are produced.

Dairy farming is a branch of agriculture. It is concerned with producing milk and milk products. Mechanization has made dairy farming much more efficient than in the past. Today, glass tubes carry the milk from the cow and the milking machine to the milk houses for processing. Feed conveyors, refrigeration, and sterilizers have lightened the tasks. Dairy farm mechanization has increased milk production. Formerly, one cow could supply milk products for every four persons in the United States. Today one cow can supply milk products for every six people in the population. See **Fig. 20-4**.

Genetic Engineering

Why do some sheep have coarse wool and others have fine wool? Why are some kernels of corn white and others yellow? The answers to these questions lie in genetics. Genetics (jen-ET-icks) is the science that studies the laws of heredity.

Fig. 20-4. Dairy farms such as this one will milk several hundred cows twice every day. The milk travels through pipes to the milk house for processing.

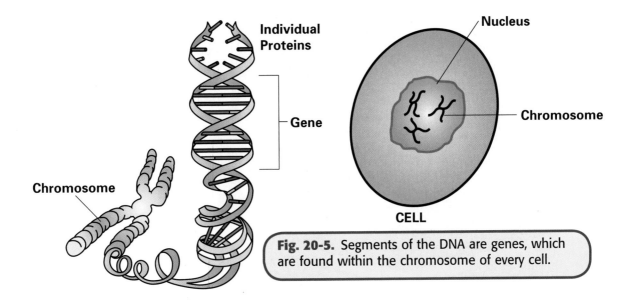

Fig. 20-5. Segments of the DNA are genes, which are found within the chromosome of every cell.

Heredity is the passing on of certain traits from parents to offspring. A trait is a physical characteristic, such as size or color.

Genes determine traits. Genes are segments of DNA. The genes are arranged on chromosomes inside the nucleus of the cells. As cells grow and multiply, they get instructions from the genes. See **Fig. 20-5**. These coded instructions produce different traits, such as coarse wool or yellow kernels of corn.

Genetic engineering is the process of inserting new genes into a cell in order to change the characteristics of an organism. Moving genes from one organism's DNA into another is called gene splicing. See **Fig. 20-6**. For more on gene splicing, see the Student CD.

Through genetic engineering, scientists have improved food production. For example, frost can destroy a potato crop. Bacteria that occur

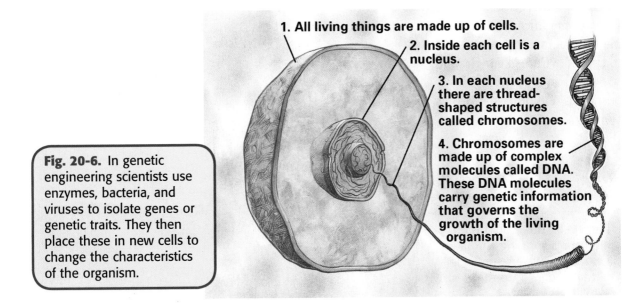

Fig. 20-6. In genetic engineering scientists use enzymes, bacteria, and viruses to isolate genes or genetic traits. They then place these in new cells to change the characteristics of the organism.

1. All living things are made up of cells.

2. Inside each cell is a nucleus.

3. In each nucleus there are thread-shaped structures called chromosomes.

4. Chromosomes are made up of complex molecules called DNA. These DNA molecules carry genetic information that governs the growth of the living organism.

Fig. 20-7. Organically grown fruits and vegetables are sold in many grocery stores. However, this represents only 30% of the produce sold in the stores.

naturally on the leaves of the potato plant freeze at 30°F. The ice crystals that form destroy the plant cells. Scientists have genetically engineered the bacteria. The bacteria cell walls now contain a protein that blocks ice crystal formation. These genetically altered bacteria are sprayed onto potato plants. The resulting potato plant won't freeze until the temperature is 23°F.

Genetic engineering has created super plants with thicker stalks and broader leaves and plants in which a larger portion is edible or usable and less is discarded as waste. Other genetic modifications include plants that are more resistant to pests, viruses, fungi, and parasites, thus requiring fewer pesticides and other dangerous sprays. Plants are also bioengineered (genetically modified) to be heartier. This means they can survive and grow in environments where weaker plants might die. Thus, more areas are available for farmland.

Today, about 70% of the processed food found in grocery stores contains ingredients made from genetically modified organisms (GMOs). Have you ever seen GMOs in a grocery store? Is there an easy way to identify specific foods as containing GMOs? See **Fig. 20-7**.

Fig. 20-8. This crop duster airplane is spraying a potato field. Crop dusters spread fertilizers and insect pesticides.

Irrigating and Fertilizing

Irrigation technology is used to supply the right amount of water to a farmer's plants and animals through the use of pipes and sprinklers. When driving past farmlands, you may see many sprinklers in use at one time.

Irrigation can be especially helpful during drought seasons. During normal weather conditions, excess water from irrigation is usually drained into nearby ponds. During droughts, the water can be pumped from the ponds to irrigate crops and be given to animals.

As crops are continuously planted and harvested, nutrients are removed from the soil. In the past, farmers always had to rotate their crops to avoid taking the same nutrients from the soil year after year. Then came the invention of fertilizer. A **fertilizer** is a chemical compound used to restore nutrients to soil. See **Fig. 20-8**. Specific fertilizers can be made for individual crops. One kind of fertilizer may be used for corn, while another kind of fertilizer may be used for beans.

Controlled Environment Agriculture (CEA)

Every kind of plant has its own growing needs. **Controlled environment agriculture (CEA)** produces the perfect growing environment so that plants can thrive. Humidity, temperature, lighting, watering, and feeding are five of the conditions controlled through this technology. The pH level of the growing environment can also be controlled. The pH level is the level of acidity or alkalinity. See **Fig. 20-9**.

Controlled environment agriculture may take place in a greenhouse or other building. Often computers are used to control temperature, humidity, lighting, pH level, and plant feeding schedules. Plants may even be on a conveyor. This conveyor moves slowly through the building as the plants grow. Artificial lights provide the energy for photosynthesis. Misters keep the humidity at perfect levels. By the time the plants reach the end of the line, they are mature.

	Beet	**Cabbage**	**Carrot**	**Radish**	**Squash**	**Tomato**
pH Preference	6.0-7.5	5.5-7.5	5.5-7.0	6.0-7.0	5.5-7.5	5.5-7.5
Light Intensity	Medium Intensity	Medium Intensity	Low Intensity	Low Intensity	Medium Intensity	High Intensity
Moisture Content	Well Drained	Damp/Well Drained	Damp/Well Drained	Damp	Damp/Well Drained	Damp

Fig. 20-9. Controlled environment agriculture (CEA) creates the perfect conditions for plants to thrive.

Flowering plants, herbs, and food products can be grown in this way. They are packaged and shipped to market from the conveyor. See **Fig. 20-10**.

Hydroponics

Did you know that a plant doesn't need soil to grow? **Hydroponics** is the process of growing plants in a soil-less environment. The word hydroponics comes from two Greek words—*hydro* meaning "water" and *ponos* meaning "work."

Hydroponics is a form of CEA. Plants are grown in a controlled environment that supplies the light, humidity, food, and water needed for rapid growth. Although plants do not need soil to grow, they do need the nutrients (food) and the support that soil normally gives. In hydroponic systems, a water/nutrient solution is fed to the plants. This is done by flooding, spraying, or pumping the solution past the roots, stems, and leaves.

Water culture systems support the plants using wood fiber, gravel, rock wool, or rice hulls. The roots of the plants are submerged in the nutrient solution. Air is bubbled through the solution so that the roots can get oxygen. See **Fig. 20-11A**.

An aggregate (AG-greg-et) is a material such as vermiculite, sand, or perlite. Vermiculite (ver-MICK-u-light) is a mineral substance that is very water absorbent. Perlite (PEARL-ite) is a lightweight volcanic glass. Aggregate systems support and grow the plant in the aggregate.

Fig. 20-10. This lettuce grows as the conveyor belt moves the plants through the hydroponic hot house.

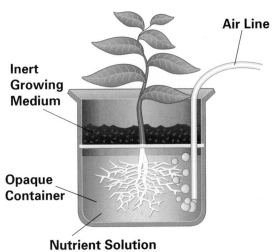

Inert Growing Medium

Air Line

Opaque Container

Nutrient Solution

Fig. 20-11A. Water culture systems support the plants using wood fiber, gravel, rock wool, or rice hulls. The roots of the plants are submerged in the nutrient solution. Air is bubbled through the solution so the plant roots can get oxygen.

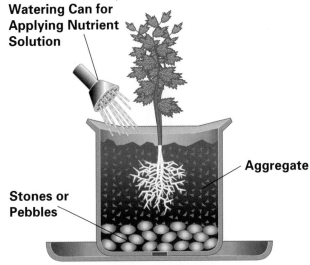

Watering Can for Applying Nutrient Solution

Aggregate

Stones or Pebbles

Fig. 20-11B. Aggregate systems support and grow the plants in a material (aggregate) such as vermiculite, sand, or perlite. The nutrient solution is flooded into the aggregate and then drained. The plant's roots feed off the nutrient-soaked aggregate.

The nutrient solution is flooded into the aggregate. When the aggregate is moist, the nutrient solution is allowed to drain out. The plant's roots feed off the nutrient-soaked aggregate. See **Fig. 20-11B**.

Aeroponic (air-o-PON-ic) systems suspend the plant in the air with the roots exposed. This allows the exposed roots to be misted with the nutrient solution. See **Fig. 20-11C**.

Advantages. Is fertile soil in short supply on Earth? Why else should farmers consider growing crops without soil? Actually, there are many advantages to soil-less agriculture.

Some areas in the world have poor soil. Such soil does not contain the nutrients or drainage plants need to thrive. Hydroponics offers an alternative to traditional agriculture for farmers in these areas. Imagine a barren desert dotted with hydroponic greenhouses growing vegetables. Such greenhouses can be found in the Middle East.

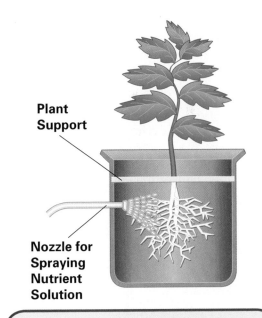

Plant Support

Nozzle for Spraying Nutrient Solution

Fig. 20-11C. Aeroponic systems suspend the plant. This allows the exposed roots to be misted with the nutrient solution.

Hydroponics can be used to grow food on a smaller scale. Metropolitan areas usually have little room for traditional agriculture. Hydroponic systems can be used to grow food in a courtyard or an alleyway. See **Fig. 20-12**. Many restaurants are even growing their own herbs using hydroponics.

Compared with traditional soil farming, hydroponics has clear advantages.

- More food can be harvested per square foot of growing area.
- A sufficient supply of nutrients is always available.
- The pH level can be maintained.
- Plenty of oxygen can reach the roots.
- The roots remain clean; the entire plant can be harvested.
- Replanting is quick.
- Crops can be changed quickly.
- The nutrient solution can be reused.
- Nutrients are not wasted; they go directly to the plant.
- The system can be automated.
- Pests are more easily controlled.
- The system is environmentally friendly.
- There is less use of pesticides.

Science Link

The Green Revolution. It might sound like the title of a horror movie, but "green revolution" actually refers to the impact of science and technology on agriculture. See if you can identify the role of science in the following developments.

- hardier crops (less sensitive to extremes of temperature, moisture levels, etc.)
- fertilizer
- natural pesticides
- more successful seed germination
- faster-growing crops
- larger produce

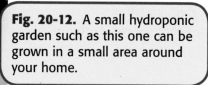

Fig. 20-12. A small hydroponic garden such as this one can be grown in a small area around your home.

Fig. 20-13. The nutrition and overall environment of the fish can be controlled in a fish farm like this one.

Reading Link

Critical Reading. Find a magazine, newspaper, or Internet article about a topic related to agricultural technology. For example, it might be an article about genetically modified plants or about fish farms. What is the author's viewpoint? What statements does the author make to support his or her viewpoint? What are the author's credentials? What sources does the author cite? Describe your own reactions to the article. Do you agree or disagree with the author? Why?

Disadvantages. Hydroponic systems also have disadvantages.
- Disease and insects can spread rapidly.
- The pH of the nutrient solution can change quickly.
- New nutrient solution is needed at all times to make up for evaporation.
- Start-up expense is high.
- A high degree of expertise is needed to manage the system.

Aquaculture

The techniques used in CEA have also been applied to aquatic (marine) plants and animals. **Aquaculture** is the raising of fish and other marine organisms in a controlled water environment.

Variables such as water temperature, water chemistry, and egg fertilization are precisely controlled. The result is larger marine organisms produced at a faster rate and in larger quantities. See **Fig. 20-13**.

Aquaculture is used to raise a variety of seafood for consumption. Tuna, trout, oysters, crabs, and abalone are a few of the farm-raised organisms that are sold throughout the world. Rice and other food plants can also be farmed using aquaculture techniques.

The aquaculture industry also produces marine organisms used by pharmaceutical and chemical companies to produce products such as vitamins, food additives, and shampoo.

From Milk to Glue

Identify a Need/Define the Problem

Bioprocessing is a bioengineering technique that uses living organisms or parts of living organisms to process or change materials. Bioprocessing is often used in agriculture. For example, casein glue is made using bioprocessing techniques. It is a very strong glue made from milk.

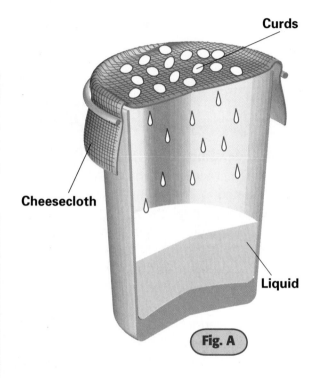

Fig. A

Labels: Curds, Cheesecloth, Liquid

Materials and Equipment

Select from this list or use your own ideas.

Glue
- 1 cup hot water
- ⅓ cup nonfat dry milk
- 3 Tbsp. vinegar
- 1 Tbsp. cold water
- ½ Tbsp. baking soda

Materials
- softwoods
- hardwoods
- fabrics
- paper
- acrylic
- ceramic

Equipment for glue
- 1 piece of cheesecloth
- measuring tools
- paper cups
- mixing sticks
- clamps

Equipment for testing
- weights
- variety of spring scales
- balance
- string

Note: To prepare smaller batches of glue, reduce the ingredients proportionally.

In this activity, you will design a bioprocessing system to manufacture casein glue. You will then test the strength of the glue on a variety of materials.

Gather Information

Casein is a protein found naturally in milk from cows. To remove the casein from the milk, bacteria are allowed to form lactic acid. This acid forms curds in the milk. The curds are then separated from the liquid. Casein glue is contained in the curds of curdled milk. Before you design your system, research casein and any alternative forms of removing this protein from milk.

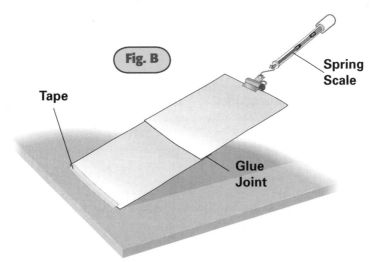

Fig. B

Tape

Spring
Scale

Glue
Joint

Develop Possible Solutions

Assume that you do not want to wait for the milk to spoil naturally. What can you do to curdle the milk? One solution is to use vinegar. By doing this, you are modeling the bioprocessing technique.

Once you have made the glue, you will need to test its strength. **Figure A** shows an example of a testing device. You will also need to develop a table to record your results.

Model a Solution

1. Mix 1 cup of hot water with ⅓ cup of non-fat dry milk.
2. Add ⅓ Tbsp. vinegar. Stir gently for three minutes. Allow the milk to curdle.
3. Using the cheesecloth, filter the curds from the liquid. See **Fig. B**. Place the curds in a paper cup. The curds contain casein.
4. Add 1 Tbsp. cold water and ½ Tbsp. baking soda to the curds. This will neutralize the vinegar. Stir to a smooth, creamy consistency. You now have your glue.

Test and Evaluate the Solution

- Glue together different combinations of test materials. Clamp them or place weights on them. Allow them to set overnight.
- Test the strength of the bonds.
- Prepare a table to record the data.
- Enter the data in the table.
- Which materials does the casein glue make the strongest bonds with? What material family do they belong to?
- If the materials were not strongly joined, what was the cause?
- Was the testing system effective?
- What changes could be made to correct any problems?

Refine the Solution

Make necessary changes to the testing system and test new samples. Compare the results to prior tests. Did your revised testing system yield better results? Why or why not?

Communicate Your Ideas

Create a portfolio documenting all of the work completed in this activity. Be sure to include your table with all relevant information, including any refinements you made to your testing system.

Growing with Hydroponics

Identify a Need/Define the Problem

Hydroponic growing systems vary in the aggregate used, plant support system, nutrient feeding system, plants, and container design. Design and build a hydroponic growing system in which you can grow flowers or food.

Gather Information

After selecting a crop to grow, gather diagrams and pictures of various hydroponic systems. See **Fig. A** for an example. Research the requirements (light, pH, temperature, nutrients) for the plants you selected to grow. Explore feeding designs and select a feeding method and aggregate if necessary. Investigate different container designs.

Safety Alert

Handling perlite and vermiculite produces dust that can irritate your eyes and lungs. Wear eye protection and a dust mask. Before using tools and equipment, look up "Safety Data Sheets" on the Student CD and prepare a data sheet for this activity. As you work on the activity, be sure to follow all safety rules.

Materials and Equipment

Select from this list or use your own ideas.

Containers
- one-liter plastic bottles
- food cans
- plastic food containers
- clay flower pots
- plastic pipe

Aggregate
- perlite
- vermiculite
- sand
- marbles
- gravel
- aquarium gravel

Seeds or plants
- lettuce
- beans
- tomatoes
- herbs
- marigolds
- rye grass
- alfalfa sprouts

Nutrients
- Purchase commercially prepared solutions. Be sure your teacher or parent helps in mixing the appropriate proportions.

General building supplies
- pine wood
- sheet metal
- plastic tubing
- fish tank aerator
- acetate for greenhouse covering

Equipment
- material processing tools and equipment

Develop Possible Solutions

Using graph paper, draw at least three unique plans for a hydroponic growing system that you think will serve your purpose.

Model a Solution

1. Choose the design you think will be the most effective.
2. Assemble the growing unit, following the design that you have chosen.
3. Develop a schedule and monitoring system to record the progress of your plants.

Test and Evaluate the Solution

- Did the hydroponic system produce healthy crops?
- If healthy crops were not produced, what was the state of the crops?

Refine the Solution

What changes might be made to correct any problems? Make any necessary changes. Check to see what effects they have on the crops.

Communicate Your Ideas

Create a portfolio of all of your work. Be sure to justify why you feel the design you chose would be more effective than the others. Also include the growth charts for the plants and any changes you made to the system.

Science Link

Chemistry and Hydroponics. In chemistry class, you learn about the different elements. How is chemistry important to hydroponics? Certain elements are needed to treat the crops in any hydroponics system. There are 16 elements which are considered essential for good plant growth. Find out what these 16 elements are and how they help plants grow. For example, why would carbon (C) be needed to help plants grow?

Fig. A

Creating a Water Filter

Identify a Need/Define the Problem

A large percentage of the drinking water in the world comes from aquifers, which are underground layers of rock and sand that hold water. The rest of the water comes from surface water, such as lakes and streams. However, surface water normally has some type of contamination and needs to be filtered and disinfected in order to be useable. In this activity, you will create a model of a filtering system for contaminated water.

Gather Information

What are some of the ways that water is naturally filtered in the earth before it ends up in the aquifer? How do water treatment plants filter water in cities where there is a large water supply system? See **Fig. A**. Do some research to find answers to these questions.

Develop Possible Solutions

Based on your research, design a way to filter water that is "contaminated" with clay and food coloring. Create at least three different designs. Make sure you document the details of each design in your portfolio.

Materials and Equipment

Select from this list or use your own ideas.

- sand of varying coarseness
- rocks
- stone
- food coloring
- clay
- 2.5-gallon spring water container with spout
- bucket
- coffee filters

Safety Alert

Do not drink any of the filtered water. It may still contain small amounts of contaminants.

Model a Solution

1. Cut out the top of a 2.5-gallon spring water container.
2. Place the filtering materials in the spring water container.
3. Create the contaminated water in a bucket by mixing water with food coloring and clay.

Test and Evaluate the Solution

- Dump the contaminated water into the filtering system.
- Place the bucket under the spout of the filter system.
- Open the spout to see the results.
- Document the condition of the filtered water.

Refine the Solution

Was your water contaminated? If so, make necessary changes to the testing system and test new samples. Compare the results to prior tests.

Communicate Your Ideas

Create a portfolio documenting all of the work completed in this activity. Demonstrate the filtering process to the class.

Math Link

Water Statistics. Throughout history, the site for choosing a settlement always involved a concern for a fresh water supply. Only about one percent of all surface water is available or acceptable for drinking.

Is one percent enough? In the year 2000, an estimated 1,688 cubic meters of water were used for each person in the population. A cubic meter of water equals about 264 gallons. If there were about 275 million people at that time, how much water was used?

Fig. A

Exploring Careers

Agricultural Engineer

ENTRY LEVEL | TECHNICAL | **PROFESSIONAL**

Agricultural engineers use their knowledge of science and engineering to improve agriculture. These engineers design agricultural machines and equipment, such as tractors. They design structures, such as shelters for animals and buildings to store crops. They also develop ways to conserve water and soil. They research ways to develop new machines and systems. Ag engineers also convey their knowledge through radio and television programs.

More than 30% of ag engineers work as consultants to farmers and farm-related businesses. Ag engineers also are employed in the crop and livestock industries, government, and agricultural manufacturing. They work in research and development, management, sales, and production.

Qualifications

An ag engineer must have a good math, science, and engineering background as well as good problem-solving skills. A bachelor's degree is required. An ag engineer would most likely take courses in engineering, math, science, administration, business and economics, computers and electronics, and communications.

Because ag engineers use their knowledge to help farmers, manufacturers, and others make better agricultural decisions, good business and communication skills are needed. Problem solving and reasoning are good traits to have.

Outlook for the Future

The job outlook for ag engineers is good. Worldwide standardization of agricultural systems and a greater interest in conservation of natural resources may increase the need for ag engineers.

Speaking Skills

In this career, it is important for you to know your audience. If you know the people's background and whether they are familiar with your topic, you will be better able to present the information they need in a way they can understand.

Researching Careers

Research the career of an ag engineer. What does the person do on the job? Has he or she invented a product or system? Is so, how has it helped agriculture? Give a brief talk to the class about what you learn.

CHAPTER 20 Review

More activities
on Student CD

Key Points

- Agriculture is a technological system that produces plants and animals for food, fiber, fuel, and other products.
- Thanks to mechanization, farmers can rely on more than just muscle power to farm crops.
- Genetic engineering is used to improve many of the foods we eat.
- Farmers use irrigation and fertilizers to help maintain their plants and animals.
- Controlled environment agriculture uses technology to create an environment for growing plants and animals.

Read & Respond

1. What five technologies have changed the nature of farming?
2. What was one of the earliest tools used to mechanize farming?
3. Define *harvesting*.
4. How is GPS used in farming?
5. Give three examples of how genetic engineering can improve the foods we eat.
6. What are GMOs?
7. How can irrigation help in times of drought?
8. What is CEA?
9. What CEA system brings nutrients directly to plants?
10. What is aquaculture?

Think & Apply

1. **Relate.** Chapter 18 discusses bioprocessing. Research how bioprocessing can be used on dairy farms.
2. **Extend.** Research aquaculture and make a list of advantages and disadvantages.
3. **Assess.** If possible, visit a nearby farm. What agricultural technologies are used? How do these make farms more productive?

4. **Organize.** Make a chart showing the evolution of agriculture technologies. Begin your chart with the invention of the plow.
5. **Assess.** Research controversy over GMOs. What are the concerns of those who oppose GMOs? What are the benefits described by supporters?

TechByte

Mystery Behind the Pop. Chemists at Indiana's Purdue University in 2005 found an answer to the age-old question, "What makes popcorn pop?" The chemists analyzed cellulose molecules in the hard seed coat of a corn kernel. They found that when heat is applied, the molecules assume a rigid crystalline form that holds in moisture. That allows pressure to build in the kernel until—pop! Food companies are researching ways to genetically engineer popcorn with lots of crystalline cellulose in the seed coat.

Appendix

Approximate Customary-Metric Conversions		
When you know:	**You can find:**	**If you multiply by:**
Length inches	millimeters	25.4
feet	millimeters	304.8
yards	meters	0.91
miles	kilometers (km)	1.6
millimeters	inches	0.04
meters	yards	1.09
kilometers	miles	0.62
Area square inches	square centimeters (cm^2)	6.45
square feet	square meters	0.09
square yards	square meters	0.84
square miles	square kilometers (km^2)	2.59
acres	square hectometers (hectares)	0.4
square centimeters	square inches	0.16
square meters	square yards	1.2
square kilometers	square miles	0.4
hectares (ha)	acres	2.5
Mass ounces	grams	28.35
pounds	kilograms	0.45
short tons	metric tons (t)	0.9
grams	ounces	0.04
kilograms	pounds	2.2
metric tons	short tons	1.1
Liquid Volume fluid ounces	milliliters	30
pints	liters	0.47
quarts	liters	0.95
gallons	liters	3.8
milliliters	fluid ounces	0.03
liters	pints	2.1
liters	quarts	1.06
liters	gallons	0.26
Temperature degrees Fahrenheit	degrees Celsius	0.6 (after subtracting 32)
degrees Celsius	degrees Fahrenheit	1.8 (then add 32)
Power horsepower	kilowatts (kW)	0.75
kilowatts	horsepower	1.34
Pressure pounds per square inch (psi)	kilopascals (kPa)	6.9
kPa	psi	0.15
Velocity miles per hour (mph)	kilometers per hour (km/h)	1.6
km/h	mph	0.62

Appendix

Polygons

A polygon is a closed figure with straight sides. It is classified by the number of sides. In a regular polygon all sides are equal in length and all angles are equal.

3 Sides = Triangle

4 Sides = Rectangle

4 Equal Sides = Square

5 Sides = Pentagon

6 Sides = Hexagon

7 Sides = Heptagon

8 Sides = Octagon

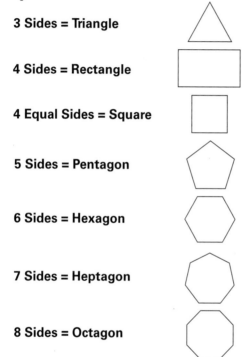

Perimeter

The perimeter of a polygon is the sum of all the sides.

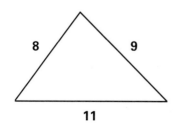

The perimeter of this triangle is:
8 + 9 + 11 = 28

Circumference

The circumference is the distance around a circle. The radius of a circle is a straight line from the center to a point on the circumference. The diameter is a straight line through the center with its ends on the circumference. The diameter is twice the radius.

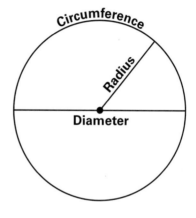

$$C = 2\pi r$$

If the radius is 7", then
$C = 2 \times 3.1416 \times 7$
$C = 43.9824"$

Here is another way to find circumference. If the radius = 7", then the diameter = 14".

$$C = \pi d$$
$C = 3.1416 \times 14$
$C = 43.9824"$

Area

The area is the number of square units covering the surface of a polygon.

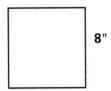

Area of a square

$A = s^2$
$A = 8^2$
$A = 8 \times 8$
$A = 64$ sq. in.

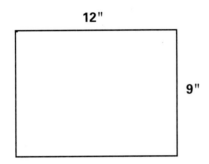

Area of a rectangle

$A = l \times w$
$A = 12 \times 9$
$A = 108$ sq. in.

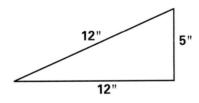

Area of a triangle

$A = \frac{1}{2}bh$
$A = \frac{1}{2} \times 12 \times 5$
$A = 30$ sq. in.

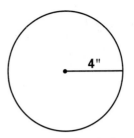

Area of a circle

$A = \pi r^2$
$A = 3.1416 \times 4^2$
$A = 3.1416 \times 16$
$A = 50.2656$ sq. in.

Volume

Volume is the amount of three-dimensional space occupied by an object.

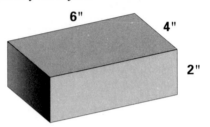

Volume of a rectangular prism

$V = lwh$
$V = 6 \times 4 \times 2$
$V = 48$ cu. in.

Volume of a cylinder

$V = \pi r^2 h$
$V = 3.1416 \times 3^2 \times 6$
$V = 3.1416 \times 9 \times 6$
$V = 169.6464$ cu. in.

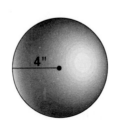

Volume of a sphere

$V = \frac{4}{3}\pi r^3$
$V = \frac{4}{3} \times 3.1416 \times 4^3$
$V = \frac{4}{3} \times 3.1416 \times 64$
$V = 268.0832$ cu. in.

Angles

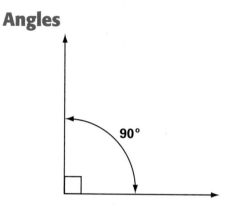

A right angle contains 90°. (Note the symbol for right angle.)

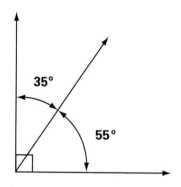

Two angles are complementary if their angles total 90°.

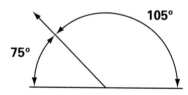

Two angles are supplementary if their angles total 180°.

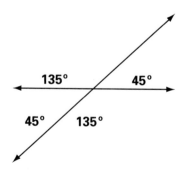

When two straight lines cross, note how the adjacent angles total 180°and how the opposite angles are always equal.

Ohm's Law

Ohm's law states that current is equal to the voltage divided by the resistance.

E = voltage
I = current
R = resistance

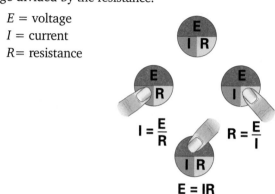

$$I = \frac{E}{R}$$

$$R = \frac{E}{I}$$

$$E = IR$$

Force

A force is a push or a pull. Force is equal to mass times acceleration.

$$F = m \times a$$

Work

Work is the application of force to make an object move in the direction of the force.

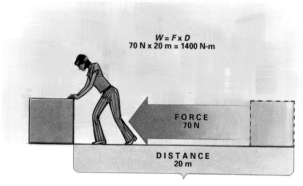

$W = F \times D$
70 N x 20 m = 1400 N-m

FORCE
70 N

DISTANCE
20 m

Glossary

A

aerodynamics (air-oh-dy-NAM-iks) The study of the forces of air on an object moving through it. (Ch. 7)

aerospace (AIR-oh-space) The study of how things fly. (Ch. 7)

AFIS Automated Fingerprint Identification System; used to help identify prints found at crime scenes. (Ch. 19)

agriculture A technological system that produces plants and animals for food, fiber, fuel, and other products; often referred to as farming. (Ch. 20)

aileron (AY-luh-ron) Changes the airflow across the wing, increasing and decreasing the amount of lift the wing creates. (Ch. 7)

air pressure A measure of the force of air pressing down on the earth's surface; also known as atmospheric pressure. (Ch. 10)

airfoil A shape designed so that air flowing around it produces useful motion. (Ch. 7)

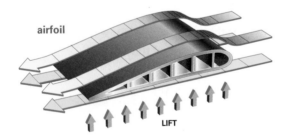

alloy A material made from combining two or more elements, at least one of which is a metal. (Ch. 4)

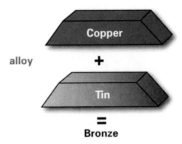

amplitude modulation (AM) Radio broadcasting signal where the amplitude (strength) of the carrier wave changes. (Ch. 12)

analog signal A continuous and variable signal. (Ch. 12)

animation The creation of simulated movement by using a series of still images. (Ch. 13)

applied physics The branch of science that applies the principles of science to solve engineering problems. (Ch. 6)

aquaculture The raising of fish and other marine organisms in a controlled water environment. (Ch. 20)

architect Someone who designs buildings and often oversees construction. (Ch. 3)

atmosphere The gaseous mixture that surrounds the earth. (Ch. 10)

automaton A mechanical device that can imitate the actions of people or animals. (Ch. 17)

B

battery A device that converts chemical energy into electrical energy. (Ch. 14)

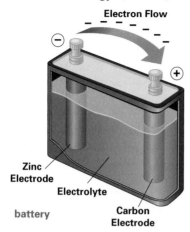

Electron Flow

Zinc Electrode
Electrolyte
Carbon Electrode

battery

Bernoulli effect States that a fast-moving fluid exerts less pressure than a slow-moving fluid (Ch. 7)

bioengineering The use of engineering to solve problems in biology, medicine, human behavior, health, and the environment. (Ch. 18)

biomass energy Energy produced from organic materials, such as trees, animal waste, and plants. (Ch. 8)

bioprocessing A bioengineering technology that uses living microorganisms or parts of organisms to change materials from one form to another. (Ch. 18)

Boyle's law Explains why gases are easy to compress. (Ch. 16)

brainstorming Process in which group members suggest ideas out loud as they think of them. (Ch. 1)

buoyancy (BOY-ann-see) is the upward force a fluid places on an object. (Ch. 9)

C

cell In electricity and electronics, a device made of two different conducting materials in a conducting solution. (Ch. 14)

charge-coupled device (CCD) A microchip inside a digital camera that converts light into an electrical signal. (Ch. 12)

circuit (SIR-cut) In electricity, the pathway through which electrons travel. (Ch. 14)

civil engineer Engineer who designs and supervises the building of structures that serve the public. (Ch. 3)

climate The average weather conditions for an area over a period of many years. (Ch. 10)

cloud A mixture of water droplets and dust particles suspended in air. (Ch. 10)

coherent light Light in which all of the light waves have the same wavelength. (Ch. 15)

combining Secondary process used to join materials together. (Ch. 5)

composite Material resulting from two or more separate materials being combined or mixed. (Ch. 4)

computer numerical control (CNC) In computer-aided manufacturing, programmed computers that perform a series of operations over and over. (Ch. 5)

computer program In robotics, a set of coded instructions the robot must follow. (Ch. 17)

computer-aided design (CAD) The process of designing on a computer. (Ch. 2)

computer-aided manufacturing (CAM) The process of using computers to operate and control many machines and processes. (Ch. 5)

computer-integrated manufacturing (CIM) Computers monitor and control every aspect of manufacturing. (Ch. 5)

computerized tomography (CT) scan Images the human body by using a rotating X-ray machine to create an image on the computer. (Ch. 18)

conditioning Secondary process that changes the internal structure of a material. (Ch. 5)

conductivity The ability of a material to allow the flow of heat or electricity. (Ch. 4)

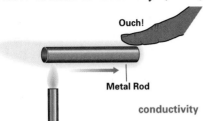

conductivity

constraints In the design process, the limits placed on the design and the designer. (Ch. 1)

continuous production Production system in which products are mass-produced, usually on an assembly line. (Ch. 5)

controlled environment agriculture (CEA) System that produces the perfect growing environment for plants. (Ch. 20)

controller In robotics, a tiny computer that acts as the robot's brain and contains the computer program. (Ch. 17)

crime scene investigator (CSI) Someone who uses a variety of tools and techniques to find and gather evidence at a crime scene. (Ch. 19)

criteria In the design process, things that the product must do or include. (Ch. 1)

current The flow of electrons. (Ch. 14)

custom production Production system in which products are made to order. (Ch. 5)

D

degree of freedom In robotics, a robot's ability to move in a particular direction. (Ch. 17)

degree of freedom

density A measure of how much mass (material) is contained in a given volume (amount of space). (Ch. 4)

design The process of creating things by planning. (Ch. 1)

diagnosis The process of examining a patient and studying symptoms to find out what illness or condition the patient has. (Ch. 18)

diesel engine An internal combustion engine that burns fuel by using heat produced by compressing air. (Ch. 9)

digital music player Different from other music players because it has a solid-state memory that requires no movement. (Ch. 12)

digital signal Separate, distinct signals in the form of binary code. (Ch. 12)

diode An electronic component that allows electrons to flow in only one direction. (Ch. 14)

directional light Light that spreads out very little compared to ordinary light. (Ch. 15)

directional light

document examiner In forensics, someone who specializes in determining the source and authenticity of documents. (Ch. 19)

Doppler radar System used to provide images that show the size and speed of a storm as well as the amount of precipitation. (Ch. 10)

drafting The process of making drawings needed so that a part or product can be manufactured or built. (Ch. 2)

drag The force of fluid friction on an object. (Ch. 7)

ductility The ability of a material to be formed and reformed. (Ch. 4)

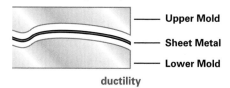
ductility

dynamic load A load that moves or changes; also called a live load. (Ch. 3)

E

elasticity The ability of a material to return to original shape after being pulled or pushed. (Ch. 4)

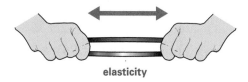
elasticity

electric vehicle A vehicle that uses electricity to turn the wheels of the vehicle. (Ch. 9)

electricity The movement of electrons from one atom to another. (Ch. 14)

electrocardiograph (ECG or EKG) A machine designed to record the electrical currents of the heart. (Ch. 18)

electromagnet A powerful magnet created by wrapping wire around an iron core and then passing electric current through the wire. (Ch. 14)

electromagnetic wave A wave produced by the motion of electrically charged particles. (Ch. 15)

electronics The study of the control of electron flow in a circuit. (Ch. 14)

element Matter made from only one kind of atom; can be combined to make compounds. (Ch. 4)

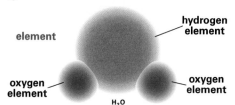
element
hydrogen element
oxygen element
oxygen element
H_2O

end effector The hand of a robot; may be in the form of a gripper, or it may hold an attachment, such as a welding torch or a paint sprayer nozzle. (Ch. 17)

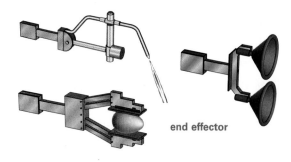
end effector

energy The ability to do work. (Ch. 8)

engineer A person who uses his or her knowledge of science, math, technology, and communication to solve technical design problems. (Ch. 1)

engineered wood Manufactured product that is a composite made by bonding together wood strands, fibers, or veneers with adhesives. (Ch. 3)

entrepreneur A person who starts a business. (Ch. 7)

ergonomics (erg-oh-NOM-icks) The study and design of equipment and devices that fit the human body, its movement, and its thinking patterns; also called human factors engineering. (Ch. 18)

ergonomics

evaporation Takes place when liquid water is turned into gas by the sun's radiant energy; explains how moisture gets into the air. (Ch. 10)

evidence In forensics, anything that helps establish the facts of the crime. (Ch. 19)

feedback control The process of sending signals, interpreting received signals, and adjusting through signals. (Ch. 17)

ferrous Any metal that contains iron. (Ch. 4)

fertilizer A chemical compound used to restore nutrients to soil. (Ch. 20)

fiber optics Thin filaments of glass through which light travels. (Ch. 15)

fluid power The use of liquids or gases under pressure to move objects or perform other tasks. (Ch. 16)

force A push or a pull that transfers energy to an object. (Chs. 3, 6, 7)

Force = mass × acceleration

or

F = m × a

forecasting In weather, making predictions based on available data. (Ch. 10)

forensics The application of science and technology to the law and the solution of crimes. (Ch. 19)

forgery The act of preparing or changing a document with criminal intent. (Ch. 19)

forming Secondary process used to change the shape of a material. (Ch. 5)

fossil fuel Energy source formed from the remains of plant and animal life over millions of years ago. (Ch. 8)

frequency modulation (FM) Radio broadcasting signal in which the frequency of the carrier wave changes. (Ch. 12)

friction Any force that opposes motion. (Ch. 6, 7)

front A line where two air masses collide; often produces unstable and potentially violent weather. (Ch. 10)

generator A device that changes mechanical energy into electrical energy. (Ch. 14)

gene splicing Moving genes from one organism's DNA into another. (Ch. 20)

genetic engineering The process of inserting new genes into a cell in order to change the characteristics of an organism. (Ch. 20)

genetics (jen-ET-icks) The science that studies the laws of heredity. (Ch. 20)

geothermal energy Energy produced under the earth's crust. (Ch. 8)

GIF An electronic file of a still image that can be animated using software; stands for Graphics Interchange Format. (Ch. 13)

global positioning system (GPS) A navigation system that uses receivers, computers, and satellites to determine the location of an object on Earth. (Ch. 20)

GMO Stands for genetically modified organism. (Ch. 20)

grain The direction, size, and appearance of wood fibers; helps determine strength of the wood. (Ch. 3)

grain

graphic communication The process of using words and images to send messages. (Ch. 11)

gravity The force of attraction that exists between two objects. (Chs. 6, 7)

gravure printing Printing process in which letters and designs are scratched into a metal plate; ink fills these grooves and is then transferred to paper. (Ch. 11)

H

hardness The ability of a material to resist denting and scratching. (Ch. 4)

hardness

holography A photographic process that uses a laser as well as lenses and mirrors to produce three-dimensional images. (Ch. 15)

humidity Amount of water vapor in the air. (Ch. 10)

hybrid vehicle A vehicle that uses a gasoline engine and electric motor to propel the vehicle. (Ch. 9)

hydraulic system Fluid power system that uses water, oil, or another liquid. (Ch. 16)

hydroponics is the process of growing plants in a soil-less environment. (Ch. 20)

hydropower Energy source produced by harnessing the mechanical energy of water and converting it to electrical energy. (Ch. 8)

I

implant In bioengineering, a prosthetic device placed inside the body. (Ch. 18)

industrial material Material that is used to make products. (Ch. 5)

inertia The tendency of an object to stay at rest or to continue to move. (Chs. 6, 7)

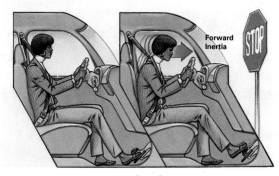

Forward Inertia

STOP

inertia

Information Age Time period that started around the mid-1800s and was brought about by the need to gather, store, and share large amounts of information. (Ch. 1)

innovation The process of improving and changing a technology that already exists. (Ch. 1)

integrated circuit A tiny piece of semiconductor material that contains miniaturized components, such as transistors and diodes, wired into minute circuits. (Ch. 14)

intermodal transportation Combining several modes of transportation to move people or products. (Ch. 9)

internal combustion engine Any engine where the fuel is burned inside the engine. (Ch. 9)

Internet A global network of computers; also called the information superhighway. (Ch. 12)

invention The process of designing new products. (Ch. 1)

irrigation Technology used to supply the right amount of water to a farmer's plants and animals through the use of pipes and sprinklers. (Ch. 20)

J

job-lot production Production system in which a specific quantity of a product is made. (Ch. 5)

K

key frame In animation, shows a beginning or ending point in an action sequence. (Ch. 13)

kinetic energy Energy in motion. (Ch. 8)

L

laser A light source that sends out light in a narrow and very strong beam. (Ch. 15)

LASIK surgery Laser surgery used to correct someone's vision by reshaping the cornea of the eye. (Ch. 18)

latent print A fingerprint that is not readily visible. (Ch. 19)

law of energy conservation States that energy can be neither created nor destroyed but can be converted from one form to another. (Ch. 8)

leading (LED-ing) In printing, the space between lines of type. (Ch. 11)

lead-through programming Method of controlling a robot by creating instructions that guide the robotic arm through a sequence of movements and then programming the computer to remember the pattern of motion. (Ch. 17)

light-emitting diode (LED) A common electronic component that emits light when connected in a circuit. (Ch. 15)

load An external force that acts on an object. (Ch. 3)

locomotive A self-propelled vehicle used to pull or push trains of rolling stock (railroad cars). (Ch. 9)

M

machine A device designed to obtain the greatest amount of force from the energy used. (Ch. 6)

maglev Trains that rise above a guideway and are propelled by magnetic fields; short for magnetically levitated. (Ch. 9)

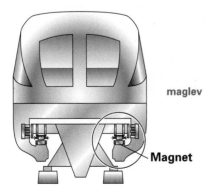

maglev

Magnet

magnetic resonance imaging (MRI) A machine that uses magnetic energy, radio waves, and computers to image the human body. (Ch. 18)

manufacturing The changing of raw or processed materials into usable products. (Ch. 5)

mass customization Production system in which standard products are produced and then modified for individual customers. (Ch. 5)

material property Describes how a material reacts under certain conditions, such as when it is heated or weight is put on it. (Ch. 4)

materials science The study of the properties and applications of materials. (Ch. 4)

matter Anything that occupies space and has mass. (Ch. 4)

mechanical advantage The number of times a machine multiplies a force. (Ch. 6)

mechanical design Engineering activity that involves designing individual parts and assemblies. (Ch. 2)

mechanization (mek-uhn-eye-ZAY-shun) The use of machines to make work easier. (Ch. 20)

medical examiner A physician who specializes in forensics. (Ch. 19)

medical imaging The process of taking pictures of the inside of the human body. (Ch. 18)

medium In physics, any solid, liquid, or gas that allows waves to pass through it. (Ch. 6)

meteorologist Someone who observes, records, and forecasts changes in the weather, based on changes within the atmosphere; also called a weather scientist. (Ch. 10)

mode A method of doing something. (Ch. 9)

modeling In animation, using computer software to create 3D computer models of characters, props, and sets. (Ch. 13)

momentum The connection between how fast an object is moving and the mass of the object. (Ch. 7)

monochromatic light Light that consists of only one color. (Ch. 15)

motion A change of position in a certain amount of time. (Ch. 6)

motion-capture animation A 3D representation of a live performance. (Ch. 13)

multimedia The combination of several forms of communication, such as text, video, photographs, spoken words, and music. (Ch. 12)

myoelectric signal Electrical signal sent through the body to help control muscles. (Ch. 18)

N

nanotechnology Engineering field that involves manipulating materials on an atomic or molecular level. (Ch. 1)

Newton's laws of motion Laws proposed by Isaac Newton that explain how force and motion work. (Ch. 6)

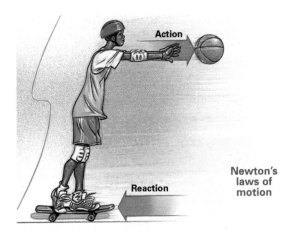

Newton's laws of motion

nonrenewable resource A resource that cannot be replaced once it is gone. (Ch. 8)

O

Occupational Safety and Health Administration See *OSHA*. (Ch. 5)

offset lithography Printing process that uses a flat surface and is based on the principle that oil and water don't mix. (Ch. 11)

Ohm's law States that that current (I) is equal to the voltage (E) divided by the resistance (R). (Ch. 14)

Current	I = E/R
Voltage	E = IR
Resistance	R = E/I

Ohm's law

OSHA Occupational Safety and Health Administration; the government agency that sets safety rules for the workplace. (Ch. 5)

P

page layout program A software program used to combine text and graphics in a document. (Ch. 11)

parallel circuit Circuit in which components are arranged in separate branches, allowing multiple paths for electrons to follow. (Ch. 14)

Pascal's principle States that when force is applied to a confined liquid, the resulting pressure is transmitted unchanged to all parts of the liquid. (Ch. 16)

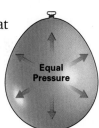

Pascal's principle

persistence of vision The blending of individual images into one image that seems to move. (Ch. 13)

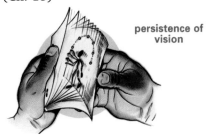

persistence of vision

phosphor A substance that gives off photons when charged with electric current. (Ch. 15)

photovoltaic cell Device that directly changes the sun's rays to electrical energy. (Ch. 8)

plywood The first engineered wood product; made by gluing thin layers of wood together. (Ch. 3)

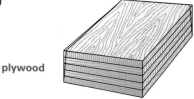

plywood

pneumatic system Fluid power system based on the use of air or another gas. (Ch. 16)

polymer Chain of molecules that forms a material; can be strong, lightweight, durable, and flexible. (Ch. 4)

positron emission tomography (PET) scan Method of imaging the human body by tracking a radioactive substance within the patient and then generating an image on a computer. (Ch. 18)

potential energy Energy that is stored or not in motion. (Ch. 8)

power The rate at which work is done. (Ch. 8)

power supply In robotics, provides power to the robot; may supply electricity, hydraulic power, or pneumatic power. (Ch. 17)

precipitation Droplets of water that fall from clouds as rain, snow, sleet, or hail. (Ch. 10)

pressure The force on a unit surface area (such as a square inch). (Ch. 16)

primary process A process that is used to change a raw material into an industrial material. (Ch. 5)

productivity In manufacturing, the relationship between how many hours are worked and the quantity of products made in that time period. (Ch. 1)

programmable logic controller (PLC) A small computer that helps control machines, such as the ones on an assembly line. (Ch. 5)

prosthesis Replacement body part for someone with a disability. (Ch. 18)

prototype A working model. (Ch. 1)

Q

quality assurance In manufacturing, making sure the product is made according to plans and meets all specifications. (Ch. 5)

quality assurance

R

radiant energy Energy that moves in waves. (Ch. 10)

radio frequency identification (RFID) A tag-and-tracking system used to track products or their containers, pallets that hold products, and the trucks and trailers that transport products. (Ch. 5)

rapid prototyping The process of using CAD data to create physical models. (Ch. 2)

raw material A material as it occurs in nature. (Ch. 5)

relief printing Printing process that uses a raised surface to put ink on paper. (Ch. 11)

rendering Surface shading used to give realism and depth to drawings. (Ch. 2)

renewable resource A resource that is plentiful and/or easy to replace. (Ch. 8)

resistance The opposition to the flow of electrons. (Ch. 14)

resolution In printing, refers to the number of dots per inch (dpi) of ink on printed images. (Ch. 11)

robot A machine that does complicated tasks and is guided by automatic controls. (Ch. 17)

robotics The design, construction, and operation of robots. (Ch. 17)

robotics

S
---

satellite radio Digital technology that makes it possible to listen to the same radio station while traveling long distances. (Ch. 12)

scanner A device that can change images such as photographic prints into an electronic form that computers can use. (Ch. 11)

screen printing Printing process that uses a stencil attached to a mesh (screen) stretched in a frame. (Ch. 11)

secondary process A process that is used to change an industrial material into a finished product. (Ch. 5)

separating Secondary process that involves the cutting of materials to specific sizes and shapes. (Ch. 5)

series circuit Circuit in which components are connected together, allowing only one path for electrons to follow. (Ch. 14)

solid modeling Three-dimensional modeling that shows the shape, area, and volume of an object. (Ch. 2)

static load A load that changes little or not at all; also called a dead load. (Ch. 3)

stop-motion animation Individual frames or models are displayed in sequence to create movement. (Ch. 13)

storyboard A series of sketches that can be used as a guide for making a show. (Ch. 13)

streaming animation Allows a movie to begin playing before the entire file has downloaded. (Ch. 13)

strength The ability of a material to resist forces such as compression and tension. (Ch. 4)

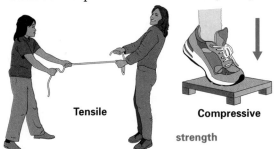

Tensile   Compressive

strength

structural engineers A civil engineer whose work focuses more on the mechanics of load-bearing structures. (Ch. 3)

structural member Building material used with similar materials to make a structure's frame. (Ch. 3)

structure Something that is constructed, or built, in order to meet a need or perform a task. (Ch. 3)

T
---

technology The practical use of human knowledge to extend abilities, satisfy needs and wants, and solve problems. (Ch. 1)

toxicology The study of how drugs, poisons, and other substances affect people. (Ch. 19)

transistor A tiny device used to control current and amplify voltage or current. (Ch. 14)

transportation The process by which people, animals, products, and materials are moved from one place to another. (Ch. 9)

typeface A set of letters, numbers, and symbols that have the same design; in computer programs, typefaces are called fonts. (Ch. 11)

U

ultrasound In bioengineering, the use of sound waves to image the inside of the human body. (Ch. 18)

universal product code (UPC) A barcode that contains information about a product and its manufacturer. (Ch. 5)

V

vessel Transports people, products, and materials over waterways. (Ch. 9)

video on demand (VOD) System that uses digital technology to make television interactive. (Ch. 12)

voltage The electric pressure that causes current to flow. (Ch. 14)

W

wave In physics, a disturbance that transfers energy from one place to another through matter or space. (Ch. 6)

weather The condition of the atmosphere in regard to temperature, moisture, wind, and clouds. (Ch. 10)

work envelope The space within which a robotic arm can move. (Ch. 17)

work envelope

work The application of force to make an object move in the direction of the force. (Chs. 6, 8)

X

X ray A type of electromagnetic wave that can pierce deep into materials like bone, tissue, fat, and muscle. (Ch. 18)

Credits

Cover Design: Squarecrow Creative Group

Interior Design: Squarecrow Creative Group

Cover Art

Getty Images/George Diebold Photography
Getty Images/Jason Reed
Nanomix, Inc.
PhotoDisc
Photo Researchers/Sam Ogden

Photo/Illustration Credits:

AETHON 392
AGStock USA/Dave Reede 460
Airbus S.A.S. 2005 ©Airbus S.A.S. 2005 163
Airbus S.A.S. 2005 ©Airbus S.A.S. 2005/exm company/P. Masclet 163
American Honda Motor Co. 390
Animazoo UK Ltd. 303
Arnold & Brown 29
Art MacDillos/Gary Skillestad 50, 65
Ken Clubb 86, 107, 110, 133, 134, 135, 137, 138, 141, 156, 204, 214, 227, 229, 230, 231, 236, 237, 241, 278, 281, 293, 295, 300, 305, 311, 313, 353, 363, 447, 465
Corbis Images 13, 178, 206, 263, 275, 320, 427, 430, 442
  S. Andreas/zefa 8, 233
  Craig Aurness 205
  Philip Bailey 146
  Morton Beebe 203, 452
  Bojan Brecelj 25
  Andrew Brookes 425
  Christie's Images 90
  Ashley Cooper 423
  Jim Craigmyle 431
  Richard Cummins 5, 22, 60
  George B. Diebold 275

Muriel Dovic/France Reportage 424
Vo Trung (NPP) Dung 429
Firefly Productions 282
Charles Gupton 275
K. Hackenberg/zefa 232
George Hall 153, 164
Richard Hamilton Smith 444
Wyman Ira 415
Ed Kashi 113
Lester Lefkowitz 10, 31, 410, 413
Michael Macor 24
John Madere 84, 383
Setboun Michel 216
John D. Norman 7, 179
Jose Luis Pelaez, Inc. 30, 338
Amet Jean Pierre 123
Carl & Ann Purcell 374
Charles O'Rear 340
Gabe Palmer 32
Louie Psihoyos 116, 369
David Pu'u 6, 94
Jose Fuste Raga/zefa 61
Roger Ressmeyer 5, 74, 75, 217
Karen Rhoden/Jim Reed 239
Joel W. Rogers 72
Reuters 193
David Sailors 69
Chuck Savage 446
Schenectady Museum 320
Alan Schein/zefa 344
H. Schmid/zefa 428
Joseph Sohm; ChromoSohm, Inc. 62
Jim Sugar 269
Svenja-Foto Zefa 145
Tim Thompson 200
William Taufic 31
Oscar White 26
Elizabeth Whiting & Associates 158
Bobby Yip/Reuters 202
Sears Craftsman 372

David R. Frazier Photolibrary, Inc. 45, 64, 65, 191, 202, 237, 277
Digital Vision 60, 84, 104, 150, 166, 198, 202
Exact Dynamics BV, The Netherlands 393
General Mills, Inc. 301
Getty Images
  AFP 11, 422
  Altrendo Images 275
  Milos Bicanski 403
  Flip Chalfant 43
  Paul Chesley 152
  Jim Cummins 261
  Robert E. Daemrich 281
  David Fairfield 382
  Hulton Archive 154
  Image Source 428
  Kaluzny-Thatcher 272
  Karen Kasmauski 63
  Lester Lefkowitz 8, 254
  Ryan McVay 309, 420
  Michael Melford/National Geographic 350
  Mason Morfit 86
  Naoki Okamoto 233
  Stephen Oliver 86, 426
  Vladimir Pcholkin 61
  Photodisc Collection 233, 269
  Ken Reid 308
  Kim Steele 252
  The Stocktrek Corp. 183
  Thinkstock 180
  Tony Ward 400
Gorman & Associates Inc. 235, 245, 248, 267, 270, 371, 372
Grant Heilmen Photography
  Gary Kreyer 448
  Authur C. Smith II 11, 445, 452
Greg Harris 128
Hasbro, Inc. MOUSETRAP® & 2005 Hasbro Inc. Used with permission 196

Index

T